MICROPROCESSOR SUPPORT CHIPS: THEORY, DESIGN, AND APPLICATIONS

MICROPROCESSOR SUPPORT CHIPS:

Theory, Design, and Applications

TJ Byers

A Micro Text/McGraw-Hill Copublication, New York, N.Y.

This book is dedicated to my wife, Twyla,
who made it all possible

Library of Congress Catalog Number: 82-060432

ISBN: 0-07-009518-3

Micro Text Publications, Inc.
McGraw-Hill Book Company
1221 Avenue of the Americas
New York, N.Y. 10020

DISCLAIMER

Although great efforts have been expended checking the accuracy of the data presented, oftentimes seeking the same information from more than one source, errors are bound to occur. In addition, the material presented here cannot be more accurate than that supplied by the manufacturer.

Therefore, the publisher and author take no responsibility whatsoever for the omission or revision of the data contained within this book. Furthermore, no responsibility is assumed for its use; nor for any infringements of patents or other rights of third parties which may result from its use. No license is granted or implied. The manufacturer reserves the right to change information without notice.

ACKNOWLEDGEMENTS

The following people contributed directly to this book: Sherri Besser, Susan Bruijnes, Susan Dunn, Mike Evans, Jim Farrell, Deiter Gessler, Denise Iwata, Glenn Lorig, Oran Marksbury, Berry Matthews, Charlie Melear, Gerry Moseley, Karen Moty, Tom Phillips, Mike Powers, Vince Rende, Marshall Rothen, Greg Shaffer, Chip Shafer, John Smith, Lothar Stern, Mark Stevens, Irwin Schwartz, Elin Thomas, Roger Woodward.

Special thanks to Irwin Schwartz and Advanced Micro Devices for supplying the microphotograph used on the cover.

NOTICE

Data has been incorporated into this book from applications information graciously supplied to Micro Text Publications, Inc. for the purposes of writing this book by the following companies: Analog Devices, Precision Monolithics, Inc., Synertek, Unitrode Corp., Datel-Intersil, NEC Electronics U.S.A., Inc., SEEQ Technology, Inc., National Semiconductor, Cybernetic Micro Systems, Motorola, Inc., Western Digital Corp., Zilog, Advanced Micro Devices, Standard Microsystems Corp., Telmos, Inc., Texas Instruments, Rockwell International, RCA, Signetics Corp., Intel, Mostek Corp.

CONTENTS

Chapter 3. Interface

Chapter 4. Control

Chapter 5. Video

Chapter 6. A/D and D/A Converters

Appendix

Preface

At the heart of every small computer beats a microprocessor. Unfortunately, all too often the microprocessor wastes much of its potential performing simple—but absolutely vital—jobs that could be more efficiently administered by another component. A growing awareness of this fact has prompted nearly all the major semiconductor manufacturers to introduce support chips that perform these mundane routines.

But, which chip is right for the application? And how is it used after you have found it?

Initiating a design can be a frustrating experience. More often than not, a data sheet is nothing more than a sketchy outline of statistics. The collection of over 100 state-of-the-art microprocessor support chips presented here will greatly facilitate the use of the new chips in current designs. All the necessary application concepts and data have been organized in a highly coherent format.

Most of what you will read pertains specifically to the hardware aspects of the device. Not only does the text list each pinout, but it also describes its function—and how to use it—in detail. When a pin is expected to perform more than one task, that is detailed. And for those times when the manufacturer expects the use of external circuitry, that too is demonstrated. If design quirks exist, they are explained and illustrated. In fact, these support chips can be seen in action by reviewing the working schematic included.

When necessary, software routines are explained. Sometimes it takes more than hardware to get a chip up and running. Each chapter, furthermore, describes the overall function of the chips contained within that group. It is hoped that this volume will provide a valuable addition to every engineering and technical library.

Canyon Country, CA, 1983

How to Use This Book

This book was organized with the user's needs in mind. It attempts to give the reader a good overview of each chip and its functions, while detailing its hardware requirements. Since many different aspects are dealt with, some explanation of the conventions of this book are necessary.

First, the book is logically laid out so that the chip of interest is completely before the reader at a glance. Each chip occupies a left and right page "spread." The text, detailing the operation of the device, is on the left. In the upper right-hand corner of this page is a pinout of the chip. On the right, are schematics representative of the device in an actual working circuit.

In the text is information pertinent to the understanding and operation of the device. A unique system of functional assignment has been established. All descriptions which actually relate to a physical pin location on the package are put in CAPITAL LETTERS, followed by the pin number in parenthesis. An example is: "The chip is enabled when the CHIP SELECT (pin 33) input is LOW." To indicate the logic levels associated with digital operation, the logic is defined by SMALL CAPITALS. A LOW equals logic "0", while a logic "1" is expressed by HIGH.

When the signal lines are a consecutive set, the numbering for the pinouts was done listing the Least Significant Bit first and the Most Significant Bit last. A good example of this is when a DATA BUS (pins 8-15) is mentioned. Pin 8 is the LSB bit and pin 15 is the MSB, with all intermediate bits in sequential order. In other words, D0 is pin 8, D1 is pin 9, etc. If the numbering appears in reverse order, such as (pins 15-8), the bit order follows along: D0 is pin 15, D1 is pin 14, etc.

Oftentimes, it was necessary to talk about a specific register within the chip to better explain the operation and interactions of the functions. Whenever a specific register is mentioned, it is written in *italics*. To distinguish between a register description and an actual software directive, all software commands are capitalized. Example: A second CPUcycle loadsthe Control Word into the *control register*. Of course, all proper names, Trademarks, and acronyms are capitalized.

Due to the complexity of a few chips described in the book, it was necessary to spread the discussion over a family of devices. Therefore, the reader must read the entire package to gain a complete picture of everything that is going on inside the chip. This gives the user a wider view of the overall performance without compromising data or becoming repetitive. References are made within the text when more than one chip is involved.

In a broader sense, chips within a certain area of interest—but not family related—have been used to tell a story. During the course of organizing the data, it became apparent even one family of chips was not enough to cover the discussion of "industry standard" information, and the details were spread over a larger group. For instance, the description of the Ethernet Local Area Network protocol is revealed by several, unrelated chips. By reading the entire Ethernet section, a fuller understanding of Ethernet operation emerges.

CHAPTER ONE TELECOMMUNICATIONS

8251A

Programmable Communication Interface

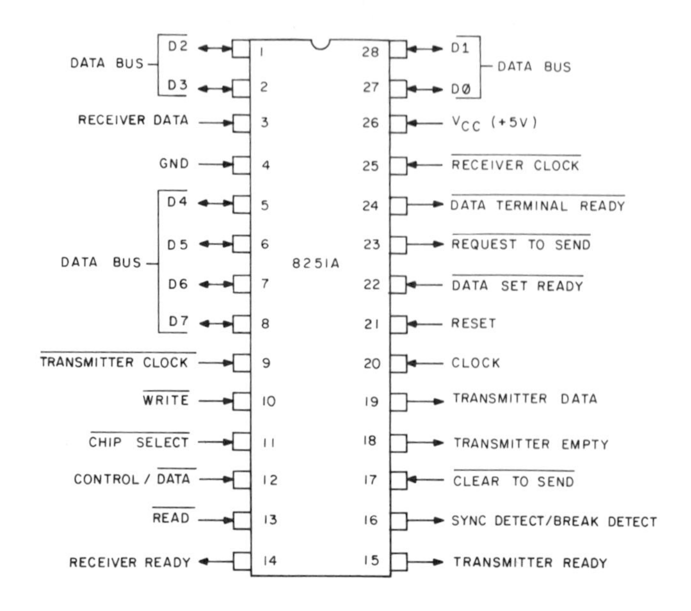

The 8251A, introduced by Intel, is a Programmable Communication Interface controller designed for use with a wide variety of microcomputers. Like other intelligent I/O devices in a microprocessor system, its functional configuration is software programmable for maximum flexibility.

The 8251A is essentially an advanced design of the industry standard USART, the Intel 8251, and operates with an extended range of CPUs while maintaining compatibility with the 8251. The controller can support most serial data techniques in use, including IBM's BiSync.

General Operation

The microprocessor, by itself, is virtually useless unless it can communicate with the outside world. In a communication environment, an interface device, such as the 8251A, must convert parallel CPU data into serial format for transmission, and convert incoming serial data into parallel form for CPU evaluation. Furthermore, the communication device should also be able to insert or delete preassigned bits or characters that are functionally unique to the specified protocol. In essence, the interface should appear transparent to the CPU, requiring a simple input or output of information.

CPU Interface

A 3-state, 8-bit DATA BUS (pins 27, 28, 1, 2, 5-8) is used to interface the 8251A to the CPU system. Whether the bus data is device control words, status information, or a data character is determined by the CONTROL/DATA input (pin 12) in association with the WRITE (pin 10) and READ (pin 13) data inputs. When the CONTROL/DATA input is HIGH, the chip is either being programmed or serviced; while the LOW state is used to transfer communication data between the 8251A and the system bus.

All data transfers occur on the LOW strobe of the CHIP SELECT (pin 11) control. When this pin is HIGH, the DATA BUS is in a high-impedence state, and neither READ nor WRITE will have an effect on the device.

Modem Interface

Most communications devices require a modulator, such as a modem, for communications interfacing. The 8251A is no exception. For this reason, the chip includes a set of input and output controls for modem interfacing.

The DATA SET READY (pin 22) is a general-purpose input that informs the communications chip that the modem is prepared to communicate data; conversely, the DATA TERMINAL READY output (pin 24) signals to the modem that the 8251A is ready to process data. When a message is assembled for transmission, the controller queries the modem with a REQUEST TO SEND (pin 23) signal, to which the modem responds with a CLEAR TO SEND (pin 17) reply. Data is subsequently transferred out of the TRANSMITTER DATA (pin 19) port. Incoming messages are received at the RECEIVER DATA (pin 3) input.

Transmitter/Receiver Controls

The 8251A controller can be used for either synchronous or asynchronous communications, with the operation of the transmitter and receiver primarily determined by the protocol specified.

The Transmitter Controls manage all matters associated with the transmission of serial data. When the transmitter is prepared to initiate operation, it signals the CPU by taking the TRANSMITTER READY output (pin 15) HIGH. Whenever the transmitter has no character to send, whether it be in standby or actively transmitting, the TRANSMITTER EMPTY (pin 18) goes HIGH.

By the same token, the Receiver Controls govern all receiver-related activities. Data reception in the Synchronous Mode begins when the receiver locates a valid Sync character, as indicated by driving the SYNC DETECT output (pin 16) HIGH. If the controller is programmed to respond to double Sync characters, as in the BiSync Mode, the SYNC DETECT goes HIGH in the middle of the second Sync character.

The RECEIVER READY output (pin 14) indicates that a character has been assembled and is ready for reading by the CPU. A subsequent read operation resets the SYNC DETECT output. Failure to read the received character prior to the assembly of the next data character sets an Overrun Error flag as the incoming data writes over the previous word, destroying it.

In the Asynchronous Mode, the SYNC DETECT becomes a BREAK DETECT output which goes HIGH whenever the RECEIVER DATA input remains LOW through two consecutive Stop bit sequences.

The transmitter and receiver baud rates are determined by the TRANSMITTER CLOCK (pin 9) and RECEIVER CLOCK (pin 25) inputs, respectively. In most communications systems, the 8251A will be handling both the transmission and reception operations of a single link. Consequently, the TRANSMITTER and RECEIVER CLOCKs will be the same, and can be tied together and driven by a single-frequency source. Although the chip is internally driven by the CLOCK (pin 20) signal, it has no effect on the baud rates or DATA BUS operations.

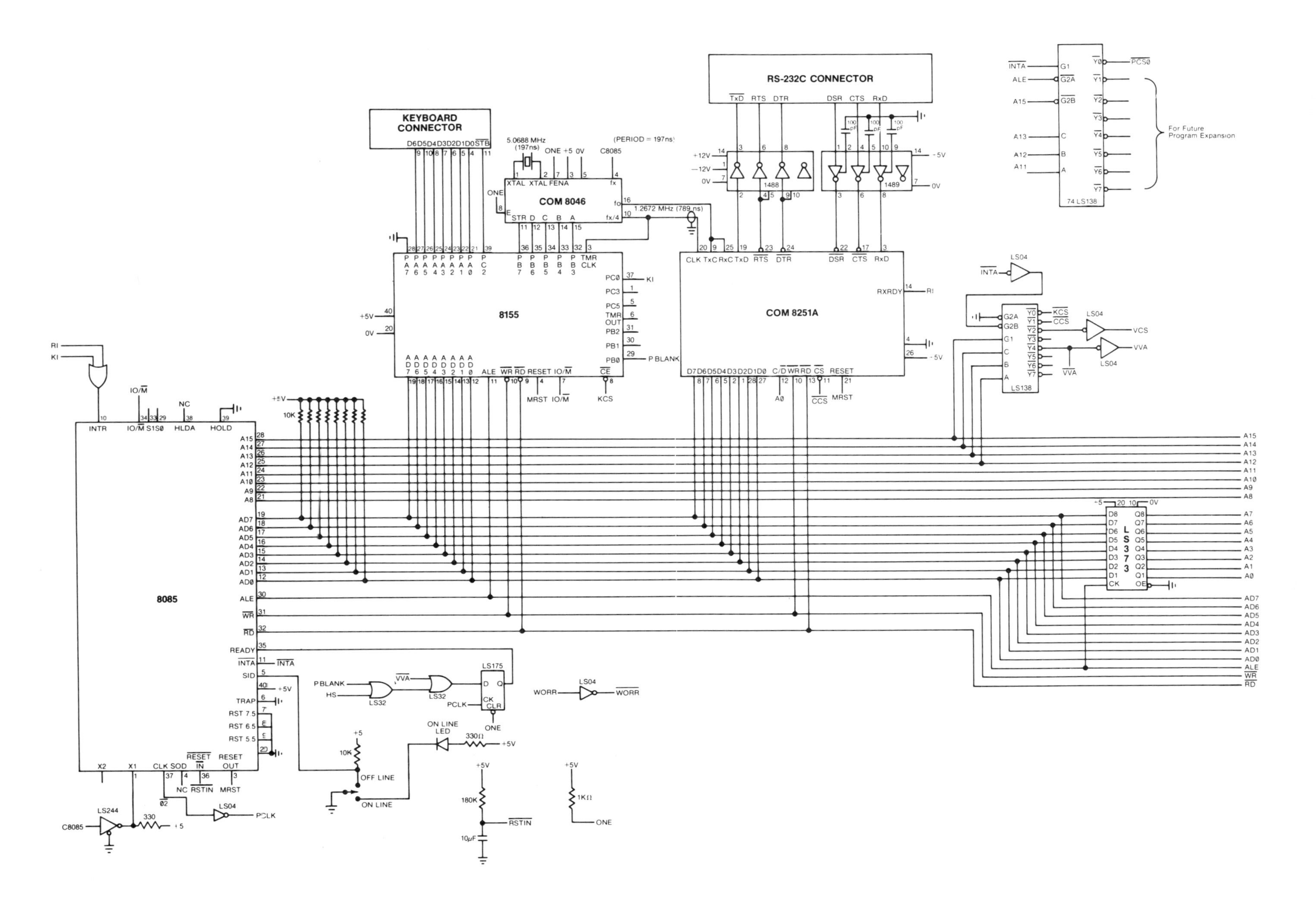

RS-232C CONNECTOR
KEYBOARD CONNECTOR
COM 8251A
COM 8046
8155
8085
74 LS138
LS138
LS373
For Future Program Expansion
1489
1488
LS175
LS04
LS32
LS244
(PERIOD = 197ns)
5.0688 MHz (197ns)
1.2672 MHz (789 ns)
ON LINE LED
OFF LINE
ON LINE

WD2501

Packet Network Interface

The WD2501, by Western Digital, is an LSI Packet Network Interface device which handles bit-oriented, full-duplex serial data communication that conform to CCITT X.25 standards. In fact, the transceiver is totally dedicated to X.25 LAP protocol, using internal firmware, and all operations except for data input and output are transparent to the user.

Bit-Oriented Communication

The 2501 communicates according to the X.25 standard, which is a bit-oriented protocol. There are several industry standards which satisfy the bit-oriented format, including SDLC and HDLC. The Level 2 protocol defined by X.25, and executed by the 2501, is essentially an outgrowth of IBM's HDLC, and is the standard adopted by the Consultive Committee for International Telephone and Telegraph (CCITT).

Messages are exchanged between stations inside a package called a frame. One frame constitutes one packet of data. Frames are bit-oriented in that each bit within the frame structure—not the characters—has a specific meaning.

Frame Format

All frames start and end with a Flag. The Flag is a unique binary pattern (01111110) which identifies the beginning of a frame. The receiver is always in constant search of a Flag, and uses it to establish synchronization and frame boundaries.

The Flag is followed by an 8-bit Address Field and an 8-bit Control Field. These two fields are directives, supplying control instructions for the communique, and are often expandable (as they are in the 2501) to include automatic addressing and loop-back testing.

Once the communication rules have been established, the Information Field is sent. This field contains the actual data packet, which may be anywhere from 5 to 8 bits in length. An important characteristic of a frame is that its contents are made code transparent by the automatic insertion and deletion of zeros. Should the message (Information Field) contain more than five consecutive logic "1" bits, the transmitter automatically inserts zeros into the data sequence. This ploy prevents the receiver from inadvertently reading the data as a Flag.

After the Information Field is sent, an error checking Frame Check Sequence is calculated and appended, and the frame closed with another Flag. Since bit-oriented protocol is basically a synchronous data link, it doesn't allow for idle frames. Therefore, when no data is available for transmission, the 2501 fills the time with contiguous Flags.

Data Transmission

Outgoing data is input to the 2501 through the DATA ACCESS LINES (pins 8-15). The packet is then assembled and serially transmitted out the TRANSMIT DATA (pin 20) output at the rate established by the TRANSMIT CLOCK (pin 19) frequency. On a local level, two stations can be interfaced using line drivers, such as EIA RS-422 or RS-232C. For longer distance communication lines, on the other hand, a modem is needed.

Two control lines direct transmitter operations. After the transmitter has assembled a frame, and is ready to send the

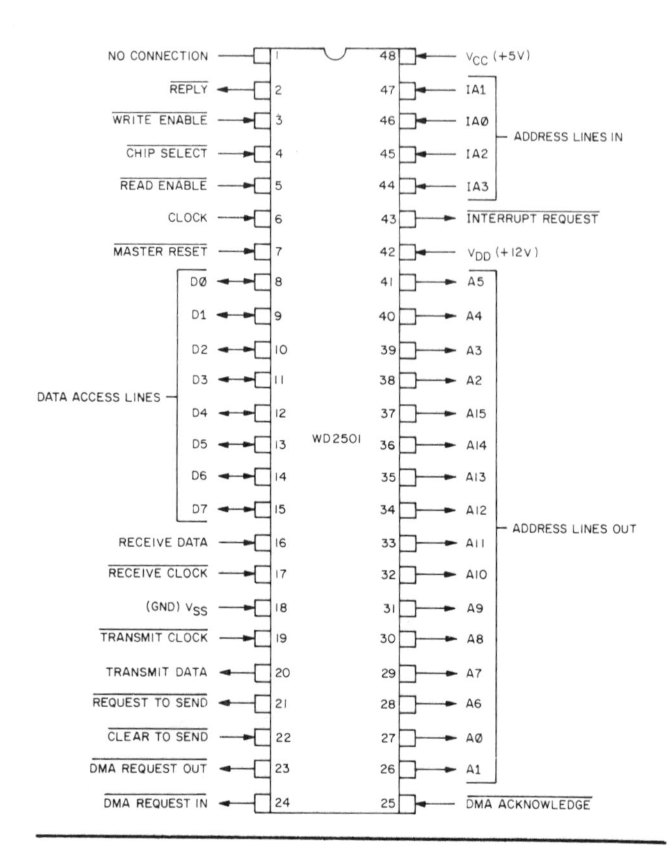

packet, it activates a REQUEST TO SEND (pin 21) open collector output. The REQUEST TO SEND output will need a pull-up resistor in most cases, since it is an open collector. The transmitter is enabled by a CLEAR TO SEND (pin 22) input.

Transmitter Abort

There are three situations that can occur which will terminate data transmission. Since all three represent an abnormal condition, the data already transmitted must be dissolved from the message. To accomplish this, an Abort command, consisting of a zero followed by at least seven "1"s (01111111...), is transmitted to terminate a frame in such a manner that the receiving station ignores the frame.

An Abort will occur whenever there is a transmitter data underrun. This happens when the data input can't keep up with the output data rate. Transmission will also terminate if a reject acknowledgement is received from the intended receiving station.

The third situation involves receiver acknowledgement. It is the responsibility of the receiving station to acknowledge the reception of a frame within a reasonable amount of time. A programmable timer within the 2501 sets this grace period between 16-ms and 16 seconds. If this timer expires while a packet is being transmitted, the packet is Aborted. A note of caution here: If a packet is longer in time than the time-out period, an Abort will always occur.

Western Digital makes two chips which comply with the two variations of the X.25 protocol. The 2501 implements LAP (Link Access Procedure) whereas the 2511 executes LAPB (Link Access Procedure Balanced). To further identify the 2501's operations, the reader should turn to the WD2511 section.

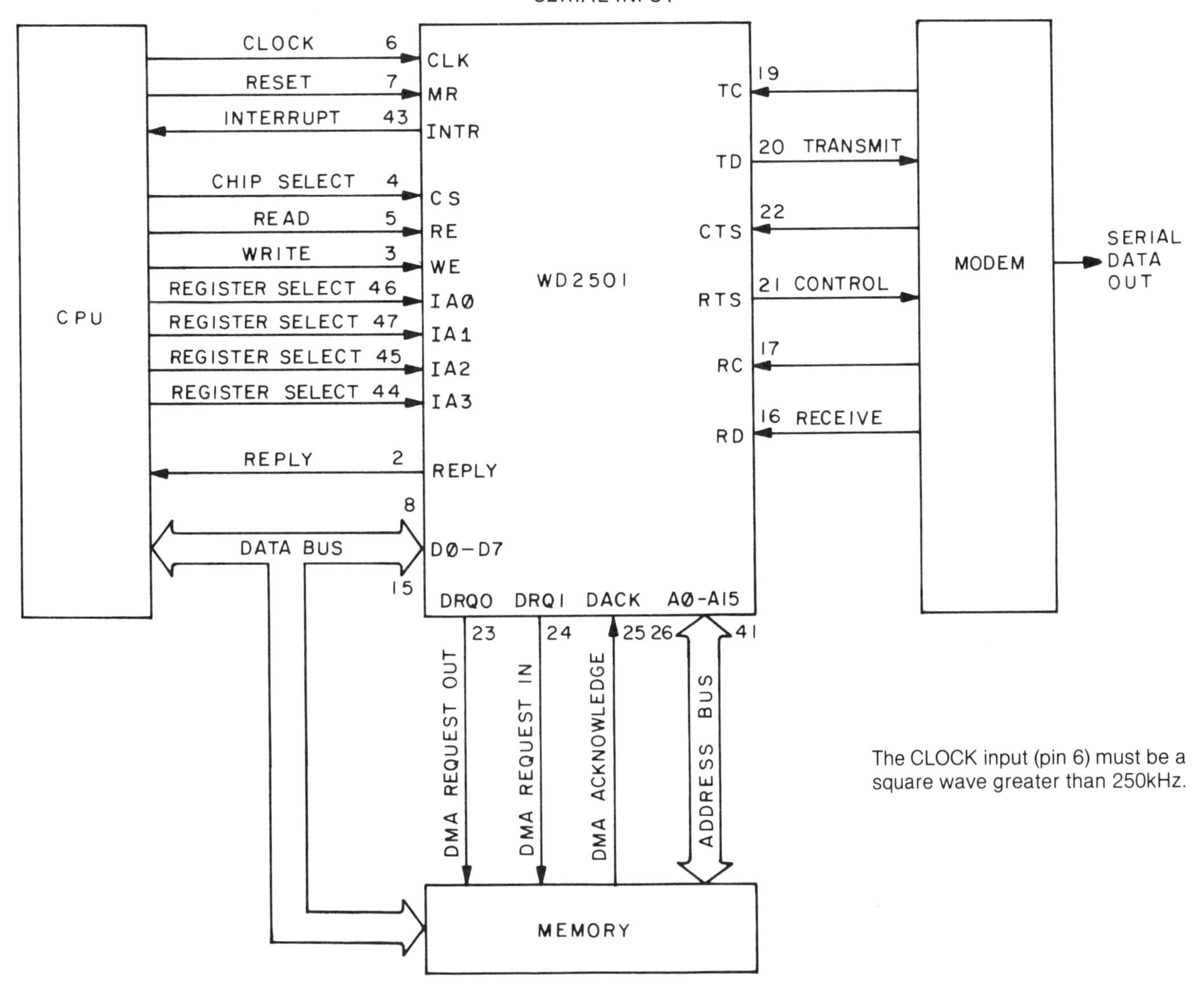

The CLOCK input (pin 6) must be a square wave greater than 250kHz.

Figure 1

The 2501 is basically a full duplex device with the receiver maintained in an always ready condition, even if the *receiver buffer* is not ready. Thus whether the received frame came from a full- or half-duplex system is of no consequence to the 2501.

WD2511

Packet Network Interface

The WD2511, by Western Digital, is an LSI Packet Network Interface device which handles bit-oriented serial data communications according to CCITT X.25 standards. The chip can also transfer information directly to or from memory using an on-chip DMA controller.

Basically, the 2511 is identical to the WD2501, also by Western Digital. However, whereas the 2501 communicates using the X.25 LAP protocol, the 2511 implements the newer X.25 LAPB.

LAPB Standards

Essentially, the X.25 protocol is an outgrowth of IBM's HDLC serial data standard. However, the LAP standard first proposed by the CCITT committee does not fully comply with HDLC, creating subtle problems. Therefore, in 1980 the committee adopted a greatly enhanced LAPB (Link Access Procedure Balanced) version of the X.25 standard. This superior protocol is scheduled to eventually phase out and replace the LAP network.

LAP and LAPB differ only in the link set-up, disconnect, reset, and receiver-not-ready procedures, all of which are in firmware and transparently performed by the chip. Therefore, the 2501 and 2511 are directly interchangeable in both hardware and software. To switch from one protocol to the other, the user simply plugs the required chip into the system.

Data Reception

Received data enters the chip through the RECEIVE DATA (pin 16) input. For data detection, the receiver must synchronize itself to the data stream using a baud rate clock. This clock may be supplied by the remote transmitter or generated by an external baud rate generator. The clock signal is input through the RECEIVE CLOCK (pin 17) pin.

Each station has a unique address, which is programmed into the receiver's *A-Field register.* After the receiver recognizes a Flag, it checks the Address Field against its own. If the two agree, the receiver is enabled; if not, the receiver continues searching for another Flag.

Once the receiver is enabled, it monitors each frame for correct Address and Frame Check Sequence. It also decodes the Control Field directive. If the frame is a packet (valid data), the Information Field is placed into an assigned register location within the chip and an INTERRUPT REQUEST (pin 43) generated. The message byte must be read from this register before the next packet arrives, or an Error Flag is set. However, up to 24 bits can accumulate before a data overrun occurs.

After a packet has been received, the receiving station must acknowledge the fact. However, with X.25, up to seven frames may be accumulated before an acknowledgement is imperative.

CPU Interface

The 2511 is controlled and monitored by 16 registers. These registers are accessed through the DATA ACCESS LINES (pins 8-15) by configuring the ADDRESS LINES IN (pins 47-44) inputs. The contents of the selected register is placed on the DATA ACCESS LINES when the READ ENABLE (pin 5) and CHIP SELECT (pin 4) inputs are LOW, and data is entered into a register when the WRITE ENABLE (pin 3) and CHIP ENABLE are

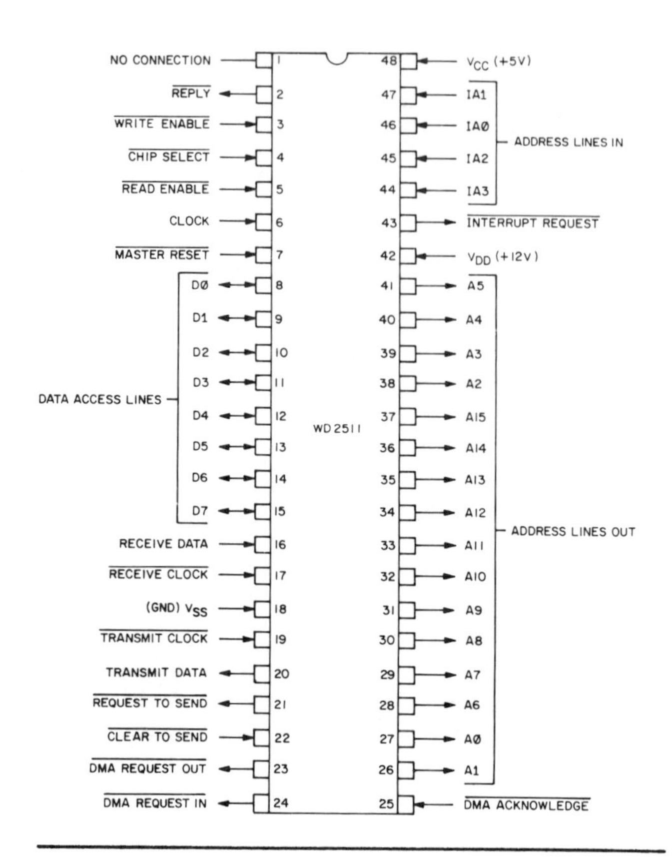

combined. Whenever a read or write exchange is taking place, the REPLY (pin 2) output goes LOW.

DMA Operation

The 2511 can also manage DMA transfers. This memory access method takes full advantage of the X.25 protocol, which allows up to seven Information Fields to go unacknowledged in each direction of the communication link. The 2511 uses two look-up tables in external memory to accomplish DMA transfers: One for transmit and one for receive.

First, the 2511 is programmed with the starting address of the external look-up tables. These locations are stored in the *look-up table register,* transmitter address first.

In the transmit mode, the 2511 will specify the first DMA location in the *look-up table register* using the ADDRESS LINES OUT (pin 26-41) outputs. The transmitter requests data by pulling the DMA REQUEST IN output (pin 24) LOW. Data is loaded on the receipt of a DMA ACKNOWLEDGE (pin 25) signal. The 2511 then automatically transmits the Flag, Address and Control Fields. Next, the DMA data byte is transmitted. At the end of the Information Field, the Frame Check Sequence are appended. The 2511 then moves on to the next DMA address and repeats the process.

If a frame needs to be repeated, the 2511 will automatically retrace the previous transmissions through the look-up table. This retransmission doesn't involve the CPU, although it is notified of the fact by setting an Error Flag bit.

The receiver reverses the process by putting data into memory, instead of extracting it. A DMA REQUEST OUT (pin 23) will place a data byte on the DATA ACCESS LINES when the DMA ACKNOWLEDGE input is pulled LOW. It must be noted that DMA transfers should not be attempted when the REPLY line is LOW.

Refer to the WD2501 section for additional details.

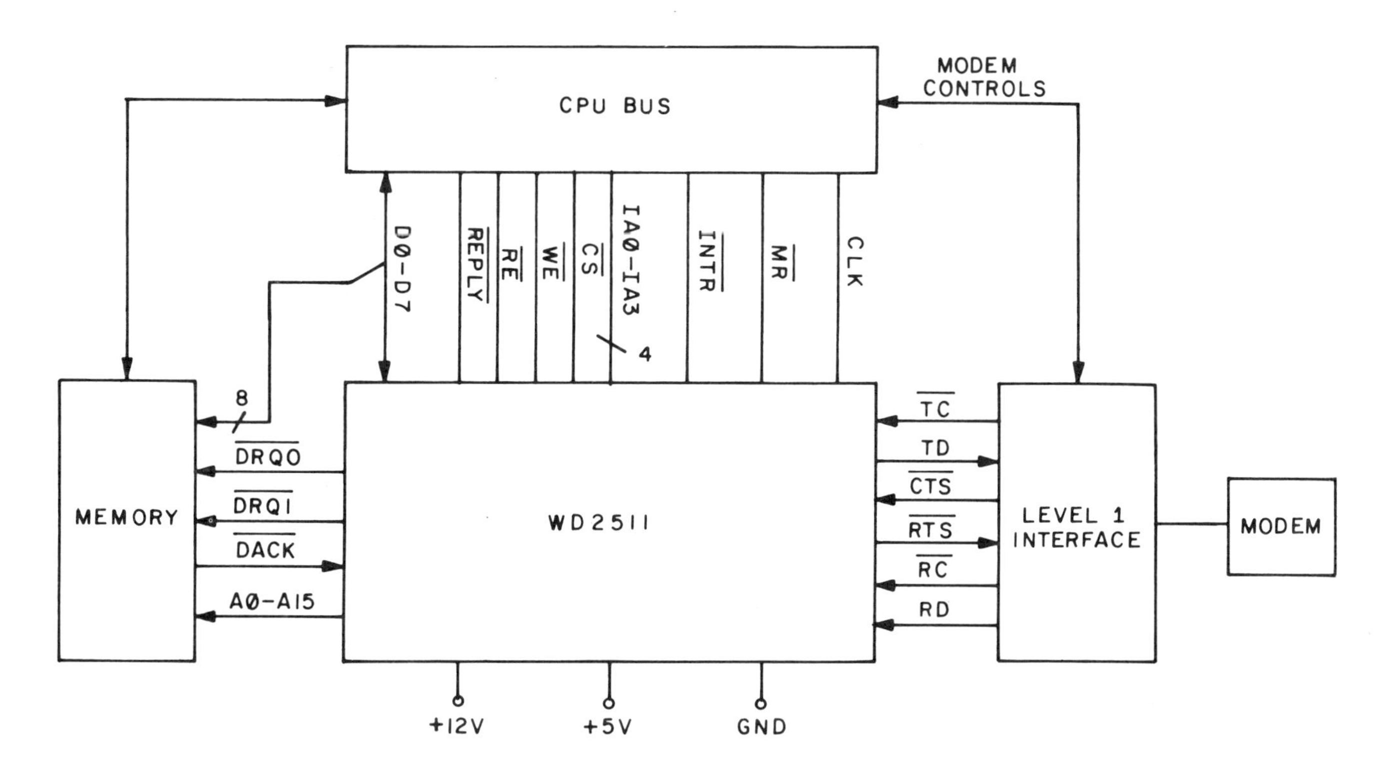

Figure 1

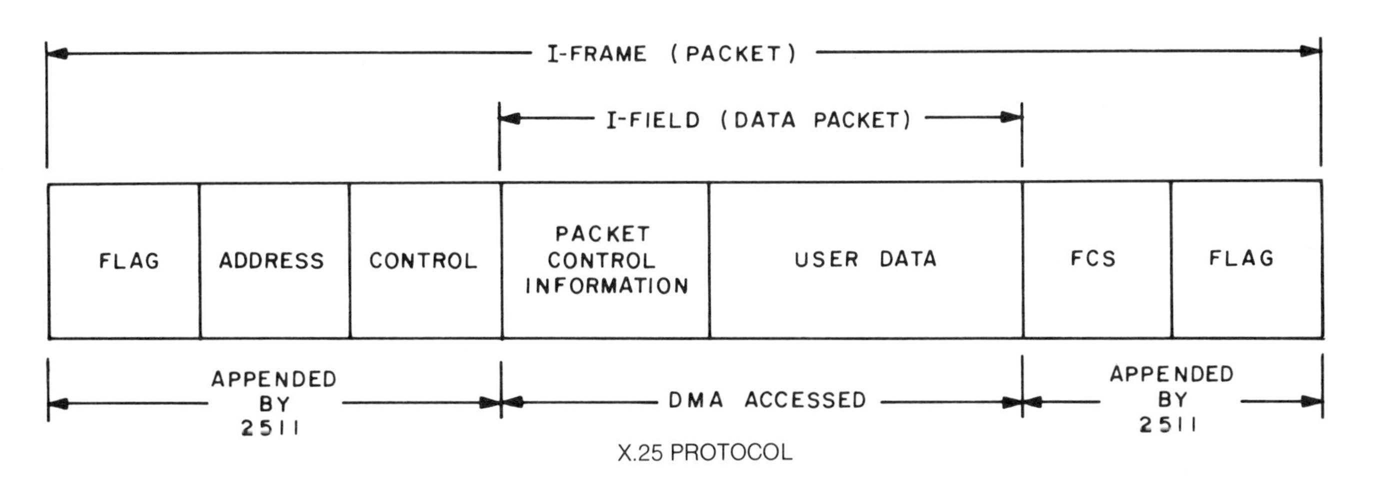

X.25 PROTOCOL

Figure 2

MC6854

Advanced Data-Link Controller

The MC6854, by Motorola, performs the complex function of CPU/data-link communications in agreement with "Advanced Data Communication Control Procedure" (ADCCP), High-Level Data-Link Control (HDLC), and Synchronous Data-Link Control (SDLC) standards. The device provides key interface requirements with improved software efficiency, and is designed for both primary and secondary stations in stand-alone, polling, and loop configurations.

Operation

The MC6854 transforms parallel 8-bit data, linked through the DATA BUS (pins 22-15), into serial NRZ or NRZI format for advanced communication applications. Data framing, word length, and protocol are all under software control.

Data transfer between the CPU and the controller begins by enabling the CHIP SELECT input (pin 9, LOW), and directing the data flow to the desired register using a combination of REGISTER SELECT inputs (pins 10 and 11) and programmed software. Writing is performed with a LOW input to the READ/WRITE pin (pin 12) and a reading is taken by forcing the input HIGH. All operations are performed under the synchronization of the ENABLE CLOCK (pin 13).

Select Data Transfer

Data is transferred over the communications lines in duplex form using individual TRANSMIT DATA output (pin 6) and RECEIVE DATA input (pin 3) ports. Neither the transmitter nor receiver data rate should ever exceed the clock frequency for accurate data transfer.

The receiver samples data on the positive edge of the RECEIVER CLOCK input (pin 4), and should be externally synchronized with the received data. The transmitter shifts its data through the *transmitter shift register* on the negative transition of the TRANSMITTER CLOCK input (pin 5), and effectively sets the output baud rate. When in the Loop Mode, the TRANSMITTER CLOCK and RECEIVER CLOCK must be of the same frequency and phase, otherwise they are separate and operate independently of each other.

Interrupt and Reset

A hardware reset is accomplished with a LOW signal to the RESET (pin 8) input. Unlike the selective reconfiguring of the software resets, this signal clears all stored status conditions, resets the transmitter and receiver, forces the REQUEST TO SEND and LOOP ON-LINE outputs HIGH, and resets the following control bits: Transmit Abort, Request To Send, Loop Mode, and Loop On-Line/DTR, with a single command.

INTERRUPT REQUEST output (pin 7) will be LOW whenever an interrupt situation exists and the appropriate Interrupt Enable has been set. The interrupt remains in effect as long as the interrupt condition exists. An output is also provided to alert the user of software flags on the FLAG DETECT output (pin 25), and is often used to initiate an external time-on counter for the Loop Mode operation.

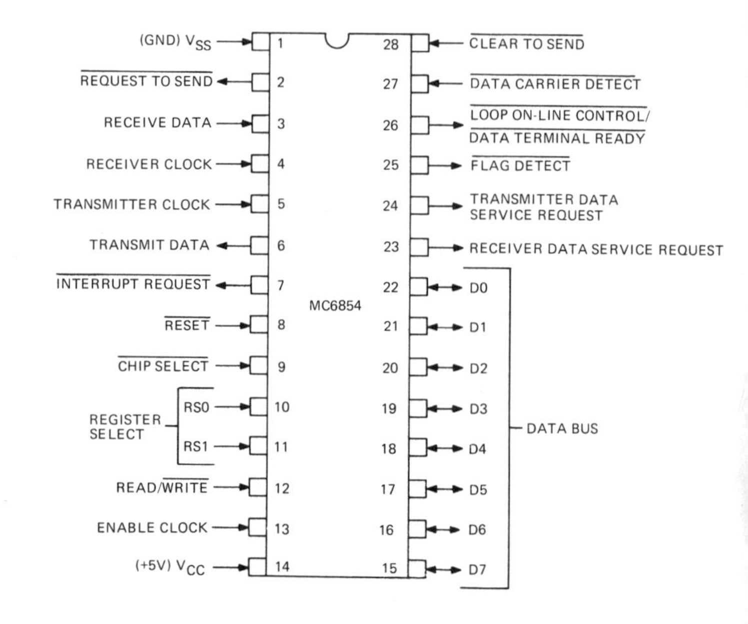

Modem and External Control

Four controls are supplied for modem and external synchronization of the communications chip. The REQUEST TO SEND output (pin 2) is controlled by the Request To Send bit in conjunction with the status of the transmitter. When this software bit goes HIGH, it forces the output pin LOW, which remains LOW until the end of the last valid frame. An Idle mark automatically resets its state.

The CLEAR TO SEND input (pin 28) provides a real-time inhibit to the Transmitter Data Register Available bit and its associated interrupt. The DATA CARRIER DETECT input (pin 27) provides real-time inhibit to the receiver section by resetting and inhibiting the *receiver register,* without disturbing the contents of the *receiver FIFO register* from the previous frame.

The LOOP ON-LINE CONTROL/DATA TERMINAL READY output (pin 26) serves as the data terminal ready output when in the non-Loop Mode, or as a loop control output in the Loop Mode.

DMA Interface

Data can also be exchanged between the MC6854 and the CPU in a block transfer form under the discipline of an external direct memory access controller. The TRANSMITTER DATA SERVICE REQUEST output (pin 24) is provided for DMA operation and when HIGH indicates that the *transmitter FIFO register* demands service, regardless of the status of the Transmitter Data Service Request Mode Control bit. Pin 24 goes LOW when the register is loaded. This output can be masked using software commands or by setting CLEAR TO SEND input HIGH.

The RECEIVER DATA SERVICE REQUEST output (pin 23) indicates that the *receiver FIFO register* is in need of service when it is HIGH, and is provided primarily for use in the DMA Mode. This output goes LOW once the register is read.

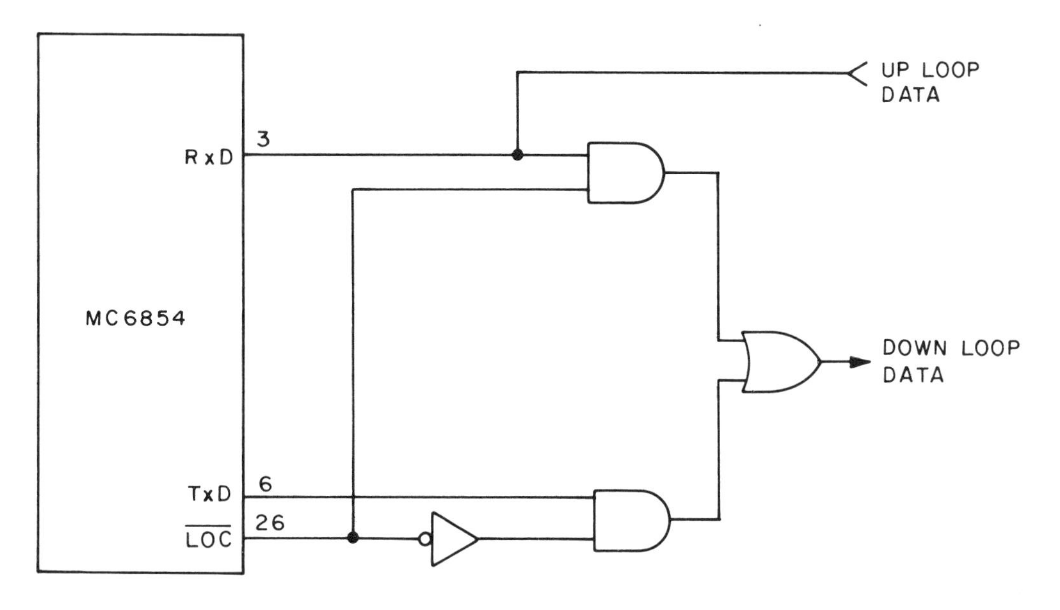

Note to users not using the MC6800 CPU: Care should be taken when performing a write followed by a read on successive enable clock pulses at high frequency rates. Time must be allowed for status changes to occur.

Figure 1

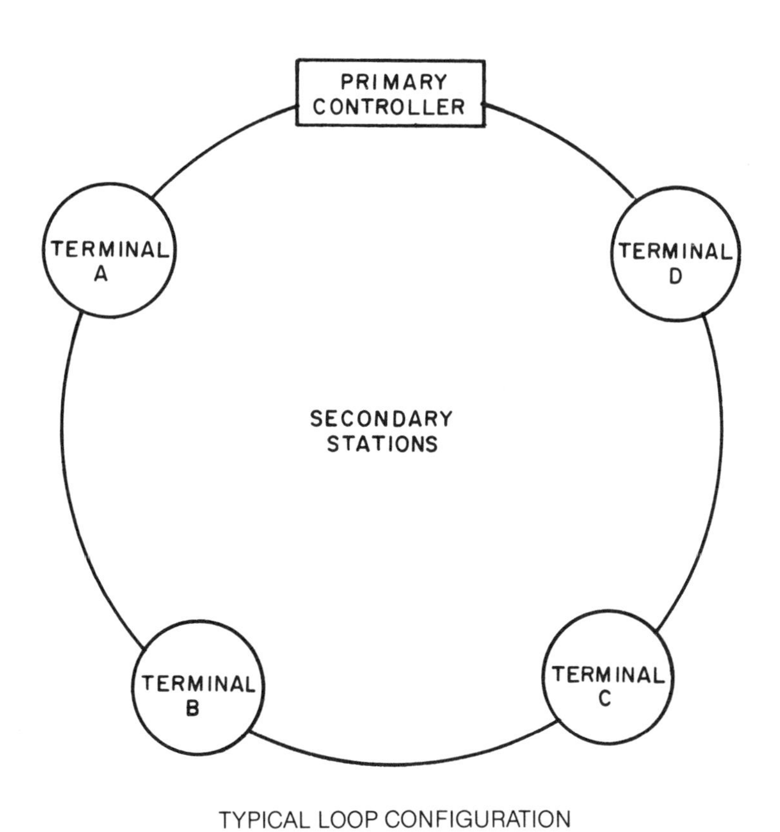

TYPICAL LOOP CONFIGURATION

Figure 2

2651

Programmable Communication Interface

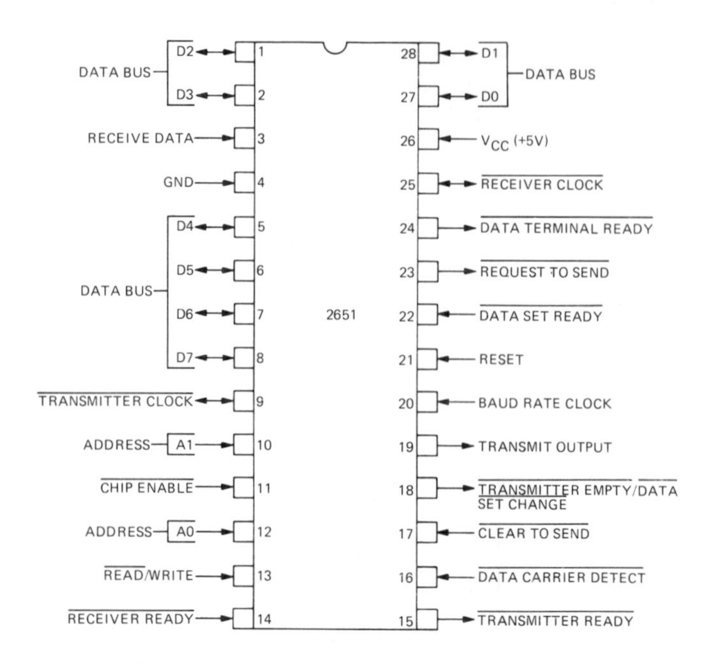

The 2651, by Signetics, is a Programmable Communication Interface device that meets the majority of asynchronous and synchronous data communications requirements. The 2651 is used as a peripheral and is programmed by the processor to communicate in commonly used serial data transmission techniques.

Operation

The 2651 interfaces the CPU to asynchronous and synchronous data communications channels by converting parallel digital data into serial data streams, and vice versa. The device requires a minimum of processor overhead to accomplish this, and can even appear transparent in several modes.

The transmitter and receiver section can operate simultaneously and independently of each other, and the 2651 will notify the CPU when it has completely received or transmitted a character and requires service. Complete device status, including data format errors and control signals, is available to the processor at any time.

CPU Interface

Prior to initiating data communications, the operational mode must be specified by performing write operations to the *mode* and *command registers.* To program a register, the ADDRESS lines (pins 12, 10) are configured and the READ/WRITE input (pin 13) is set HIGH. Conversely, reading the contents of a register requires that the READ/WRITE input be pulled LOW. The transfer takes place on the LOW strobe of the CHIP ENABLE (pin 11) input. When CHIP ENABLE is HIGH, it places the DATA BUS (pins 27, 28, 1, 2, 5-8) in a high-impedence state. These registers are cleared when a RESET (pin 21) input is applied.

Synchronous Receiver Operation

After programming, the 2651 is ready to perform the desired communications functions. The receiver is conditioned to receive serial data at the RECEIVE DATA input (pin 3) when the DATA CARRIER DETECT input (pin 16) is LOW and the Receiver Control bit is set.

In the Synchronous Mode, the receiver first enters the hunt phase of operation. During the hunt phase, data is shifted into the *receiver shift register* one bit at a time. The contents of this register is then compared to the sequence stored in the *Syn1 register.* If the two agree, the hunt mode is terminated and character assembly begins. If the 2651 has been programmed to recognize double Sync characters, the second character must match the contents of the *Syn2 register,* otherwise the receiver continues searching.

Once synchronization has been achieved, the receiver assembles the incoming characters and transfers them to the *receive data holding register.* Parity checks are made on the characters if so specified, and each time a valid character is assembled, the RECEIVER READY output (pin 14) is asserted. If the holding register is not read before a new character is ready for transfer, an Overrun Error bit is set, and the contents of the register is considered invalid.

Synchronous Transmission Operation

The transmitter is enabled when the Transmit Control bit is set and the CLEAR TO SEND input (pin 17) is LOW. In the Synchronous Mode, the TRANSMIT DATA output (pin 19) is held HIGH and the TRANSMITTER READY output (pin 15) is LOW, indicating that it can accept a character for transmission.

This condition exists until the first character to be transmitted — usually a Syn character — is loaded into the *transmit data holding register.* Data is transferred from the holding register to the *transmit shift register,* after which a continuous stream of characters is transmitted out the TRANSMIT DATA output. Once the holding register is cleared, the TRANSMITTER READY output again asserts itself, and the next character is loaded.

No extra bits (except parity, when used) are generated by the 2651 when in the Synchronous Mode other than those presented by the CPU—unless the CPU fails to input a new character to the 2651 by the time the transmitter has completed sending the previous character. Since synchronous communication doesn't allow for gaps between characters, the transmitter must fill the gap by inserting Idle or Syn characters (according to the protocol specified). Normal transmission of the message resumes when a new character is available in the *transmit data holding register.*

Baud Rate

The 2651 contains a baud rate generator which can generate sixteen commonly used baud rates, any one of which may be selected for full duplex operation. The BAUD RATE CLOCK input (pin 20) accepts the base frequency (5.0688-MHz) for division to the baud rate. Both transmitter and receiver baud rates can be monitored at the TRANSMITTER CLOCK (pin 9) and RECEIVER CLOCK (pin 25), respectively. When an external baud rate generator is used, these pins become inputs.

Refer to the 2661 section for further details.

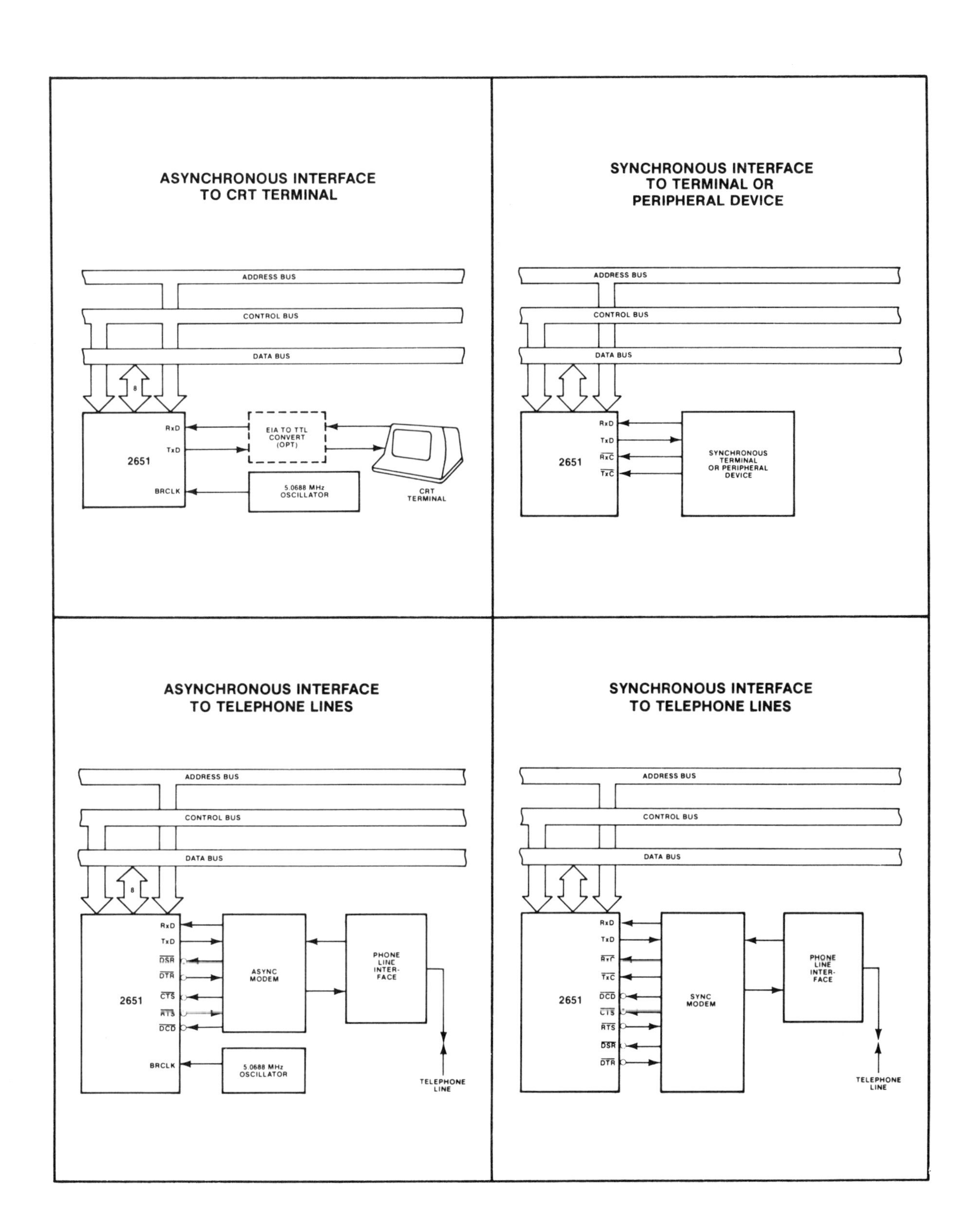
ASYNCHRONOUS INTERFACE
TO CRT TERMINAL
ADDRESS BUS
CONTROL BUS
DATA BUS
8
RxD
TxD
2651
BRCLK
EIA TO TTL
CONVERT
(OPT)
5.0688 MHz
OSCILLATOR
CRT
TERMINAL
SYNCHRONOUS INTERFACE
TO TERMINAL OR
PERIPHERAL DEVICE
ADDRESS BUS
CONTROL BUS
DATA BUS
RxD
TxD
2651
RxC
TxC
SYNCHRONOUS
TERMINAL
OR PERIPHERAL
DEVICE
ASYNCHRONOUS INTERFACE
TO TELEPHONE LINES
ADDRESS BUS
CONTROL BUS
DATA BUS
8
RxD
TxD
DSR
DTR
2651
CTS
RTS
DCD
BRCLK
ASYNC
MODEM
PHONE
LINE
INTER-
FACE
5.0688 MHz
OSCILLATOR
TELEPHONE
LINE
SYNCHRONOUS INTERFACE
TO TELEPHONE LINES
ADDRESS BUS
CONTROL BUS
DATA BUS
RxD
TxD
RxC
TxC
2651
DCD
CTS
RTS
DSR
DTR
SYNC
MODEM
PHONE
LINE
INTER-
FACE
TELEPHONE
LINE

2661

Enhanced Programmable Communication Interface

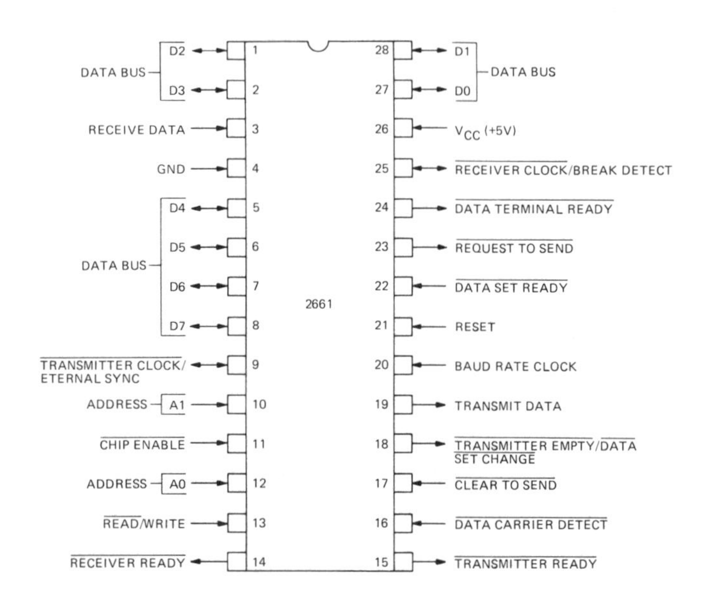

The 2661 Enhanced Programmable Communication Interface, by Signetics, is a CPU peripheral device that meets the majority of asynchronous and synchronous data communications requirements by interfacing parallel digital systems to serial data communications channels. The 2661 is a pin compatible version of the 2651 which has been enhanced to provide greater flexibility and improved software, including special support for the BiSync protocol.

CPU Interface

Prior to initiating data communications, the 2661 must be programmed with the desired protocol. To program the device, the program registers are accessed by configuring the ADDRESS inputs (pins 12, 10) while the READ/WRITE input (pin 13) is held HIGH. The READ/WRITE control directs the flow of data through the DATA BUS ports (pins 27, 28, 1, 2, 5-8) that interface the CPU and 2661. A HIGH input writes the registers, a LOW signal reads them.

Data is transferred on a LOW pulse of the CHIP ENABLE (pin 11) input. In its HIGH state, the CHIP ENABLE places the DATA BUS in a high-impedence mode. The registers are cleared when a RESET (pin 21) is asserted.

Asynchronous Transmitter Operation

In the Asynchronous Mode, transmission begins when the CLEAR TO SEND input (pin 17) is LOW, the Transmit Control bit is set, and the first full character enters the *transmit data holding register.* The contents of the holding register is then transferred to the *transmit shift register,* forcing the TRANSMITTER READY output (pin 15) LOW to signify the holding register is empty and ready to accept the next character.

The transmitter automatically sends a Start bit out the TRANSMIT DATA port (pin 19), followed by the programmed number of data bits. Each character can be programmed for a byte length between 5- and 8-bits. The transmitter then appends an optional parity bit and the programmed number of Stop bits.

If, following transmission of the data bits, no new character is available in the *transmit data holding register,* the TRANSMIT DATA output remains in the marking (HIGH) condition and the TRANSMITTER EMPTY (pin 18) output goes LOW. If desired, the transmitter output can be forced into a break condition (continuous LOW) during this phase by setting the Send Break command bit HIGH. Transmission resumes when the CPU loads a new character into the holding register.

Asynchronous Receiver Operation

The 2661 is conditioned to receive data when the DATA CARRIER DETECT input (pin 16) is LOW and the Receiver Control bit is enabled. The receiver first looks for the market to space (HIGH to LOW) transition of the Start bit on the RECEIVE DATA (pin 3) input line. When a transition is detected, the input is sampled again one-half bit later. If the RECEIVE DATA input is still LOW after this interval, a valid Start bit is assumed and the incoming bits are assembled. If not, the search continues.

The data is transferred to the *receive data holding register* and the RECEIVER READY output (pin 14) is asserted. When the character length is less than 8-bits, the high order unused bits are set to zero. The character is then checked for parity, framing, and overrun, and their corresponding status bits are strobed into the *status register* on the positive going edge of the RECEIVER CLOCK (pin 25). Shoiuld a break condition occur, where the RECEIVE DATA input remains -LOW for an entire character and Stop bit, the BREAK DETECT output (pin 25, 2661 only) is forced HIGH and the holding register is loaded with all zeros. After a break condition, the RECEIVE DATA input must return HIGH before the Start bit search can resume.

Modem Interface

The 2661 is well suited for modem applications, and has three inputs and three outputs which are used for handshaking and status indicators between the two. After the device has been initialized, the DATA TERMINAL READY output (pin 24) is asserted. When the transmitter is ready to send data, it indicates so with a REQUEST TO SEND (pin 23) output. If the modem is ready to accept data, it answers with a CLEAR TO SEND signal that enables the transmitter.

The DATA SET READY input (pin 22) is used as a Ring Indicator for the receiver section, and when it changes state, the DATA SET CHANGE output (pin 18) goes HIGH. Data reception begins when the DATA CARRIER DETECT input is activated.

Baud Rate

The 2661 contains a baud rate generator which is virtually identical to the one in the 2651. However, the 2661 chip comes in three versions—each with a different pattern of baud rates that span the range from 45.5-baud on up to 38400-baud. In the Asynchronous Mode, the baud rate selections are limited to 1X, 16X, and 64X the BAUD RATE CLOCK (pin 20).

The reader may refer to the 2651 chip for further details.

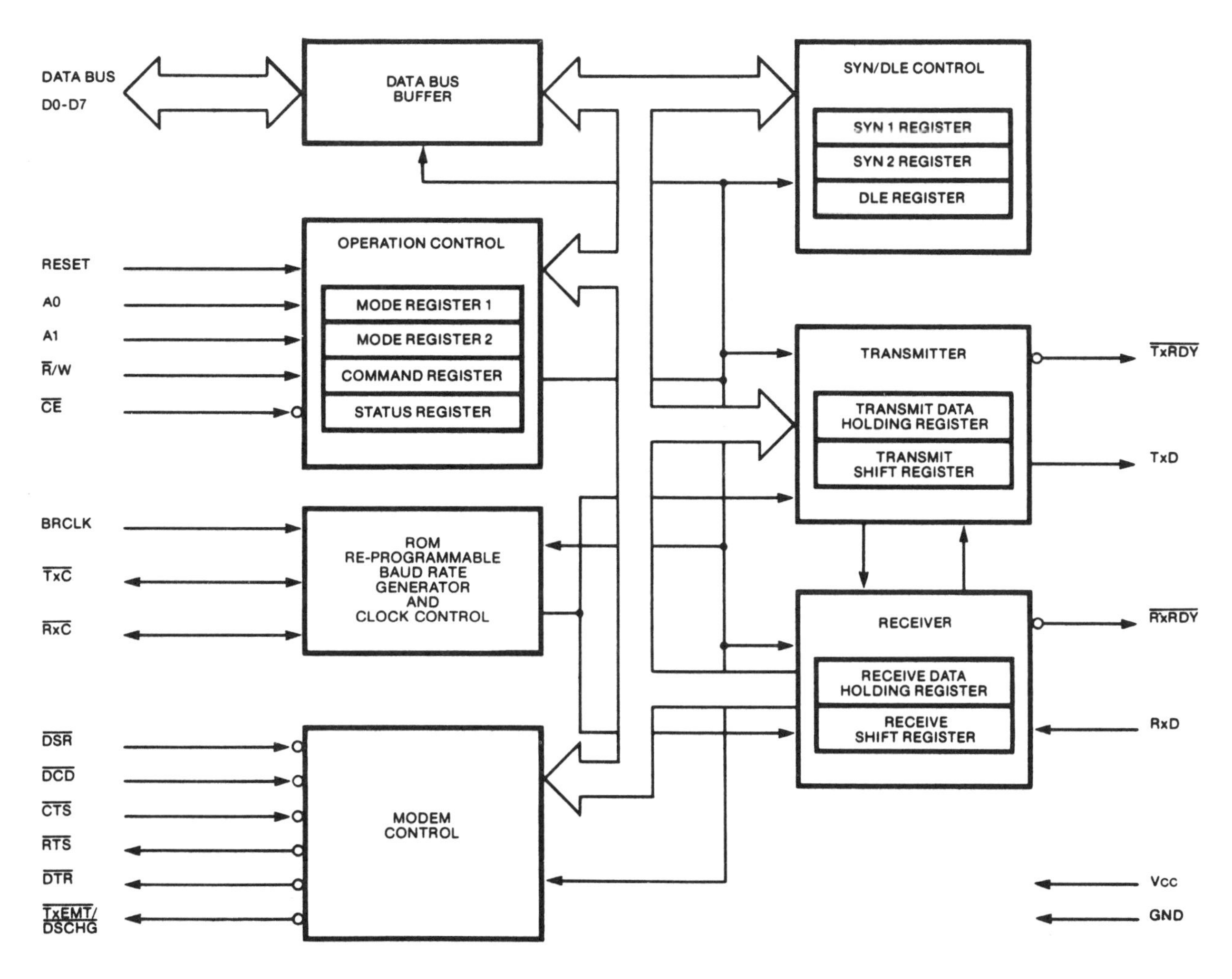

COM 2661 ORGANIZATION

Figure 1

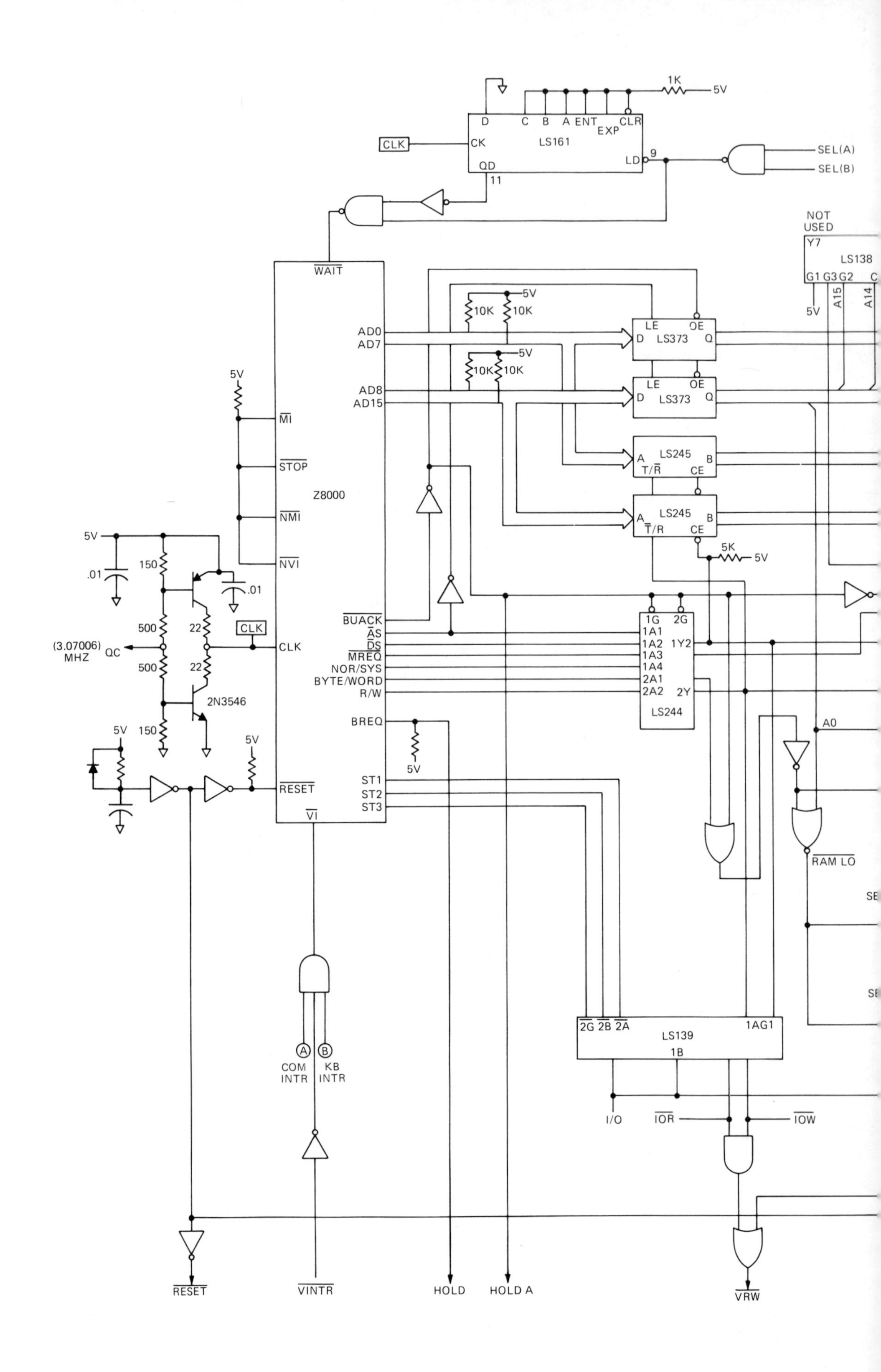

1K
5V
D
C
B
A
ENT
EXP
CLR
CLK
CK
LS161
LD
9
QD
11
SEL(A)
SEL(B)
NOT USED
Y7
LS138
G1 G3 G2
C
5V
A15
A14
WAIT
5V
10K
10K
AD0
AD7
LE
OE
D
LS373
Q
5V
10K
10K
AD8
AD15
LE
OE
D
LS373
Q
5V
MI
STOP
Z8000
NMI
NVI
A
LS245
B
T/R
CE
A
LS245
B
T/R
CE
5K
5V
5V
.01
150
.01
BUACK
AS
DS
MREQ
NOR/SYS
BYTE/WORD
R/W
1G
2G
1A1
1A2
1Y2
1A3
1A4
2A1
2A2
2Y
LS244
500
22
CLK
CLK
(3.07006) MHZ
QC
500
22
2N3546
BREQ
A0
150
5V
5V
5V
RESET
ST1
ST2
ST3
VI
RAM LO
COM INTR
KB INTR
2G
2B
2A
1AG1
LS139
1B
I/O
IOR
IOW
RESET
VINTR
HOLD
HOLD A
VRW

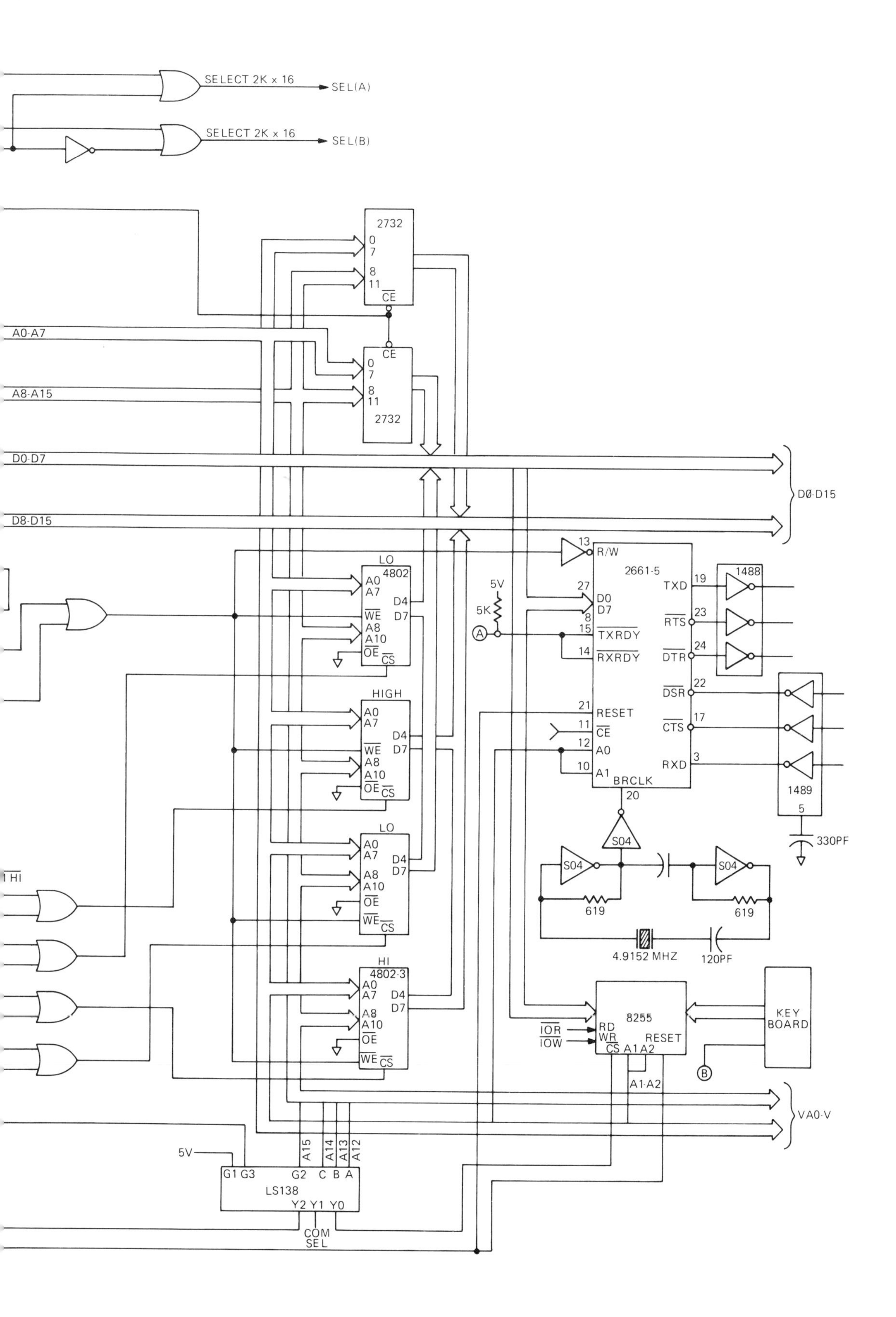

SELECT 2K x 16
SEL(A)
SELECT 2K x 16
SEL(B)
2732
CE
A0-A7
A8-A15
2732
D0-D7
D8-D15
DØ-D15
LO
4802
HIGH
LO
HI
4802-3
2661-5
R/W
TXD
RTS
DTR
DSR
CTS
RXD
RESET
TXRDY
RXRDY
BRCLK
1488
1489
5V
5K
330PF
S04
619
4.9152 MHZ
120PF
8255
KEY BOARD
IOR
IOW
A1-A2
VA0-V
LS138
COM SEL

Z8030

Serial Communications Controller

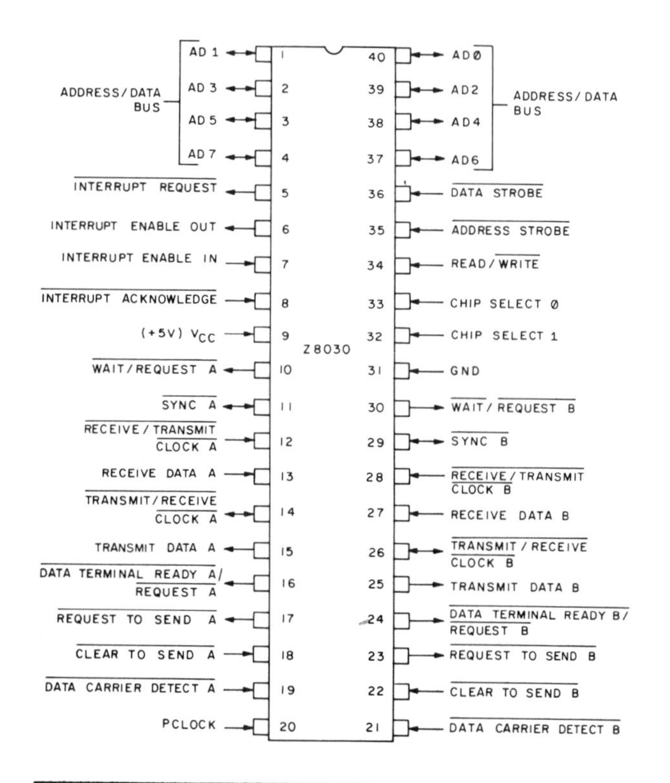

The Z8030, by Zilog, is a dual-channel, multi-protocol Serial Communications Controller designed for use with the Zilog Z-Bus. This versatile chip supports virtually all serial communications protocols in Synchronous and Asynchronous modes, including BiSync. The chip also contains a variety of sophisticated internal functions, such as an internal PLL data separator, that dramatically reduces the need for external circuitry.

Z-Bus Interface

Specifically designed for operation with Zilog's Z-Bus, the Z8030 uses an eight-bit multiplexed ADDRESS/DATA BUS (pins 40-37, 1-4) CPU interface. Both data and register addresses pass through this bidirectional interface.

The controller recognizes the ADDRESS/DATA BUS word as a register address when the ADDRESS STROBE (pin 35) is pulsed LOW. After the address has been latched into the chip, the register is accessed by pulling the DATA STROBE input (pin 36) LOW. If the ADDRESS STROBE and DATA STROBE coincide, it is interpreted as a reset.

The direction of the data flow is determined by the level present on the READ/WRITE (pin 34) input. In its LOW state, a WRITE operation is performed; a READ cycle requires a HIGH signal to this pin. Two inputs are used to enable the device: CHIP SELECT 0 (pin 33, LOW) and CHIP SELECT 1 (pin 32, HIGH).

Programming

The Z8030 contains 13 write registers for each of the two communications channels, which are separately programmed to configure the functional personality of that channel. The CPU first issues a series of commands to initialize the basic network mode. This is followed by a set of commands that qualify conditions within the selected mode.

For example, the Asynchronous Mode bit, character length, clock rate, number of stop bits, and parity might constitute the first order of commands. This could be followed by interrupt definition and allocation, and finally, the receiver or transmitter enable command.

Asynchronous Communications

Asynchronous communications can be established independently for each channel. Character length (5 to 8 bits), parity, CRC verification, and start/stop bits can be configured to reflect virtually any existing serial asynchronous protocol. The Z8030 also operates in an Auto Mode, relieving the CPU of the burden, and the following discussion will assume the device to be programmed for the Auto Mode.

The transmitter section receives parallel data characters through the ADDRESS/DATA BUS, arranges them according to protocol, and serially shifts them out the TRANSMIT DATA (pin 15 for Channel A; 25 for Channel B) port. In the Auto Mode, pulling the CLEAR TO SEND input (pin 18, A; 22, B) LOW enables the transmitter. After the transmitter has sent its message, and the buffers are clear of data, the REQUEST TO SEND output (pin 17, A; 23, B) goes LOW. This output can be used for a number of functions, including a CPU interrupt.

The receiver accepts a serial stream of data via the RECEIVE DATA (pin 13, A; 27, B) input when enabled by a DATA CARRIER DETECT (pin 19, A; 21, B) signal. The DATA CARRIER DETECT input often monitors the presence of a carrier frequency when the controller is operating with a modem, terminating communications in its absence. The SYNC pin (pin 11, A; 29, B) can also be programmed to respond to CLEAR TO SEND and DATA CARRIER DETECT signals when in the Auto Mode

When the Auto Mode is not specified, the above operations are controlled by software, and the control lines (CTS, RTS, & DCD) become bit-programmable general-purpose I/O ports.

Baud Rate Generator

The Z8030 has been designed to operate with either an internal or external baud rate clock. For each channel, the TRANSMIT/RECEIVE CLOCK (pin 14, A; 26, B) pin feeds the external baud frequency to the transmitter. Likewise, the RECEIVE/TRANSMIT CLOCK (pin 12, A; 28, B) pins supply the baud rate to the receiver. In the Asynchronous Mode, the external receiver clock may be in multiples of 1X, 16X, 32X, or 64X the actual baud rate.

Although these pins have been given transmitter or receiver designations, they are fully interchangeable. In other words, pin 14 can be exchanged for pin 12 by simply stating so in software. The only restriction is that Channel A pins are applicable only to Channel A, and vice versa.

Each channel also contains a programmable internal baud rate generator. Each generator consists of two 8-bit *time constant registers* that form a 16-bit time constant from which the baud rate is derived. This time constant may be changed at any time, but the new value does not take effect until the next loading of the counter. The output of the internal generator may be connected to the transmitter or receiver, or both, and can be externally output, but only through its respective TRANSMIT/RECEIVE CLOCK pin.

Refer to the Z8530 section for further discussion.

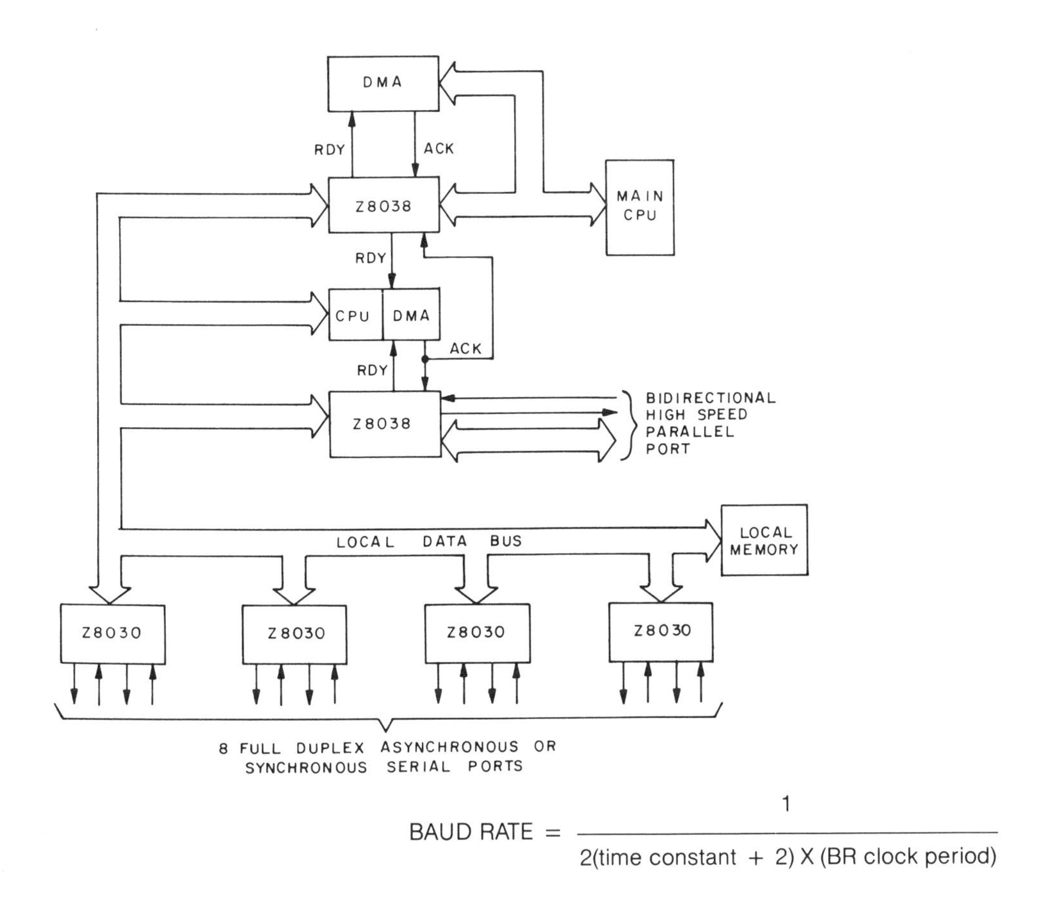

$$\text{BAUD RATE} = \frac{1}{2(\text{time constant} + 2) \times (\text{BR clock period})}$$

Figure 1

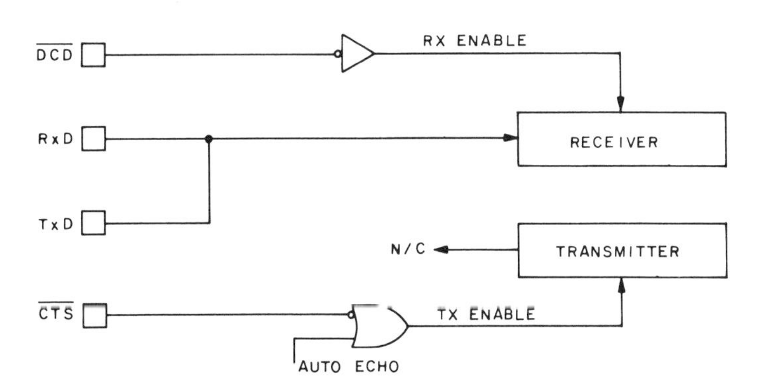

AUTO ECHO MODE

Figure 2

Z8530

Universal Serial Communications Controller

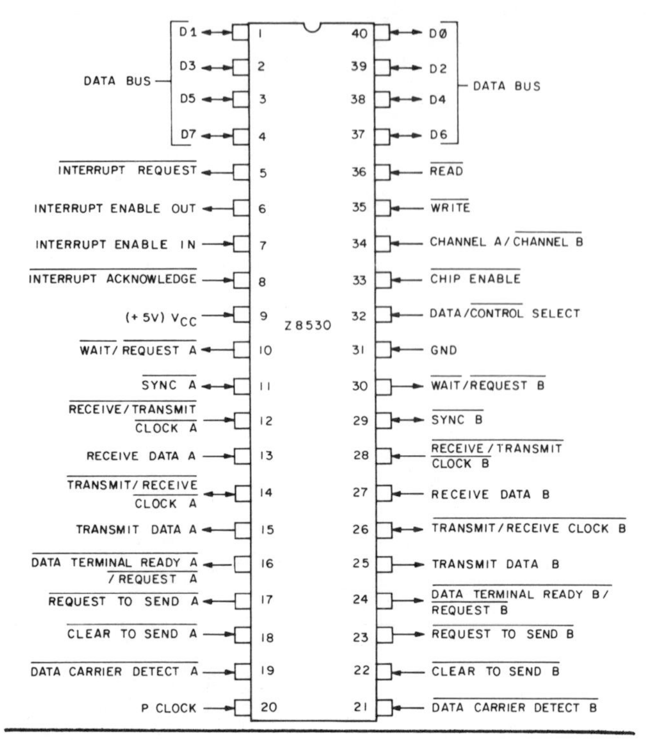

The Z8530, by Zilog, is a Serial Communications Controller which contains two independent full-duplex communications channels, each with its own baud rate generator. Basically, the Z8530 is identical in operation to the Z8030. However, the CPU interface has been reconfigured for the more universal non-multiplexed data bus, making the controller compatible with most existing microprocessors.

CPU Interface

The Z8530 interfaces to the processor through eight bidirectional DATA BUS (pins 40-37, 1-4) lines. This three-state bus is enabled when the CHIP SELECT input (pin 33) is LOW.

The DATA BUS timing signals are controlled by the READ (pin 36) and WRITE (pin 35) inputs. As their nomenclature implies, they direct the flow of traffic to and from the chip. Strobing both these inputs simultaneously resets the device.

Programming

The 8530 contains 13 write registers in each channel that are programmed by the CPU to configure the operational mode of that channel. Unlike the Z8030, however, register addressing is direct for the *data registers* only. In all other cases, two operations must be performed to access a register.

Programming begins by engaging the desired communications channel, using the CHANNEL A/CHANNEL B select (pin 34) input, and placing the device in the Control Mode by setting the DATA/CONTROL SELECT input (pin 32) LOW.

Once the chip is locked into the Control Mode, a WRITE operation will latch the address of the register to be programmed into the three LSB positions of the *write register 0.* A second WRITE command loads the specified register with the Control Word on the DATA BUS. If the CPU wishes to read a register, instead, the address cycle is followed by a READ operation, whereupon the register's contents is placed on the DATA BUS for examination. Following the second operation (READ or WRITE), the address bits in the *write register 0* are automatically cleared in preparation for the next address cycle. Notice that the two operations must be performed in sequence at all times in order to maintain correct programming order.

Synchronous Communications

The Z8530 supports both byte-oriented (MonoSync, BiSync, etc.) and bit-oriented (SDLC, HDLC, etc.) synchronous communications. Automatic bit insertion, deletion, and framing is performed in accordance with the specified protocol. Virtually any format or operational mode is possible, including Local Loop and Auto Echo. Two standard CRC equations are available for error detection and/or correction.

In the byte-oriented mode, the device can operate with either internal or external sync. When the receiver recognizes a Sync character while in the internal sync mode, it activates the SYNC (pin 11, A; 29, B) output. During external sync operations, however, the SYNC pin becomes an input that must be driven LOW two PCLOCK (pin 20) periods after the last bit in the Sync character is received.

In the bit-oriented mode, the SYNC pins act as outputs, and are valid upon receipt of a Flag. For all modes, the Z8530 generates an interrupt after an incoming character has been assembled. Several interrupt options are open to the user.

Interrupts

There are three types of interrupts available under program control: Transmit, Receive, and External/Status. The Transmit and Receive interrupts, obviously, deal with the smooth flow of data through the communications device, as shown above. All three types of interrupts can activate the INTERRUPT REQUEST (pin 5) output. An INTERRUPT ACKNOWLEDGE (pin 8) input places the interrupt vector on the DATA BUS.

The interrupts may also be programmed to generate DMA block transfers. To a DMA controller, the REQUEST output (pin 10, A; 30, B) indicates that the Z8530 is ready to transfer data to or from memory. The DATA TERMINAL READY/REQUEST line (pins 16, A; 24, B) is interchangeable with the REQUEST output, and allows full-duplex operation under DMA control.

Serial Data Formats

The Z8530 may be programmed to encode and decode the serial data in four different ways, making it applicable to cassette, diskette, tape drive, and modem interfacings. The four formats are: NRZ, NRZI, FM (bi-phase mark) and Bi-Phase Space. In addition, the receiver can decode Manchester data by using an internal digital phase-locked loop.

The PLL oscillator is crystal controlled. When programmed in the crystal oscillator mode, a high-gain amplifier is connected between the RECEIVE/TRANSMIT CLOCK (pin 12, A; 28, B) and SYNC (pin 11, A; 29, B) pins, across which a crystal—sixteen times the baud frequency—is placed. The output of this oscillator is available at the TRANSMIT/RECEIVE CLOCK (pin 14, A; 26, B) pin.

Zilog has recently introduced a Z8531 version of this chip, which has all its attributes but operates in the Asynchronous Mode only. Refer to the Z8030 section for more details.

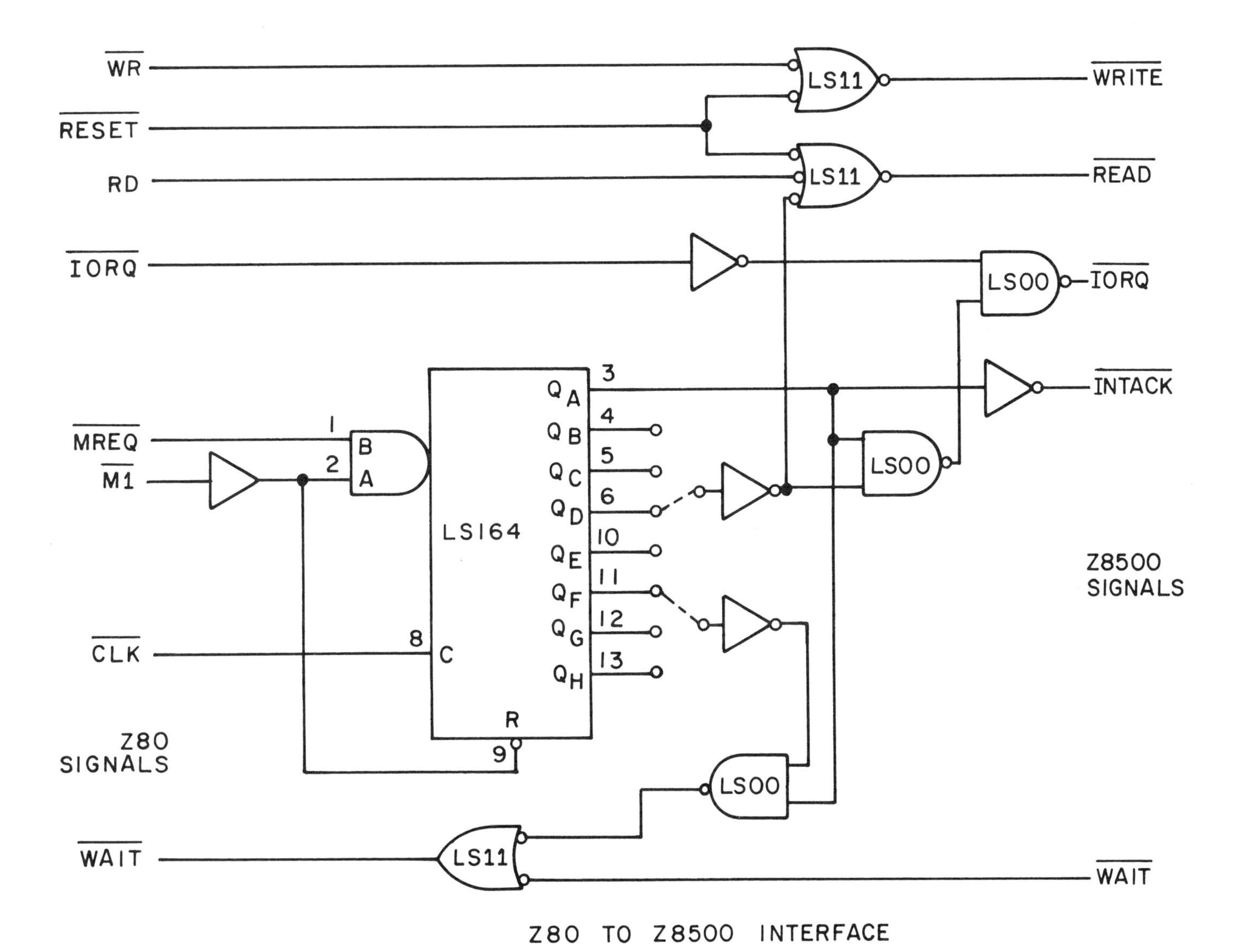

Figure 1

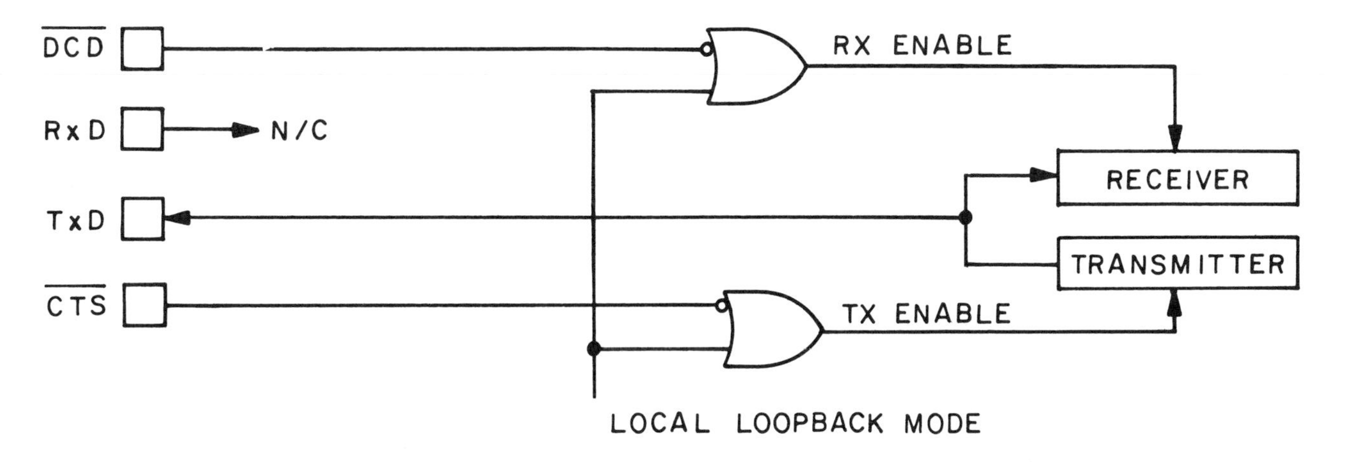

Figure 2

8274

Multi-Protocol Serial Controller

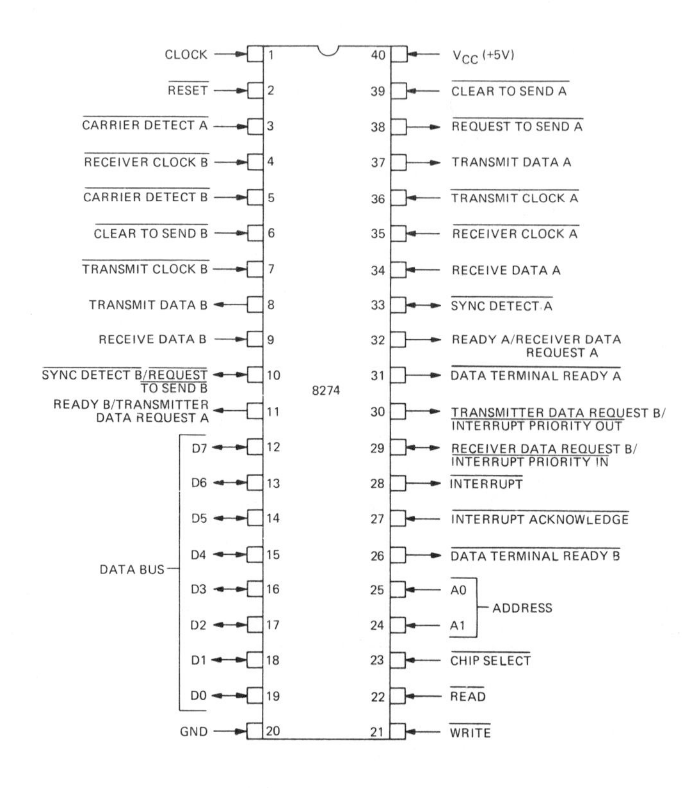

The 8274 Multi-Protocol Serial Controller, by Intel, is a microprocessor communications device which supports Asynchronous, Byte Synchronous, and Bit Synchronous protocols. It supports UART capabilities plus MonoSync, BiSync, HDLC, and SDLC standards. The 8274 also endorses several CPU interface options: Polled, Wait, Interrupt, and DMA-driven.

CPU Interface

The 8274 contains two separate communications channels, each with a complete transmitter and receiver section which can communicate independently in full duplex. The selection of either channel is governed by the A0 ADDRESS (pin 25). When this input is LOW, channel A is interfaced to the CPU; when HIGH, channel B is actively engaged.

The chip interfaces to the CPU through a bidirectional DATA BUS (pins 19-12). After the channel is selected, data is written by pulling the WRITE line (pin 21) LOW and strobing the CHIP SELECT (pin 23) input. The A1 ADDRESS (pin 24) selects between data or command information transfer, and accesses the program registers when it is HIGH.

Asynchronous Operation

In the Asynchronous Mode, the controller is basically a UART, which uses start and stop bits to frame the message, and is well suited for modem applications. Although the following description pertains to both communications channels, the pinouts will be given for channel A only.

For asynchronous operation, the registers must be initialized with the desired protocol. Once initialized, the DATA TERMINAL READY output (pin 31) goes HIGH. The transmit function begins when the Transmit Enable bit is set and the first data byte is entered, thus forcing the REQUEST TO SEND output (pin 38) LOW. The modem signals when it is ready to accept the data byte by pulling the CLEAR TO SEND (pin 39) input LOW, an operation that automatically adds the Start bit and serially shifts the data out the TRANSMIT DATA (pin 37) port on the falling edge of the TRANSMIT CLOCK (pin 36). The baud rate is programmable to 1, 1/8th, 1/32nd, and 1/64th the TRANSMIT CLOCK frequency.

The receiver function begins when the Receive Enable bit is set. If the Auto Mode option is selected, the receiver will hold off operations until it recognizes a LOW input to its CARRIER DETECT (pin 3) input. The data is received at the RECEIVE DATA input (pin 34), and is sampled at mid-bit time on the rising edge of the RECEIVER CLOCK (pin 35). The receiver first searches for a valid Start bit, and assembles the character following its detection. The CPU then reads the character by enabling the READ (pin 22) control, while strobing the CHIP SELECT. Data transfers between the 8274 and the CPU are synchronized by the READY (pin 32) line. If the CPU fails to read a data character after more than three characters have accumulated, the Receive Overrun bit is set and the third character is discarded and replaced by the fourth.

Synchronous Operation

There are two synchronous communications modes and four protocols available with the 8274: Byte Control (MonoSync and BiSync) and Bit Oriented (HDLC and SDLC). Both channels are basically identical, but we will limit the following discussion to channel B only.

With the controller in the Byte Mode, the transmitter initiates communications with a Sync character; in the Bit Mode, a Flag sequence is generated instead. Data is loaded into the *transmit buffer*, transferred to the *transmit shift register*, and serially output through the TRANSMIT DATA (pin 8) pin at the rate set by the TRANSMIT CLOCK (pin 7).

An interrupt is generated each time the *transmit buffer* becomes empty. The interrupt is subsequently satisfied by writing another character into the transmitter. DMA transfers are performed using the TRANSMITTER DATA REQUEST (pin 30) interrupt. If the data character is not loaded by the time the *transmit shift register* is emptied, an underrun condition exists. The 8274 has two programmable options for solving underrun: it can insert Sync characters, or it can send the CRC characters generated so far (Frame Check Sequence).

After the receiver is initialized, it hunts for a Sync character or Flag sequence, whichever the protocol specifies. The recognition of either will activate the SYNC DETECT output (pin 10) and enable the receiver. Data may be transferred with or without interrupts. When a DMA request for data transfer is made, the RECEIVER DATA REQUEST output (pin 29) is taken HIGH. Channel A differs from B in the respect that a transmitter or receiver DMA transfer can occur in both the Byte and Bit modes, whereas in channel B it can only be performed in the Bit Mode.

The assembly of RECEIVE DATA (pin 9) continues until the 8274 is RESET (pin 2) or, in the case of the Auto Mode, the CARRIER DETECT (pin 5) signal is lost. Unlike the Asynchronous Mode, however, the baud rate is not internally programmable, and is set by the RECEIVER CLOCK (pin 4) frequency.

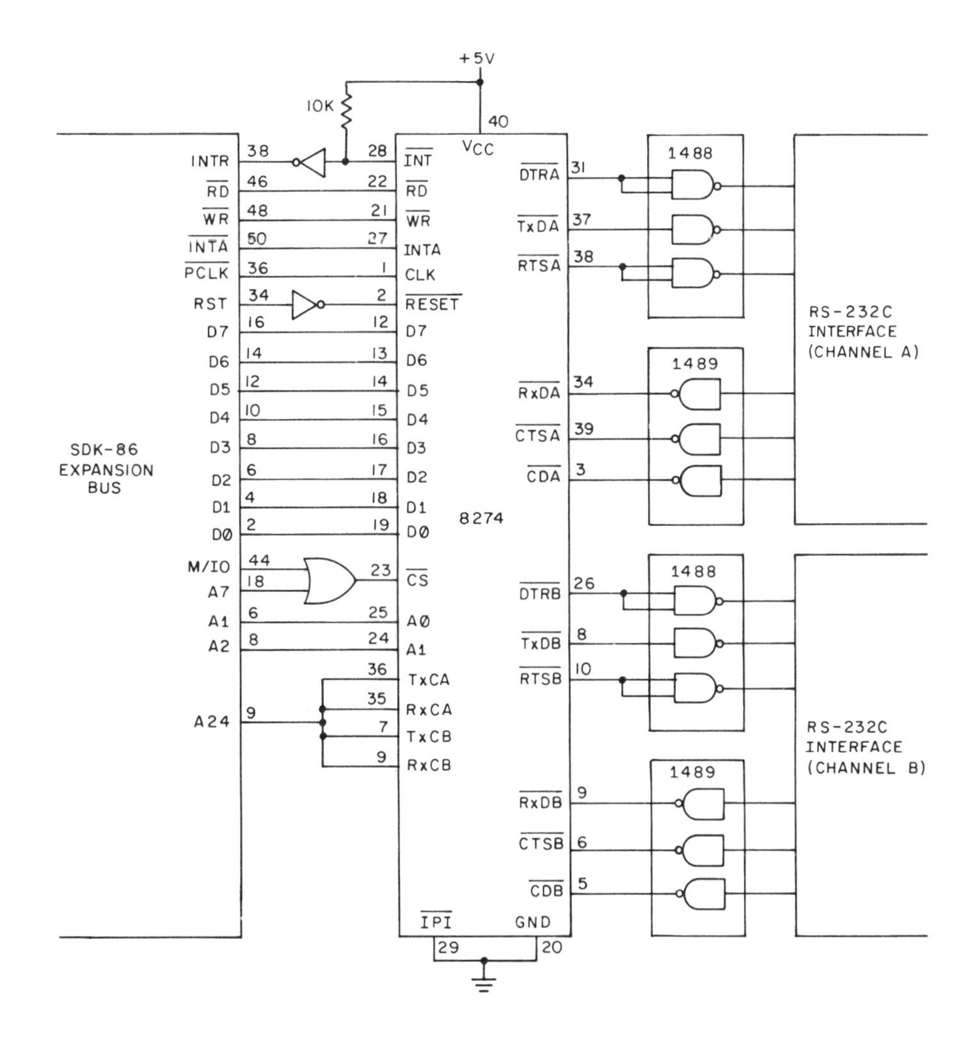

Figure 1

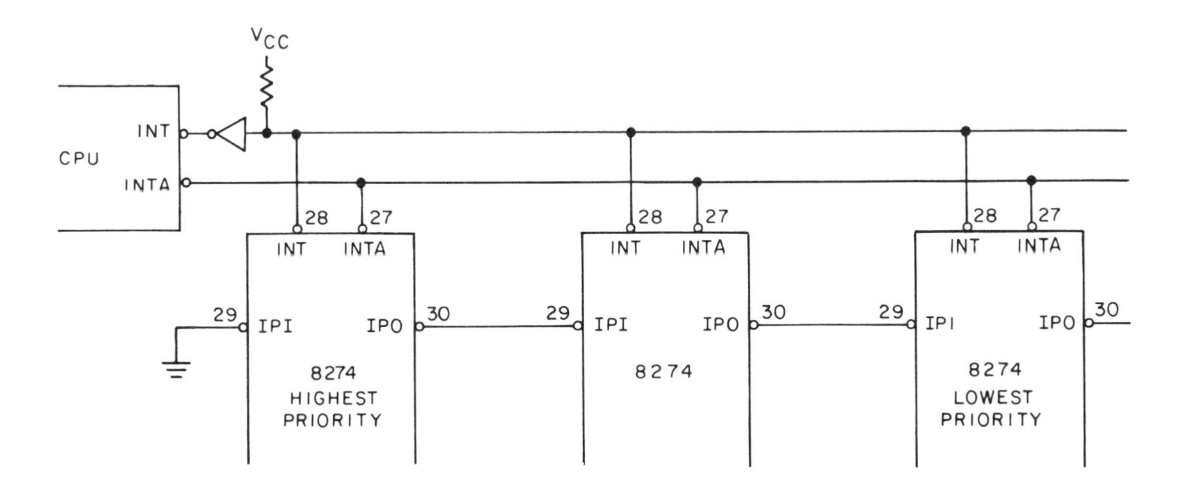

Figure 2

SC2652

Multi-Protocol Communications Controller

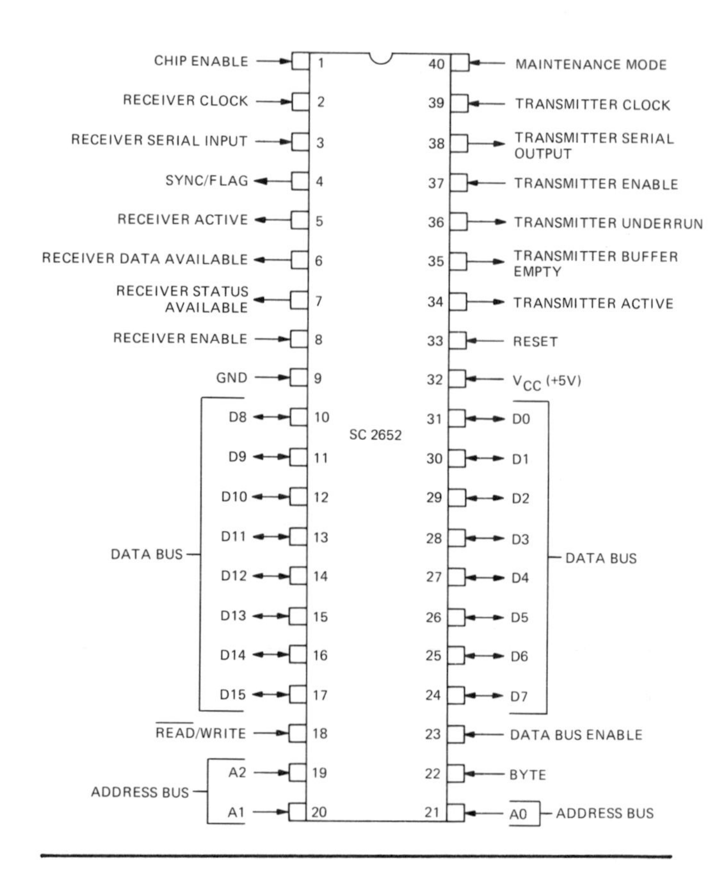

The SC2652 Multi-Protocol Communications Controller, by Signetics, is an LSI communications circuit that formats, transmits, and receives synchronous serial data while supporting bit-oriented and byte-control protocols. These protocols include: SDLC, ADCCP, and HDLC bit-oriented along with DDCMP and BiSync byte-control standards.

Data framing, word length, and protocol selection are software functions, and are programmable through either 8-bit or 16-bit wide data bus interfacings.

Data Bus

The DATA BUS consists of two 8-bit bidirectional data ports, D0 to D7 (pins 31-24) and D8 to D15 (pins 10-17). When interfacing to a 16-bit bus, each DATA BUS port connects to its respective data line and the BYTE input (pin 22) is asserted LOW. For 8-bit DATA BUS configurations, however, DATA BUS lines D0 through D7 must be WIRE ORed to their corresponding HIGH and LOW order bits on the D8 to D15. That is, D0 parallels with D8, D1 with D9, etc. The BYTE input is now shifted HIGH.

Programming

Prior to initiating data transmission or reception, the *parameter control register* and *parameter control sync/address register* must be loaded with control information from the processor. The CPU selects the register by configuring the register ADDRESS BUS (pins 21-19). The CHIP ENABLE (pin 1) and READ/WRITE control (pin 18) must be established before a DATA BUS transfer can take place.

During a write operation, where the READ/WRITE input is HIGH, the DATA BUS will place its contents into the addressed register on the leading edge of a DATA BUS ENABLE (pin 23) pulse. A read operation requires the READ/WRITE input to be forced LOW; upon receipt of a DATA BUS ENABLE strobe, the addressed register will place its contents on the DATA BUS.

The organization of the ADDRESS BUS depends upon the DATA BUS interfacing used. If the DATA BUS is operated in the 8-bit mode, all three ADDRESSes are required to select the register. Using 16-bit bytes, however, only requires two ADDRESS lines, A1 (pin 20) and A2 (pin 19); the status of A0 (pin 21) is immaterial.

Receiver Operation

After initializing the parametric control registers, the RECEIVER ENABLE (pin 8) input is set HIGH to enable the receiver. Received serial data is input to RECEIVER SERIAL INPUT (pin 3), synchronized, and shifted into an 8-bit *control character shift register* on the rising edge of the RECEIVER CLOCK (pin 2). The RECEIVER CLOCK input must be at the same baud rate as the incoming data, and is normally supplied by the remote transmitter.

The contents of the *control character shift register* is examined bit-by-bit until a Flag (bit-oriented protocol) or Sync (byte-control protocol) sequence is found. Once a match is confirmed, the SYNC/FLAG (pin 4) output is driven HIGH for one RECEIVER CLOCK cycle.

The detection of a Flag in the Bit-Oriented Mode indicates a data frame is upcoming. The characters are assembled and transferred to the *receiver data/status register* for presentation to the CPU. At that time, the RECEIVER DATA AVAILABLE output (pin 6) is forced HIGH, compelling the CPU to read the character within one RECEIVER CLOCK cycle, otherwise an Overrun bit will assert itself and the succeeding characters will be lost.

The first character following the Flag is the Secondary Station Address. If the controller is configured as a secondary station, this address is internally compared to the station's address. If they coincide, then the data is intended for that station and the RECEIVER ACTIVE (pin 5) interrupt is taken HIGH. No match indicates that another station is being addressed, and the receiver searches for the next Flag. If the device is programmed as a primary station, no address check is made, and the RECEIVER ACTIVE output is activated immediately following a Flag detection.

Data that follows is assembled into specific character lengths and read by the CPU. As before, the RECEIVER DATA AVAILABLE is asserted each time a character is transferred into the *receiver data/status register* and is cleared when the register is read.

In the Byte-Control Mode, the receiver initially searches for two successive Sync characters that match the protocol specified by the receiver. The next received character sets the RECEIVER ACTIVE output and enables the receiver data path. Should the character length exceed 8-bits, the RECEIVER STATUS AVAILABLE (pin 7) goes HIGH, informing the CPU that both halves of the *receiver data/status register* need servicing. This output is also used to indicate an error in the event that an error control has been ordered.

The 2652 has been designed to the same specifications as the COM 5025, and although they are not totally identical, they are hardware (and basically software) compatible. Additional information concerning their operations is presented in that section.

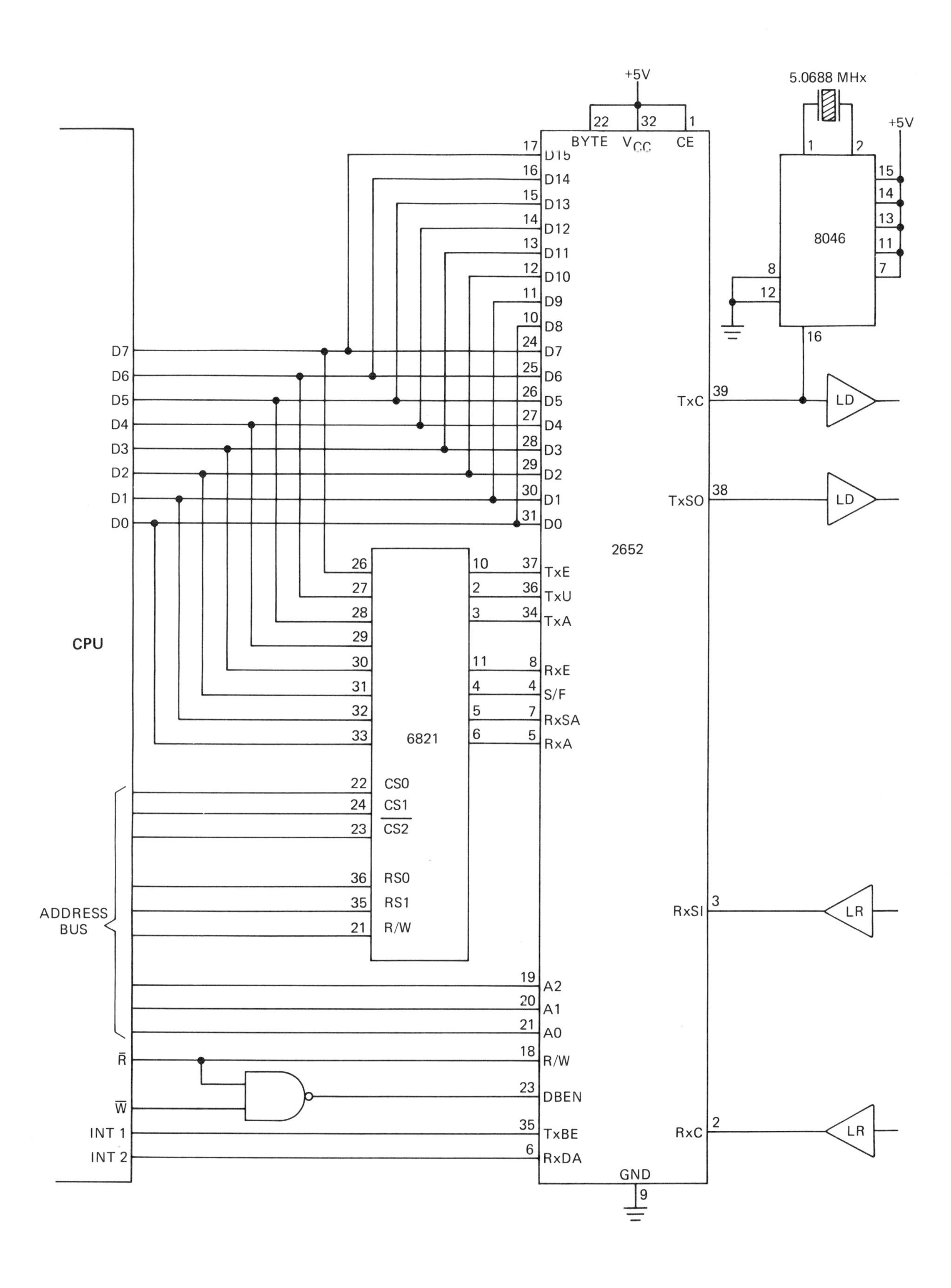

+5V
5.0688 MHz
+5V
BYTE
VCC
CE
8046
2652
6821
CPU
ADDRESS BUS
D15
D14
D13
D12
D11
D10
D9
D8
D7
D6
D5
D4
D3
D2
D1
D0
TxE
TxU
TxA
RxE
S/F
RxSA
RxA
CS0
CS1
CS2
RS0
RS1
R/W
A2
A1
A0
R/W
DBEN
TxBE
RxDA
GND
TxC
TxSO
RxSI
RxC
LD
LD
LR
LR
R̄
W̄
INT 1
INT 2

COM 5025

Multi-Protocol Universal Synchronous Receiver/Transmitter

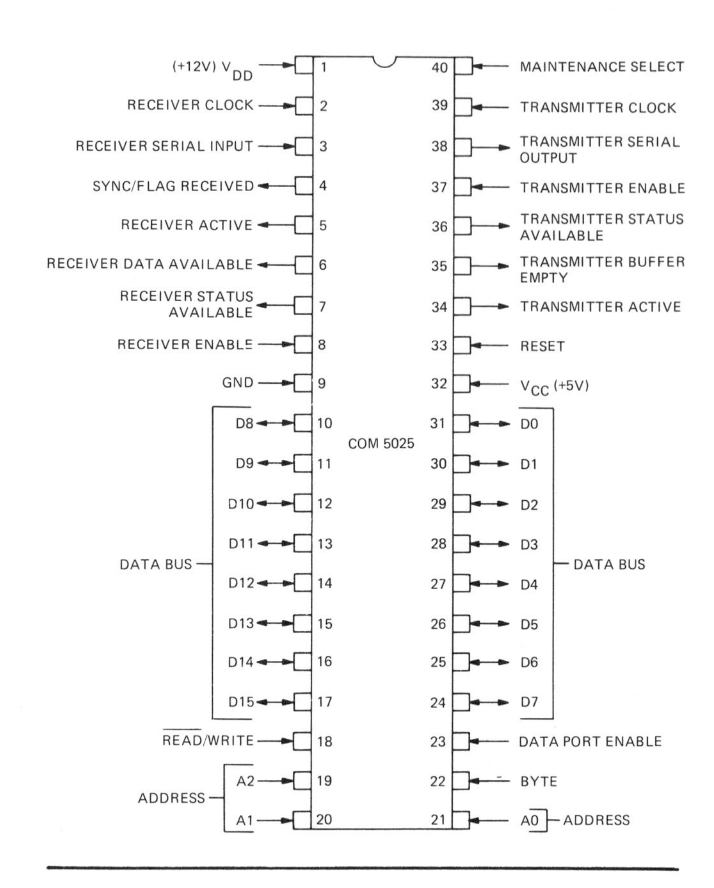

The Multi-Protocol COM 5025, by Standard Microsystems Corp., is a Synchronous Receiver/Transmitter controller that meets synchronous communications requirements according to Bi-Sync, SDLC, HDLC, ACDDP, and DDCMP protocols. The the 5025 can be configured to interface to an 8- or 16-bit data bus, and is capable of operating at 1.5-MHz.

CPU Interface

The DATA BUS consists of two 8-bit bidirectional data ports, D0 to D7 (pins 31-24) and D8 to D15 (pins 10-17). The DATA BUS may be configured for either 8-bit or 16-bit interface, depending upon the system requirements, using the BYTE (pin 22) control. When interfacing to a 16-bit bus, each data port connects to its respective data line and the BYTE input is pulled LOW. For 8-bit DATA BUS configurations, however, the lower order lines must be WIRE ORed to the higher order lines (D0 is ORed to D8, etc.) and the BYTE control is set HIGH.

Programming

Prior to initiating data transmission or reception, the *data length select register* and *mode control register* must be programmed with control information. The CPU selects the registers by addressing the register ADDRESS (pins 21-19) inputs. The organization of these ADDRESSes will depend upon the configuration of the DATA BUS.

Altogether, there are four registers in the 5025 controller: three are read/write and one is read only. Each register is 16-bits in length, and is programmed with instructions on the DATA BUS. When the DATA BUS is in the 16-bit mode, each line represents one bit of the register and only two ADDRESS inputs (A1 & A2) are needed to locate any register.

When operating in the 8-bit mode, however, the controller must use a double fetch operation to program a register. The significance of the input data, whether it be HIGH or LOW ordered, is determined by the A0 (pin 21) input. A HIGH input to this pin activates DATA BUS lines D8 through D15, while a LOW signal asserts the D0 to D7 ports.

The direction of the data flow through the DATA BUS is controlled by the READ/WRITE (pin 18) input. A write operation requires a HIGH input to this pin; a read operation is performed when it is driven LOW. Data transfer is initiated on the leading edge of the DATA PORT ENABLE (pin 23) strobe.

Transmitter Operation

Once the *data length select register* and *mode control register* have been initialized, a Transmit Start of Message bit is set and the TRANSMITTER ENABLE (pin 37) control is raised HIGH. Transmitter operations then begin according to the protocol specified in software.

In the Bit-Oriented Mode, a Flag is sent immediately following the activation of the transmitter, as indicated by the TRANSMITTER ACTIVE (pin 34) output. The Flag is used to synchronize and frame the message that follows. When the Flag sequence begins, the 5025 drives the TRANSMITTER BUFFER EMPTY (pin 35) output HIGH, after which the CPU loads the first character of the frame into the controller. Flag generation will continue as long as the Transmit Start of Message bit remains set, and this bit should be cleared at the same time the first character is loaded, or immediately thereafter.

All consecutive characters are entered whenever the TRANSMITTER BUFFER EMPTY output goes HIGH. If an underrun occurs, that is to say the CPU is not keeping up with the transmitter, the TRANSMITTER STATUS AVAILABLE output is driven HIGH and an error flag is set. To retransmit the message once this condition has occurred, the CPU must perform another start of message sequence.

Data is serially shifted out the TRANSMITTER SERIAL OUTPUT (pin 38) at the TRANSMITTER CLOCK (pin 39) baud rate. As each character is transmitted, a Frame Check Sequence is generated as specified by the Error Control Mode.

When operated in the Byte-Control Mode, a string of Sync characters is transmitted when the Transmit Start of Message bit and the TRANSMITTER ENABLE are set HIGH. The TRANSMITTER ACTIVE output is driven HIGH, and the TRANSMITTER BUFFER EMPTY output signals a ready condition to the CPU. The register is now loaded with a message byte at every TRANSMITTER BUFFER EMPTY interrupt.

Should the CPU create an underrun condition, the TRANSMITTER STATUS AVAILABLE output is forced HIGH and a Sync sequence is issued. The CPU must then reiterate the Transmit Start of Message command and retransmit the message to recover. However, this procedure does not comply with IBM's Bi-Sync protocol, and the user must not underrun when supporting this format.

The 5025 has been designed to the same specifications as the SC2652, and although they are not totally identical, they are hardware (and basically software) compatible. Additional information concerning their operations is presented in that section.

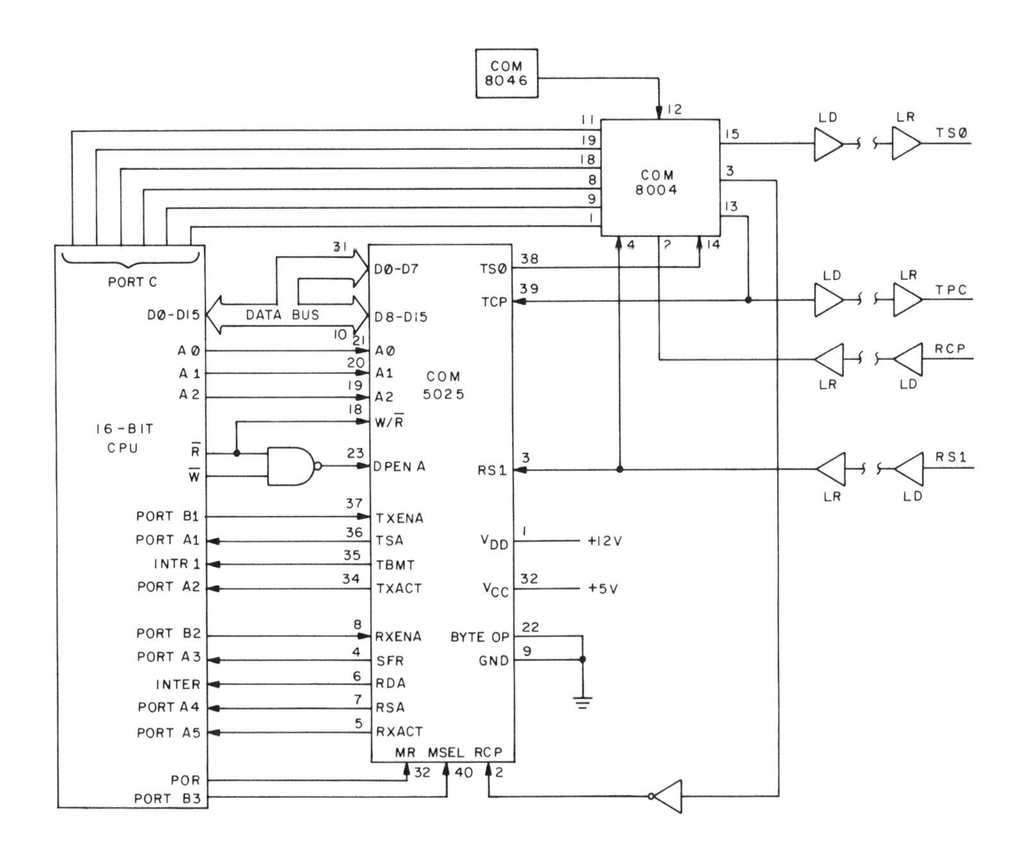
COM 8046
COM 8004
COM 5025
16-BIT CPU
PORT C
DØ-D15
DATA BUS
DØ-D7
D8-D15
AØ
A1
A2
W/R̄
R̄
W̄
DPEN A
PORT B1
PORT A1
INTR 1
PORT A2
PORT B2
PORT A3
INTER
PORT A4
PORT A5
POR
PORT B3
TXENA
TSA
TBMT
TXACT
RXENA
SFR
RDA
RSA
RXACT
MR MSEL RCP
TSØ
TCP
RS1
VDD
VCC
BYTE OP
GND
+12V
+5V
LD
LR
TSØ
TPC
RCP
RS1

Figure 1

WD2840

Local Network Token Access Controller

The WD2840 is a Western Digital device intended for Local Network applications using Token Access protocol. The WD2840 autonomously performs all local network communications tasks, including network initialization, addressing, data transmission, acknowledgements, and diagnosis. In other words, the user sees the data link as transparent, and needs to be concerned with only the input and output of data.

Token Access Protocol

Efficient data processing often requires the interconnection of one or more devices. Utilizing a shared common bus, the Token Access protocol approaches the problem in a unique way.

First, every station sharing the bus is assigned an ID number. This can be any number between 1 and 254, but no two should be the same. Next, the first station is given a *token* which allows that station access to the communications bus. The first station is then free to send its message.

After the message is sent, the token is passed to the next station. The address of the token-receiving node is contained within the passing node, and the address does not have to be contiguous. For example, if station 19 has the token, it may pass it along to station 53.

Assuming station 53 gains possession of the token, it now has the option of using the communications line. If it has no message to send, it simply forwards the token to the next station on the list. The token then proceeds around the network in circular fashion, giving everyone an equal opportunity at the medium.

The node possessing the token may take advantage of the bus, but with some reservation. When it has a message to send, it addresses the intended receiver and transmits a data frame. After the data packet has been sent, a timer is started (if specified) and the transmitter waits for a response from the receiver.

If an acknowledgement is received before the timer expires, the message is understood to have arrived intact, and the transmitter sends the next data packet. If, however, no response is made by the time-out period, a second attempt is made. Should the timer expire again, the transmitter assumes the receiver is off-line, and the frame is skipped.

The transmitter then addresses another receiver or passes the token. To prevent filibustering by one station, a time limit is placed on its activities. When its time is up, the token must be relinquished to the next station.

WD2840 Operation

The 2840 is basically similar to the WD2501 and WD2511 devices. In fact, they are very nearly hardware compatible, and one can more or less be exchanged for the other. Since the pinouts have been described in detail in those sections, the following is a brief summary of pin functions.

Host Interface

Sixteen registers control and monitor the 2840's operations. These registers are accessible through the bidirectional DATA ACCESS LINES (pins 8 to 15) by configuring the ADDRESS LINES IN (pins 44-47). The direction of data flow is determined by the WRITE ENABLE (pin 3) or READ ENABLE (pin 5) input in conjunction with the CHIP SELECT (pin 4).

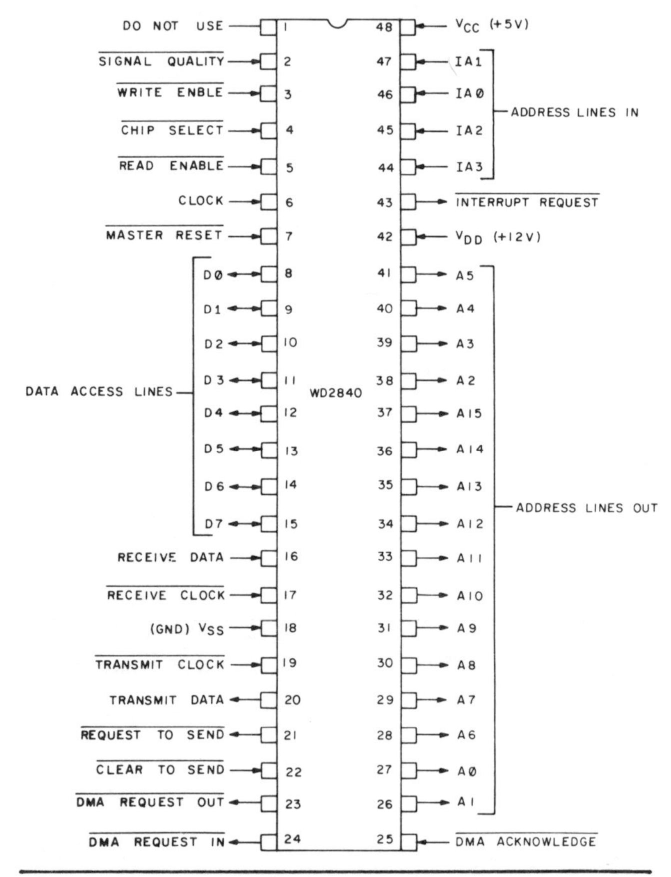

Network Interface

The 2840 is designed to logically interconnect 2 to 254 devices over a shared communications medium. Examples of expected mediums include coax cable, twisted pairs, RF, and CATV. The receiver accepts data through the RECEIVE DATA (pin 16) input; the RECEIVE CLOCK (pin 17) sets the baud rate. Since most applications will require the use of a line transceiver, like a modem, the SIGNAL QUALITY input (pin 2) disables the receiver when external monitoring circuitry decides the received signal level is below a reliable threshold.

Data is transmitted out the TRANSMIT DATA (pin 20) port at the baud hate set by the TRANSMIT CLOCK (pin 19). Data is present when the REQUEST TO SEND output (pin 21) is LOW; a CLEAR TO SEND (pin 22) response from the modem enables the transmitter.

Memory Interface

Although the memory control signals are identical to those used by the WD2501 and WD2511, the memory configuration is a little different. The 2840 uses a buffer chaining scheme to allow efficient memory utilization, while minimizing CPU involvement.

Basically, in order for the controller to meet the network's data rate and processing delay requirements, the data is stored in buffers that are directly accessed by the 2840. Up to 16 buffers may be used for either transmit or receive, or both, in a chain architecture. It is the responsibility of the controller to respond to real-time network obligations by extracting messages from the buffers, while the CPU has the less demanding task of removing used buffers from the transmit chain, or visa versa for the receiver.

The memory is addressed by the 2840 using sixteen ADDRESS LINES OUT (pins 26-41) lines. Data is transferred to memory on a DMA REQUEST OUT (pin 23) or from memory on a DMA REQUEST IN (pin 24) signal when the DMA ACKNOWLEDGE (pin 25) input is LOW.

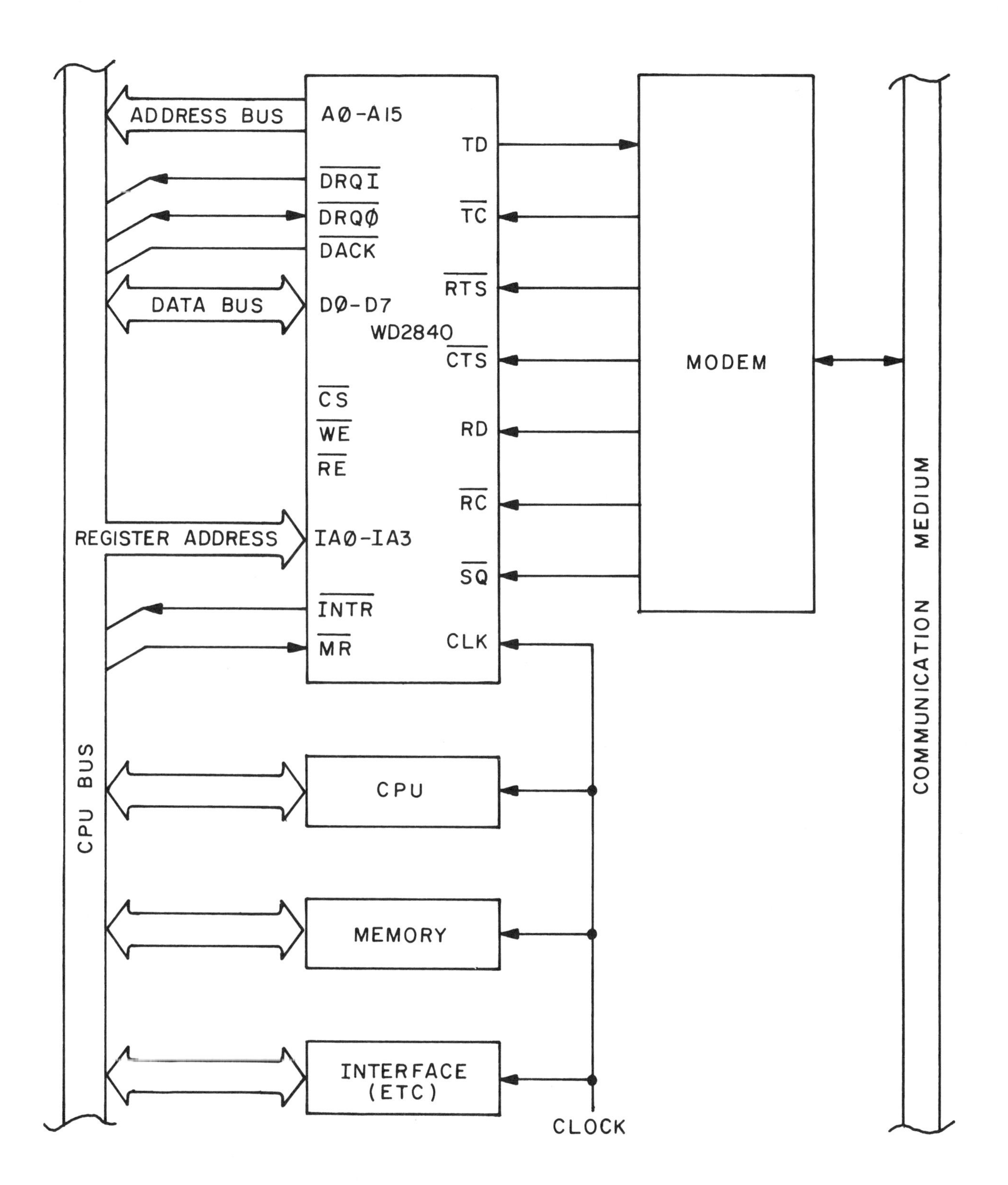
ADDRESS BUS
A0-A15
TD
DRQI
DRQ0
TC
DACK
RTS
DATA BUS
D0-D7
WD2840
CTS
MODEM
CS
WE
RD
RE
RC
REGISTER ADDRESS
IA0-IA3
SQ
INTR
MR
CLK
CPU BUS
CPU
MEMORY
INTERFACE
(ETC)
CLOCK
COMMUNICATION MEDIUM

Figure 1

COM 9026

Local Area Network Controller

The COM 9026, by Standard Microsystems Corp., is a special-purpose communications adapter for interconnecting processors and intelligent peripherals using the ARCNET local area network. ARCNET is a self-polling "modified token passing" network operating at a 2.5-Mbit rate, and is the protocol adoped by Tandy for use with their Radio Shack TRS-80 systems.

Processor Interfacing

The 9026 is a command-oriented device, with just eight commands performing all network operations. Commands are executed according to the configuration of the AD0 address (pin 28) and READ/WRITE (pin 7) inputs. A typical command cycle goes as follows.

To enter a command, the I/O REQUEST (pin 8) is pulled LOW, the READ/WRITE input set, and the command address placed on the multiplexed ADDRESS/DATA BUS (pins 28-21). Due to the critical timing requirements of bus multiplexing, though, the ADDRESS/DATA BUS must be buffered through an external input latch, as shown in fig. 1. Asserting the ADDRESS STROBE (pin 10) latches the address into the buffer and the I/O REQUEST status into the chip. The CPU then reconfigures its bus line to represent the intended command.

At this point, the network controller assumes a waiting state. Once the data bus is valid, the CPU activates the DELAYED WRITE (pin 6) input, thus initiating a series of events that places the command into the chip. First, the controller forces the ADDRESS/DATA INPUT ENABLE output (pin 14) LOW. This transfers the AD0 address from the buffer into the 9026. Next, an INTERFACE LATCH ENABLE (pin 18) command gates the CPU's data word into the 9026, completing the cycle. A total of eight CLOCK (pin 19) half cycles are required to achieve this.

Reading from the *status register* involves the same procedure, but with the READ/WRITE input held HIGH. However, a separate latching buffer is used for data outputting, making the state of the DELAYED WRITE of no consequence. The data is simply transferred to the buffer immediately following its request, and read at the CPU's convenience.

Memory Interface

The CPU does not communicate data directly with the 9026 chip. Instead, an external RAM buffer is used to temporarily hold data packets for either CPU or controller processing. As a result, the RAM buffer must behave as a dual-ported memory . . . which means interface bus arbitration and control is necessary if this RAM is to be a standard part.

For processor RAM buffer access, the MEMORY REQUEST (pin 9) line is set LOW, placing the ADDRESS/DATA BUS into a high-impedence state. While latching the RAM memory address request into the input buffers, the CPU begins its access cycle on the trailing edge of the ADDRESS STROBE. However, these cycles run completely asynchronous with respect to the 9026, and the controller immediately puts the CPU into a wait state by asserting the WAIT (pin 12) output. When the 9026 has synchronized the CPU's access cycle to its own rhythm, the WAIT signal is removed and the processor is allowed to complete its cycle.

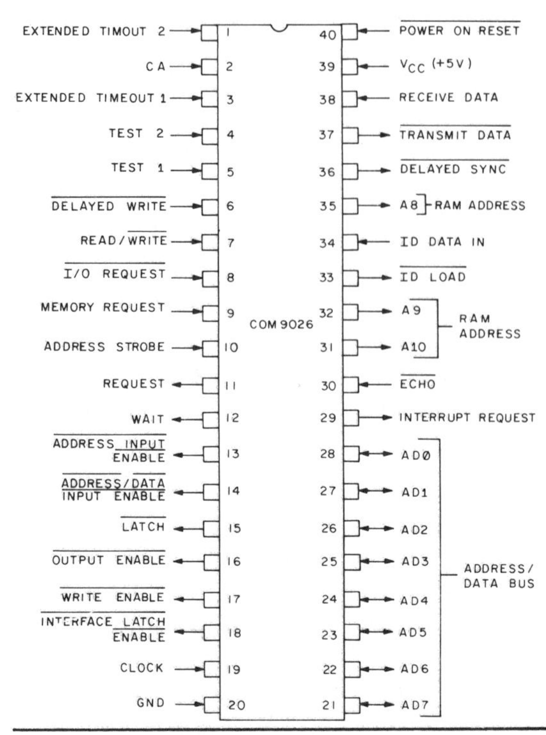

To read the RAM, the CPU must first set the READ/WRITE input HIGH and specify a memory location. Once the address is latched into the input buffer, it is transferred to the ADDRESS/DATA BUS by an ADDRESS/DATA INPUT EN-

To read the RAM, the CPU must first set the READ/WRITE input HIGH and specify a memory location. Once the address is latched into the input buffer, it is transferred to the ADDRESS/DATA BUS by an ADDRESS/DATA INPUT EN-

ABLE strobe. The upper three RAM address bits (A8, A9, A10) are placed on line by the ADDRESS INPUT ENABLE (pin 13). Immediately following this sequence, the LATCH output (pin 15) locks the lower eight addresses into an external latch driving the buffer RAM.

Now that the RAM has been accessed, the address inputs are removed from the bus and the RAM data is placed on line using an OUTPUT ENABLE (pin 16) command from the 9026. This data is stored in the output buffer and read by the CPU.

To write into memory, the procedure is very much the same. The memory address is again strobed into the input buffers and placed on line by the ADDRESS/DATA INPUT ENABLE and ADDRESS INPUT ENABLE signals. After the LATCH output has secured the RAM address, the line is cleared and a coincidence of WRITE ENABLE (pin 17) and INTERFACE LATCH ENABLE (pin 18) outputs writes data from the CPU into the buffer memory.

The 9026 controller chip manipulates the ADDRESS and DATA lines to its own advantage when transferring data between the buffer RAM and the 9026. In this case, however, the buffer latches are not engaged and the CPU never sees the activity, making all operations transparent.

Station Configuration

On the other side of the chip is the network interface. The 9026 serially communicates with other stations on the line through the TRANSMIT DATA (pin 37) output and RECEIVE DATA (pin 38) input using token passing protocol. Should any node fail to respond to a token passing within a predetermined time, the chip initiates a Network Reconfiguration. The response time is adjustable through the EXTENDED TIMEOUT 1 and 2 (pins 1 and 3, respectively) inputs.

The station's ID address is set by mechanical switches, rather than in software. The ID address is placed into an external shift register and serially shifted into the chip through the ID DATA IN (pin 34) input on an ID LOAD (pin 33) output signal. This address is entered at the beginning of every POWER ON RESET (pin 40) cycle.

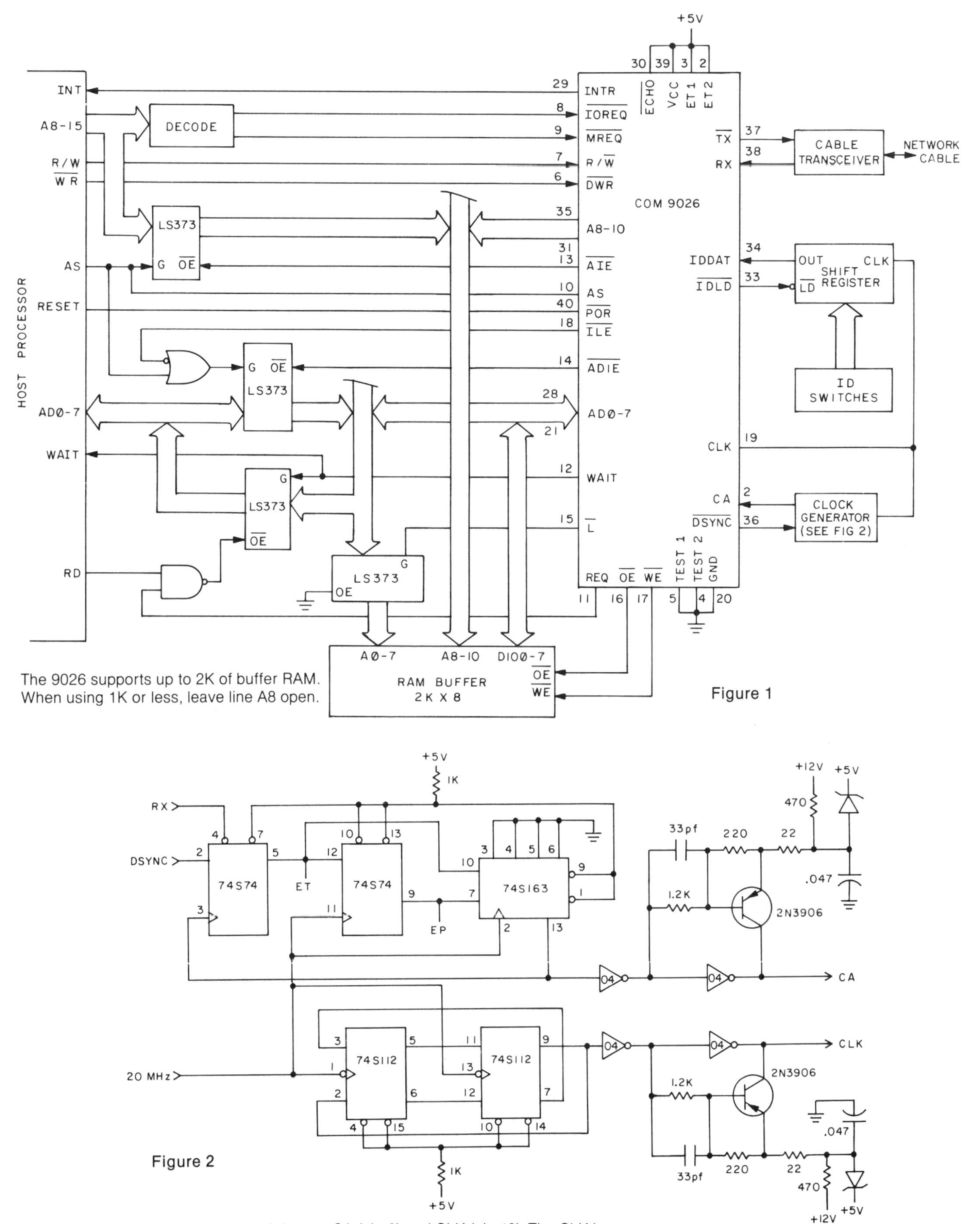

The 9026 supports up to 2K of buffer RAM. When using 1K or less, leave line A8 open.

Figure 1

Figure 2

The 9026 uses two separate clock inputs, CA (pin 2) and CLK (pin 19). The CLK input is a free-running 5-MHz signal, while the CA clock is a gated input. Figure 2 illustrates a typical CA clock input.

MC6860

Digital Modem

The MC6860, by Motorola, is a Digital Modem chip specifically designed to implement a serial data communications over a voice-grade channel, including existing telephone lines.

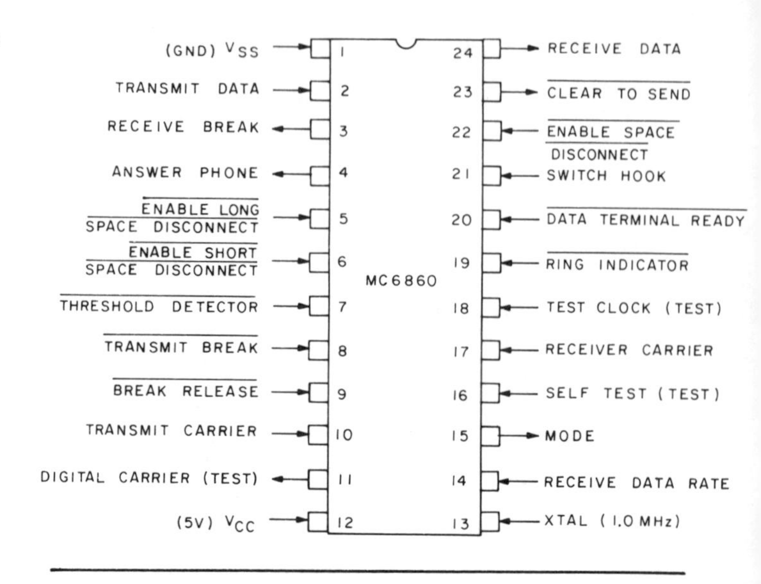

Operating Modes

The MC6860 modem is a full-duplex device which uses two carrier frequencies to establish a two-way data link through a single communication line, and is capable of operating in two modes. In the Originate Mode, the modem initiates and establishes the data link between the stations, while the Answer Mode designates the chip as the receiving station. Mode selection is determined by the sequencing of external events performed while gaining access to the communication line. The MODE (pin 15) output indicates which mode the chip is currently using.

Answer Mode

To initiate data reception, the RING INDICATOR input (pin 19) is pulled LOW, putting the modem in the Answer Mode. The chip will recognize a 51-ms signal from a CBS interface or at least 20-cycles of the 20-47-Hz ringer from a CBT decoder to this input as a receiver enable signal. If the DATA TERMINAL READY (pin 20) line is LOW, indicating the communication terminal is available to send or receive data, the ANSWER PHONE output (pin 4) is asserted and 2225-Hz carrier signal transmitted.

At the far end of the link, the originating modem senses this 2225-Hz signal and responds with a 1270-Hz carrier of its own. An external circuit monitors the incoming carrier for signal level and consistency, driving the THRESHOLD DETECTOR input (pin 7) LOW when all conditions are met. If the carrier signal remains steady for 150-ms, the RECEIVE DATA output (pin 24) is unclamped.

The FSK signal from the remote modem is now received through the telephone lines and filtered to remove extraneous signals, such as the chip's own 2225-Hz transmit tone. This filtering can be either a bandpass filter which admits only the desired frequency or a notch filter that rejects the known interference. The data signal is then limited, to preserve the axis crossings, and fed to the RECEIVER CARRIER (pin 17) demodulator input where the data is recovered from the FSK carrier and output through the RECEIVE DATA port.

The receiver has been optimized for two common baud rates: 300-bps and 600-bps. To gain full benefit, the RECEIVE DATA RATE (pin 14) must be strapped LOW for 600 baud and tied HIGH for 300 baud.

In order to maintain data flow through the chip, the input to the THRESHOLD DETECTOR must go LOW for at least 20-us out of every 32-ms. If the THRESHOLD DETECTOR signal is lost for more than 51-ms, disconnect procedures are placed into effect.

Automatic Disconnect

Should the receiver detect a space of 150-ms or longer, the modem clamps the RECEIVE BREAK output (pin 3) HIGH. This condition will exist until a BREAK RELEASE (pin 9) command is issued. In lieu of a BREAK RELEASE order, the modem will automatically release itself from the line. Grounding the ENABLE SHORT SPACE DISCONNECT (pin 6) input hangs up the modem after receiving a 0.3-second space; grounding the ENABLE LONG SPACE DISCONNECT (pin 5), instead, stretches that time to 1.5-seconds.

Originate Mode

Upon receipt of a SWITCH HOOK (pin 21) command, the modem is placed in the Originate Mode. If the DATA TERMINAL READY is enabled, the modem will commence a search for the 2225-Hz signal from the remote answering station. It will continue to look for this signal all the while the HOOK SWITCH is engaged, and for 17 seconds after release. Disconnect occurs if it isn't located.

Reception of the 2225-Hz tone for at least 450-ms at an acceptable level, on the other hand, prompts the transmitter to respond with a 1270-Hz carrier. After an additional 300-ms of steady carrier (750-ms in all), the CLEAR TO SEND (pin 23) output is taken LOW and FSK modulated data can be transmitted out the TRANSMIT CARRIER (pin 10) port, as well as receiving data at the RECEIVER CARRIER input. As before, external filters remove the local carrier signal. Data for output and encoding is fed into the chip through the TRANSMIT DATA (pin 2) input.

At the completion of the call, a 34-ms pulse to the TRANSMIT BREAK (pin 8) input directs the transmitter to initiate hang-up procedures. The transmitter complies by sending a steady space tone (1270-Hz unmodulated). This space tone is sent for a period of 3 seconds, after which the modem hangs up the phone, provided the ENABLE SPACE DISCONNECT input (pin 22) is LOW. Loss of RECEIVER CARRIER any time during this period automatically terminates contact immediately.

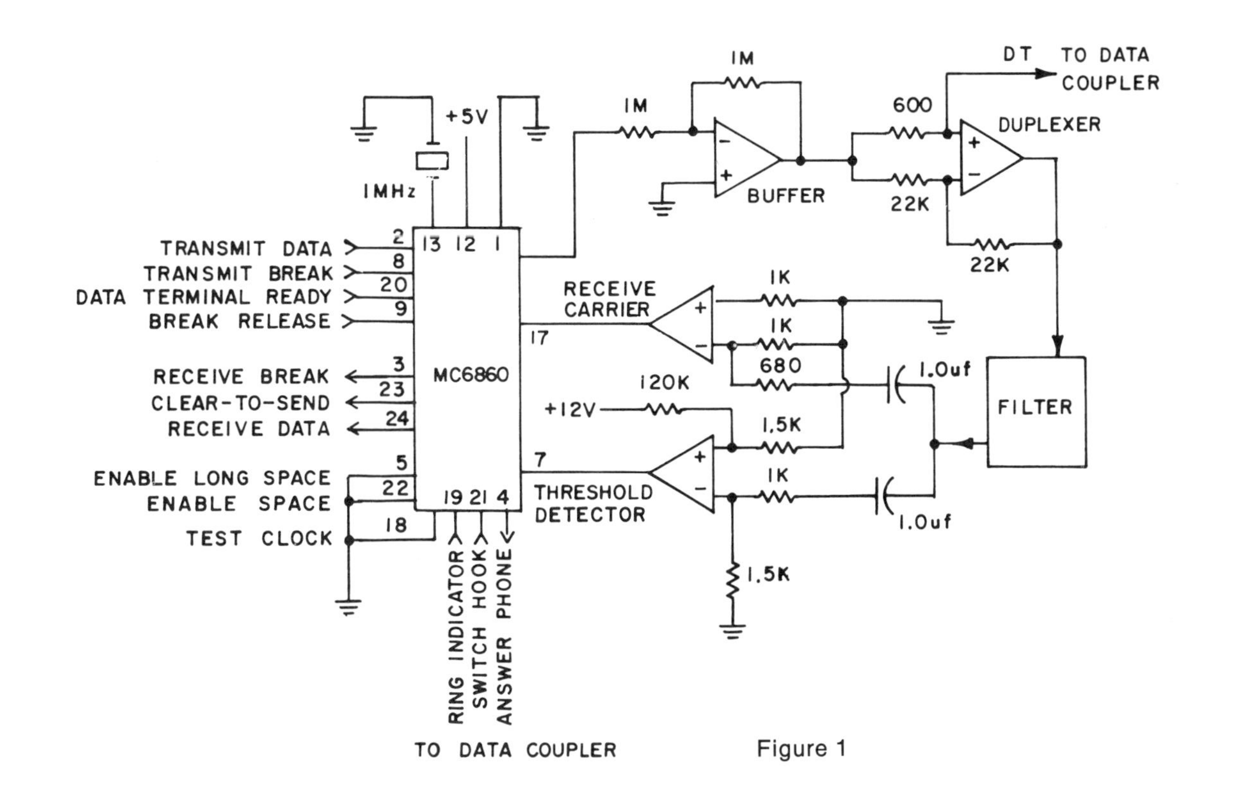

Figure 1

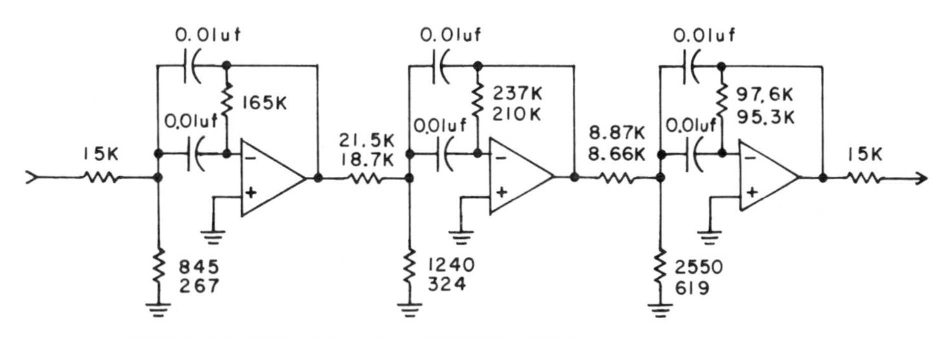

Figure 2

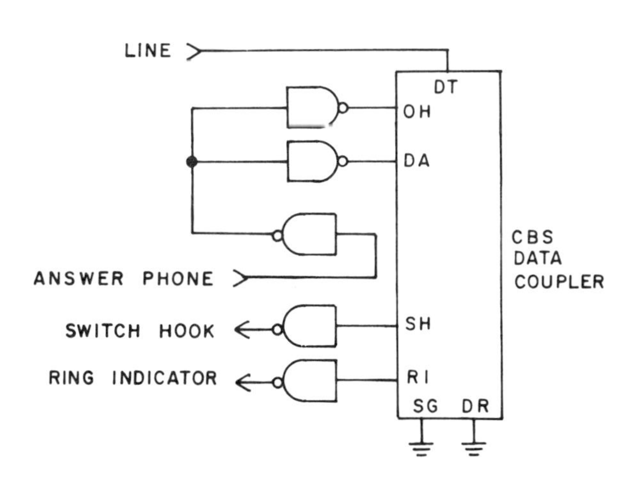

Figure 3

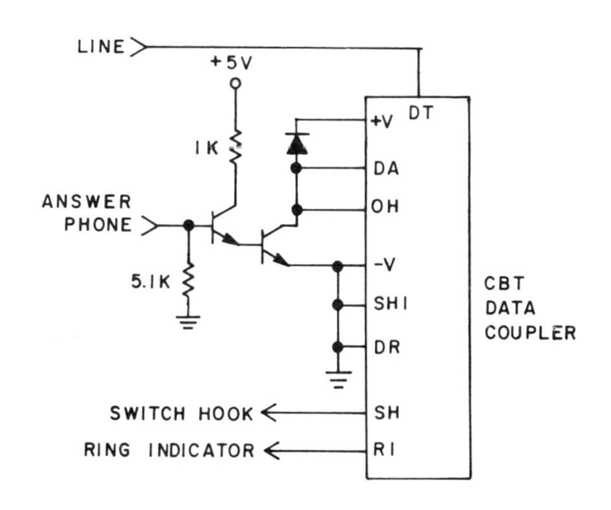

Figure 4

TMS 99532

300 BPS FSK Modem

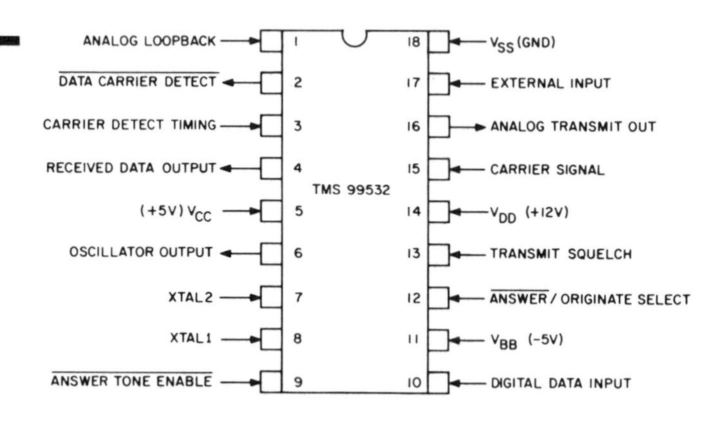

The TMS 99532 FSK Modem, by Texas Instruments, is a telecommunications device that transmits and receives serial data over the switched telephone network using frequency-shift keyed (FSK) modulation. It provides all the necessary modulation, demodulation, and filtering required to implement a serial, asynchronous communication link.

The modem will communicate at any rate up to 300 bits per second (baud) in simplex, half duplex, or full duplex capacities. The filters are of the switched capacitor design and require no external components or special signal conditioning for operation.

Receive

The operation of the modem is determined by the ANSWER/ ORIGINATE MODE SELECT (pin 12) control input. When this input is forced LOW, the modem is in the Answer, or Receive, Mode. This means that it receives 1270-Hz serial data from a remote modem, and responds with a 2225-Hz reply.

The 1270-Hz carrier is received at the CARRIER SIGNAL (pin 15) input, where it's processed by internal filters and demodulated. The resulting binary word is serially outputted to the RECEIVED DATA OUTPUT (pin 4).

To prevent erroneous data interpretation due to carrier signal dropout, the carrrier is monitored for consistency, as reported by the DATA CARRIER SELECT (pin 2) output. A LOW output on this pin indicates the carrier signal is greater than −43-dBm, and suitable for acquisition. Once the carrier has been locked in, though, it may drop to as little as −53-dBm before the DATA CARRIER DETECT output will indicate a loss of signal by driving pin 2 HIGH.

When no valid carrier is present, the RECEIVED DATA OUTPUT pin is clamped HIGH. A simple RC network on the CARRIER DETECT TIMING (pin 3) pin determines the response time of the carrier detect circuit.

Transmit

To get in the Transmit, or Originate, Mode, the ANSWER/ ORIGINATE MODE SELECT input is set HIGH. In this mode the modem sends data at 1270-Hz and listens to a 2225-Hz signal, the reverse of the Receive Mode.

The serial input data from the CPU is fed into pin 10, the DIGITAL DATA INPUT. The binary bits are FSK modulated and the analog output passed through the ANALOG TRANSMITS OUT port (pin 16) for interfacing with the telephone communications lines. This signal is fully buffered and compatible with a broad range of low-speed direct-connect interface devices or acoustical couplers.

The ANALOG TRANSMIT OUT signal is controlled by the TRANSMIT SQUELCH (pin 13) circuit. When this input is HIGH, the signal presented to the DIGITAL DATA INPUT will not appear on the ANALOG TRANSMIT OUT output.

The TMS 99532 modem chip also has provisions for transmitting analog signals (tones) in conjunction with the digital information. The analog signal is input to the EXTERNAL INPUT (pin 17) where it passes through the transmitter anti-alias (rejection) filter before being output on the ANALOG TRANSMIT OUT pin. This maneuver prevents the modem from rejecting the EXTERNAL INPUT as unwanted interference. The EXTERNAL INPUT is typically used in DTMF (touch tone dialing) applications. It should be noted that the TRANSMIT SQUELCH function has no control over this signal and won't disable it from exiting the ANALOG TRANSMIT OUT port.

Analog Loopback

The TMS 99532 includes a special test function called Analog Loop Back. Basically, it is a self test function which places the signal processing filters, both transmit and receive, on the same frequency when the ANALOG LOOPBACK input (pin 1) is raised HIGH.

Furthermore, the procedure connects the two analog signals together internally. Any signal presented to the DIGITAL DATA INPUT will now transverse the entire signal path of the chip and emerge on RECEIVED DATA OUTPUT. The analog output pins are disabled during the test to prevent the signals from entering the phone lines. The test can be performed in either Answer or Originate Modes.

Clock

Timing pulses for chip operation are generated by an internal 4.032-MHz oscillator. An external crystal is connected across pins 7 and 8, the XTAL2 and XTAL1 inputs, for stable frequency control.

However, the modem can also be clocked by an external TTL/CMOS generator of the same frequency. In this configuration, the signal is input to XTAL1 input only, leaving XTAL2 unconnected. The oscillator frequency is available at the buffered OSCILLATOR OUTPUT (pin 6) pin.

Power Supply

Three power supplies are required to operate the chip. Two bipolar voltages are required to supply the +5-volts and −5-volts needed on pins 5 and 11, respectively. In addition, a +12-volt supply must be connected to pin 14. Pin 18 is the common ground.

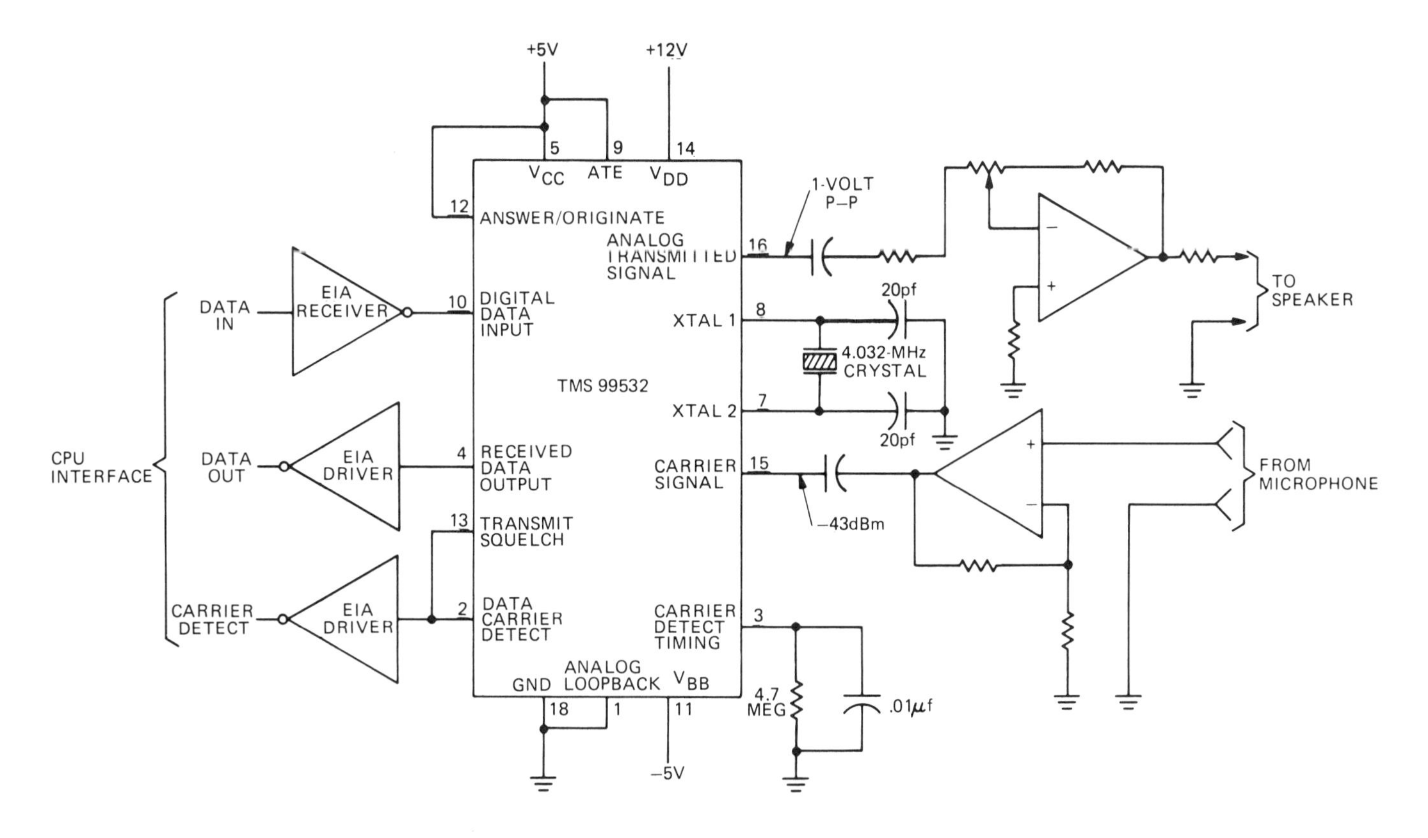

Figure 1

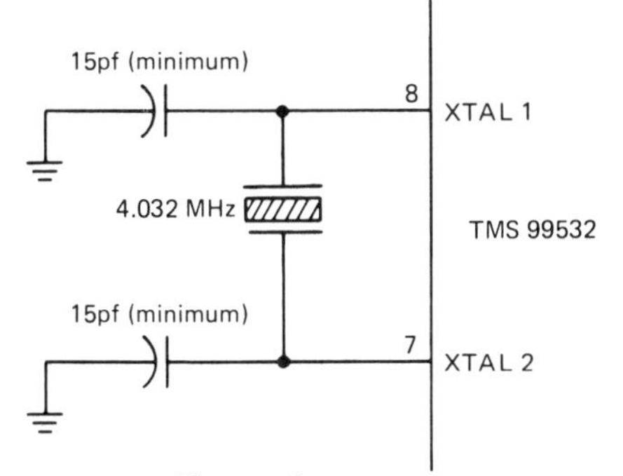

Figure 2

Pin 9 generates a 2100-Hz tone when the ANSWER TONE ENABLE (pin 9) is pulled LOW. This tone is used to signal when a modem link has been established, although it is seldom used in the U.S.

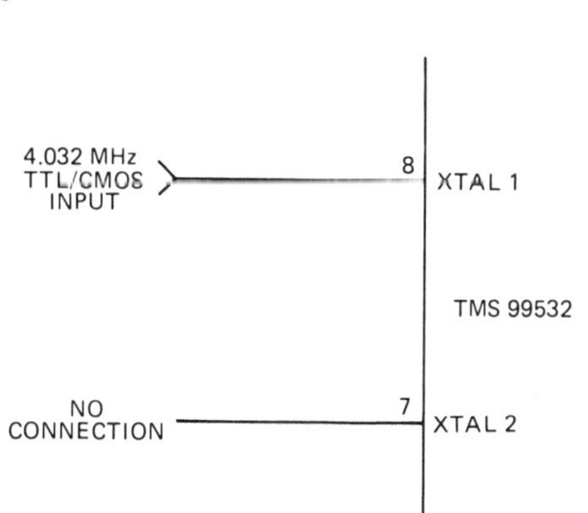

Figure 3

DP8340

Serial Bi-Phase Transmitter/Encoder

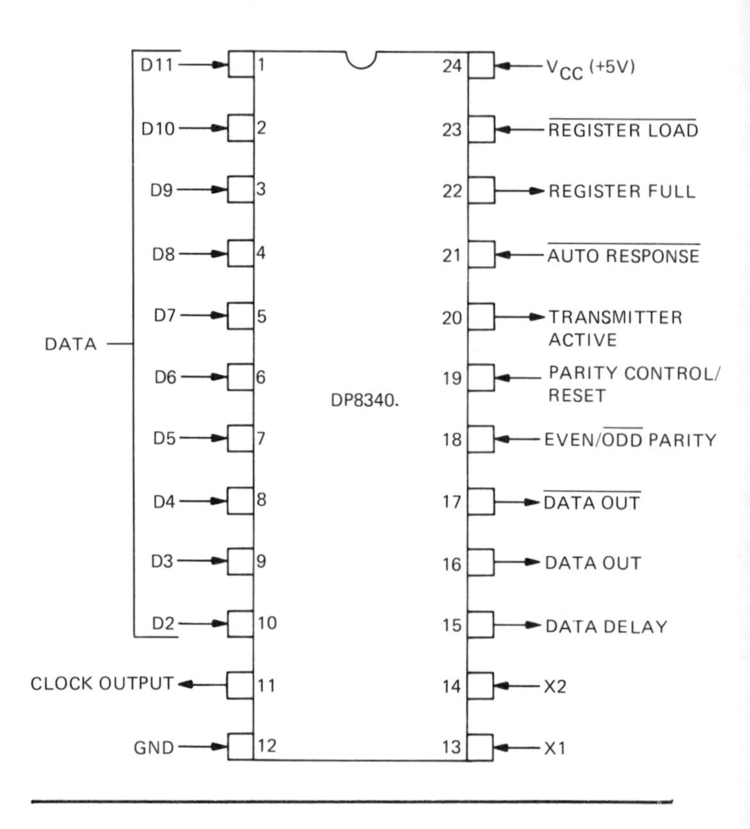

The DP8340, made by National Semiconductor, is a Serial Bi-Phase Transmitter/Encoder that generates a complete encoding of parallel data conforming to the protocol as defined by the IBM 3270 Information Display System Standards into high-speed serial transmissions. Although the device has been optimized to communicate through coaxial lines using the IBM format, it also adapts to general high-speed serial data communications over other than coax, at frequencies either higher or lower than the IBM standard.

When used in conjunction with its complementary chip, the DP8341 Receiver/Decoder, a flexible communications system is formed.

Operation

Operation of the DP8340 is such that all protocol functions are automatic. Upon receipt of parallel input data, the chip generates a Line Quiesce Pattern, Code Violation, and Sync Bit in accordance with IBM 3270 standards. If the message is a multi-byte communique, the internal format logic will modify the message data format for multibyte transmission as long as the next byte is loaded into the *input holding register* before the last data bit of the previous data byte is transferred out of the internal *output shift register*. After the message is completed, the encoder initiates an Ending Sequence and returns to an idle state.

CPU Interface

Messages are loaded into the DP8340 through ten DATA input (pins 10-1) ports. Data is transferred into the *input holding register* on the trailing edge of a REGISTER LOAD (pin 23) strobe; data must be valid during the LOW state of this control signal. The word is then encoded and placed into the *output shift register* for transmission.

The next byte will be loaded into the *input holding register* and held until the contents of the *output shift register* are cleared. When both registers are filled, a REGISTER FULL (pin 22) flag goes HIGH to inform the CPU that the transmitter is accepting no additional messages. This flag is cleared when the *input holding register* is emptied.

Data Control

The encoded message must contain a certain amount of user input to be valid. This involves the setting of the parity. Parity is determined by the EVEN/ODD PARITY (pin 18) input—when this pin is HIGH, even parity is generated; when it is LOW, odd parity is encoded into the word.

Oftentimes a data word will only consist of 8 bits instead of 10, in which case an additional parity bit is required following the word in the bit 10 position. The bit generated is odd parity on the previous eight bits of data, and is established by the PARITY CONTROL/RESET (pin 19) input.

With the PARITY CONTROL input is in the LOW state, the DATA bit 10 is ignored and odd parity of the previous data is placed in the normal bit 10 position, while overall parity (bit 12) is even or odd as controlled by the EVEN/ODD PARITY input.

If the PARITY CONTROL/RESET input is driven to a voltage that exceeds the power supply V_{CC} (pin 24) input (typically 7.5-volts), the device is reset.

In conformance with IBM protocol, the receiving unit must respond to a completed message with a Clean Status pattern, indicating that it was properly received and understood. The AUTO RESPONSE (pin 21) function generates this sequence automatically when pulled LOW.

Output

After encoding, the formatted message is shifted out of the *output shift register* in serial form over the transmission line. When the transmitter is about to transmit or is in the process of transmitting data, the TRANSMITTER ACTIVE output (pin 20) asserts itself HIGH. Serial data is available at three output ports for convenient application.

The DATA (pin 16) and $\overline{\text{DATA}}$ (pin 17) outputs are a direct bit representation of the Bi-Phase data. As you can imagine, the DATA signal is a true logic output, while the $\overline{\text{DATA}}$ bits are an inversion—or negative logic—signal. Either may drive the transmission line directly using a suitable interface.

However, these two outputs add flexibility to the DP8340 by allowing it to transmit high-speed differential output signals. Since the DATA output is a true complement of the DATA signal, the two are always in mutual opposition and a differential receiver can easily sort the digital information from background noise, even when the link is through a pair of twisted wires.

The DATA DELAY output (pin 15) provides the necessary increment to clearly define the four DC levels of the pulse, and is often used in the translation logic.

Clock

The internal oscillator is controlled by an external crystal connected between X1 (pin 13) and X2 (pin 14). The CLOCK OUTPUT (pin 11) is a buffered signal derived from the crystal oscillator, and is designed to drive the DP8341 Receiver/Decoder chip.

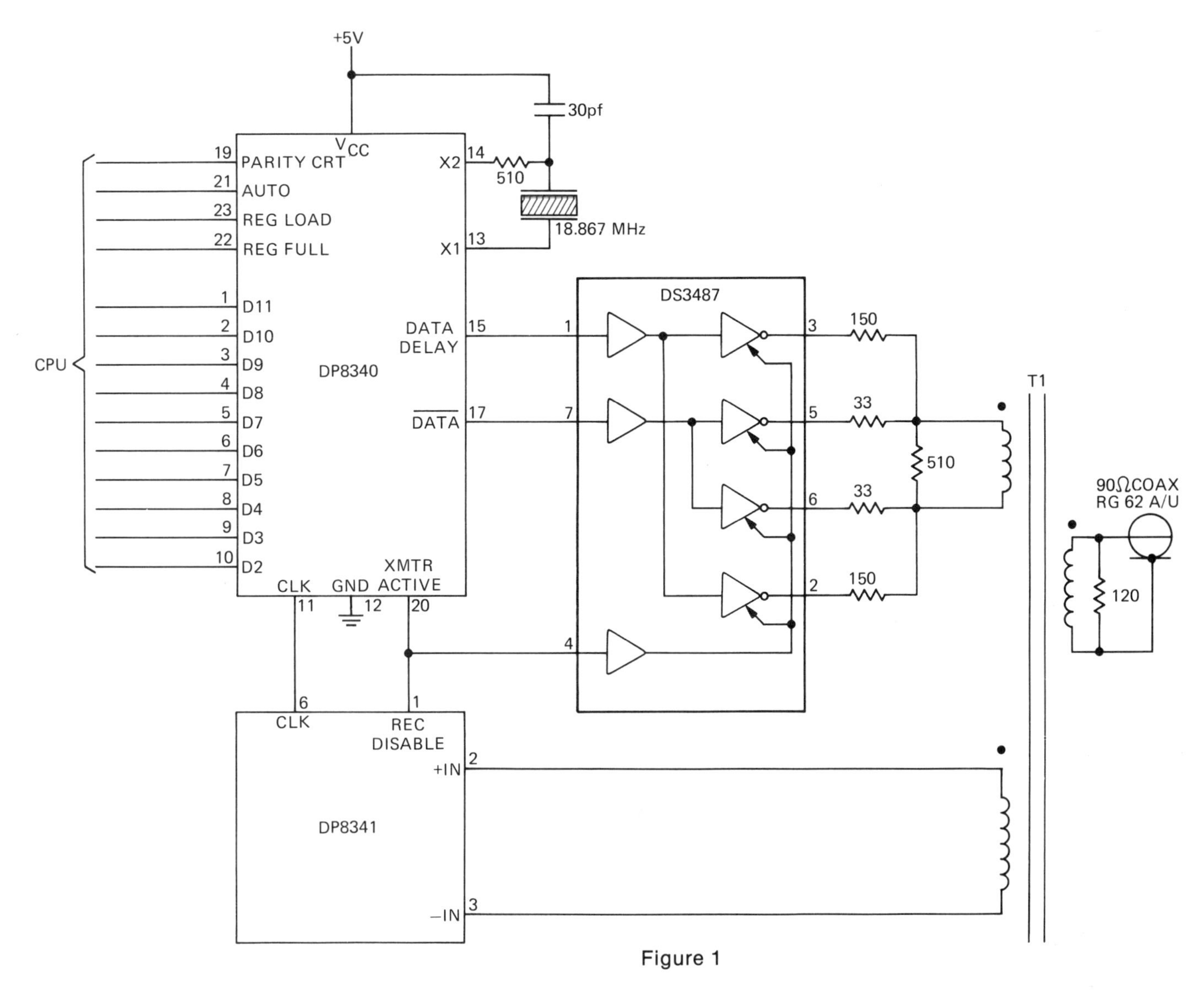

Figure 1

T1 is a 1:1:1 pulse transformer.
Pulse Engineering No. 5762
Technitrol No. 11LHA

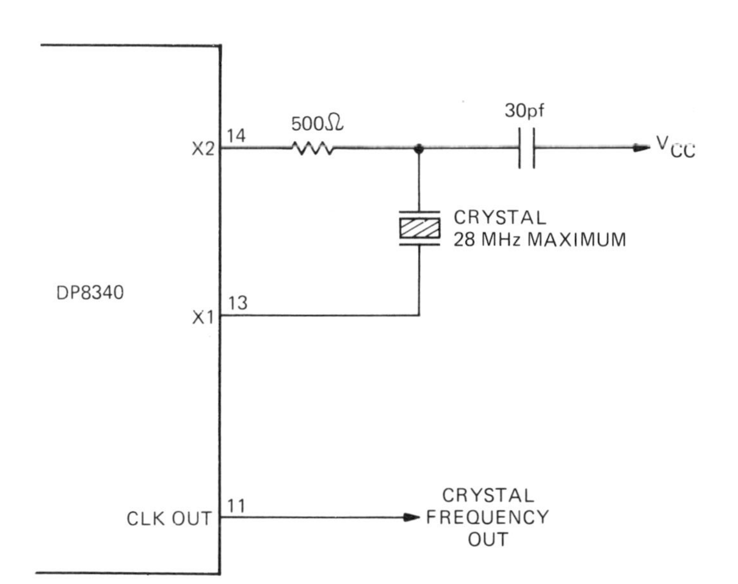

Figure 2

The master clock runs at a frequency of eight times the baud rate, which is 18.867-MHz for the IBM 3270 protocol.

DP8341

Serial Bi-Phase Receiver/Decoder

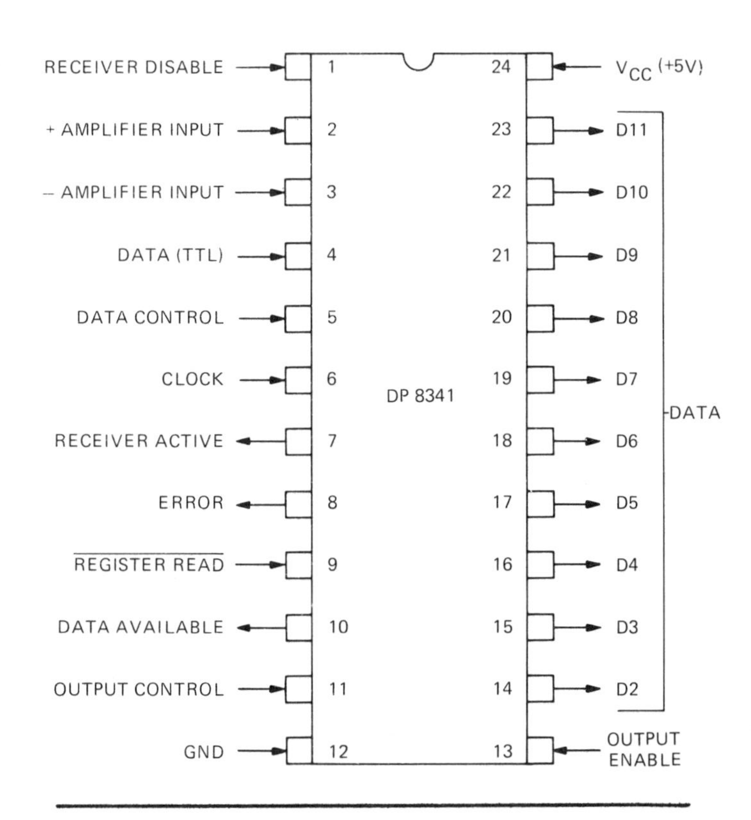

The DP8341, by National Semiconductor, is a Serial Bi-Phase Receiver/Decoder that provides complete decoding of input serial data according to IBM 3270 Information Display System Standards.

The DP83441 receiver and its companion chip, the DP8340 transmitter, are designed to provide maximum flexibility in system designs. The separation of transmitter and receiver functions permits the addition of receivers to a system without the necessity of adding unused transmitters. This is particularly advantageous in control units where data is typically multiplexed over several Bi-Phase lines and the receivers generally outnumber the transmitters.

Data Input

The DP8341 has two types of serial data input ports. One is a differential amplifier receiver, which is the primary input, that may be directly connected to a transformer-coupled coax line. Furthermore, the + AMPLIFIER (pin 2) and – AMPLIFIER (pin 3) complementary inputs, with their high common mode rejection ratio, allow the receiver to successfully detect the presence of a differential signal even when buried in a background of noise. This is a definite advantage in noisy environments.

The DATA input (pin 4), on the other hand, is a TTL compatible input that can serve as an alternate data input port when driven by a suitable signal. Using this pin as an alternate input also allows self testing of the peripheral system without disturbing the transmission line.

The DATA CONTROL (pin 5) selects which of the inputs is used for data entry. When this pin is LOW, the DATA input is actively engaged; when HIGH, the differential AMPLIFIER INPUTS are receiving. Both input functions are disabled when the RECEIVER DISABLE (pin 1) control is HIGH. This input is normally tied to the transmitter to prevent the transmitted data from feeding back through the receiver, and is basically a transmit/receive switch.

Data Output

As the ten bits of data are shifted into the *input shift register* from the demodulator circuit, the receiver will verify that even parity is maintained on the data bits and the sync bit. After one complete byte is assembled, it is transferred to the *output holding register*, and the DATA AVAILABLE output (pin 10) is set HIGH. Data is read from the *output holding register* through the tri-state DATA output (pins 14-23) buffers. The register is read by first enabling the output drivers with a HIGH input to the OUTPUT ENABLE (pin 13) input, and then strobing the REGISTER READ (pin 9) input LOW.

The next word byte enters the *output holding register* upon the return of the REGISTER READ input to HIGH, and can be subsequently read. After the last byte is read, however, and no new bytes are available to fill the output register, the last received byte will remain in the register until new information is received—even though the register has been read. The only way to clear the *output holding register* is to replace its contents with a new word.

After the detection of a starting sequence, the RECEIVER ACTIVE (pin 7) output is forced HIGH. This output will remain HIGH until the receipt of a valid Ending Sequence or the detection of an error.

Error Detection

All incoming data is checked very carefully to ensure it conforms to the IBM 3270 protocol. Should an error be spotted, the ERROR (pin 8) output is pulled HIGH, and the RECEIVER ACTIVE signal drops LOW.

Moreover, the DP8341 has an intrinsic error identification logic circuit which informs the user as to what kind of error occurred. This error code is available at the DATA output pins when the OUTPUT CONTROL (pin 11) line is forced LOW. In normal operation, this input remains HIGH to output the correct message byte received. Once an error has been detected, and the receiver drops out of operation, driving this input LOW will display the error bit on the DATA output. Altogether, there are seven errors the receiver will recognize. A table listing them and their respective data bits is shown in fig. 2 on the schematic.

The ERROR output returns LOW after the *error register* has been read, using a READ REGISTER operation in association with the OUTPUT CONTROL, or when the next valid starting sequence is received.

Clock

A clock is used to synchronize all internal operations and sets the baud rate. The CLOCK (pin 6) frequency must run at eight times the desired baud (18.87-MHZ for IBM 3270 protocol) and is normally derived from the DP8340 transmitter CLOCK OUTPUT pin.

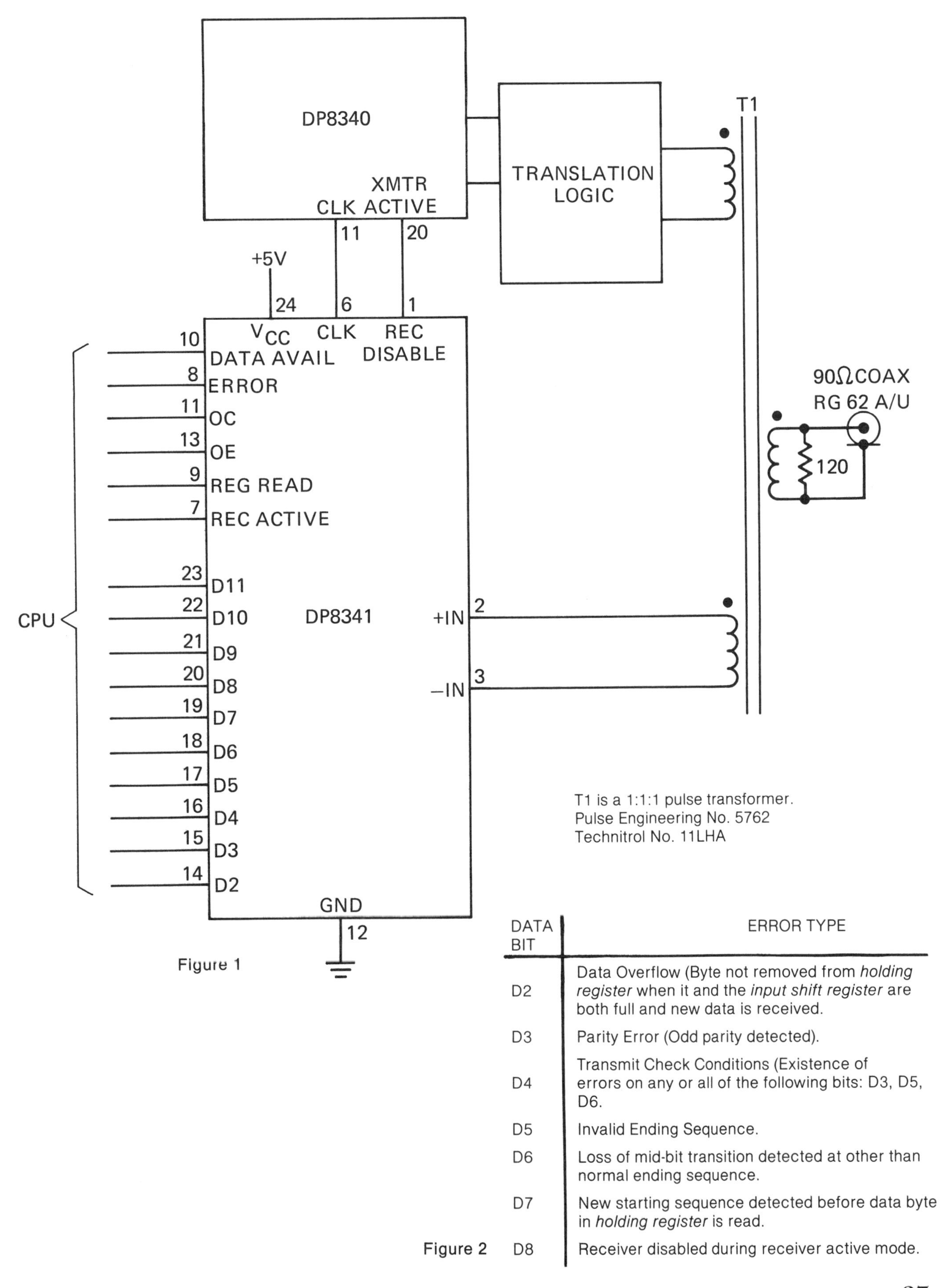

Figure 1

DATA BIT	ERROR TYPE
D2	Data Overflow (Byte not removed from *holding register* when it and the *input shift register* are both full and new data is received.
D3	Parity Error (Odd parity detected).
D4	Transmit Check Conditions (Existence of errors on any or all of the following bits: D3, D5, D6.
D5	Invalid Ending Sequence.
D6	Loss of mid-bit transition detected at other than normal ending sequence.
D7	New starting sequence detected before data byte in *holding register* is read.
D8	Receiver disabled during receiver active mode.

Figure 2

COM 9004

Compatible COAX Receiver/Transmitter

The COM 9004 Compatible COAX Receiver/Transmitter, by Standard Microsystems Corporation, is designed to implement a communication interface between IBM compatible control units and terminals following the IBM 3270 protocol. The chip is specifically designed for use with coaxial cable transmission lines.

The device consists of a receiver and transmitter section that processes Manchester II phase encoded data as specified by IBM protocol. The transmitter and receiver sections are separate and may be used independently of each other.

Transmitter

The COM 9004 can process both 8- and 10-bit words in the IBM format. The first 8 bits are entered in parallel form on the bidirectional TRANSMIT/RECEIVE DATA BITS (pins 6-13) lines, while bits 9 and 10 have dedicated inputs: TRANSMIT BIT 9 (pin 38) and TRANSMIT BIT 10 (pin 39), respectively.

When communicating with 8-bit words, the TRANSMIT BIT 9 SELECT (pin 4) input is driven LOW. In this mode, the last two bits are masked and bit 9 represents the parity bit. Parity is determined by the TRANSMIT PARITY (pin 3) input: a LOW input causes odd parity and a HIGH level generates an even parity.

In the 10-bit mode, the TRANSMIT BIT 9 SELECT input is set HIGH. This unmasks the final two bits and shifts the parity bit from bit 9 over to bit 11. Bit 9 now contains word information.

The transmit sequence is initiated by strobing the TRANSMIT DATA STROBE (pin 36) LOW. This transfers the data from the data bus into the *transmit holding register* and forces the TRANSMIT BUFFER EMPTY (pin 37) output LOW. Pin 16 is more or less the equivalent of a chip enable function, and has an effect on the receiver status lines as well.

The TRANSMIT BUFFER EMPTY output remains LOW until its contents are accepted by the *transmit shift register,* after which it returns HIGH and new data can be entered. The data is then encoded and serially shifted out of the chip via the TRANSMIT DATA (pin 33) port.

If a new word is loaded into the *transmit holding register* before the parity bit of the previous word is transmitted, a Sync bit is generated and the new word transmitted. If not, the last word transmitted initiates the Ending Sequence code. During the entire time that valid data is present, beginning with the line Quiesce character, the TRANSMIT GATE output (pin 32) is LOW. This output recovers to its HIGH state when the Ending Sequence is started, and can be used to operate external transmit circuitry.

Receiver

The receiver portion interfaces with the COAX through pin 30, the RECEIVE DATA input, and all incoming signals are checked for a Code Violation. Once a line Quiesce, Code Violation, and Sync Bit are detected, in that order, the status of the CODE VIOLATION DETECT (pin 25) output goes HIGH, initiating the receiver decoding process. The received word is first checked for parity. Parity is governed by the RECEIVER PARITY (pin 17) input: when HIGH the receiver will recognize even parity, when

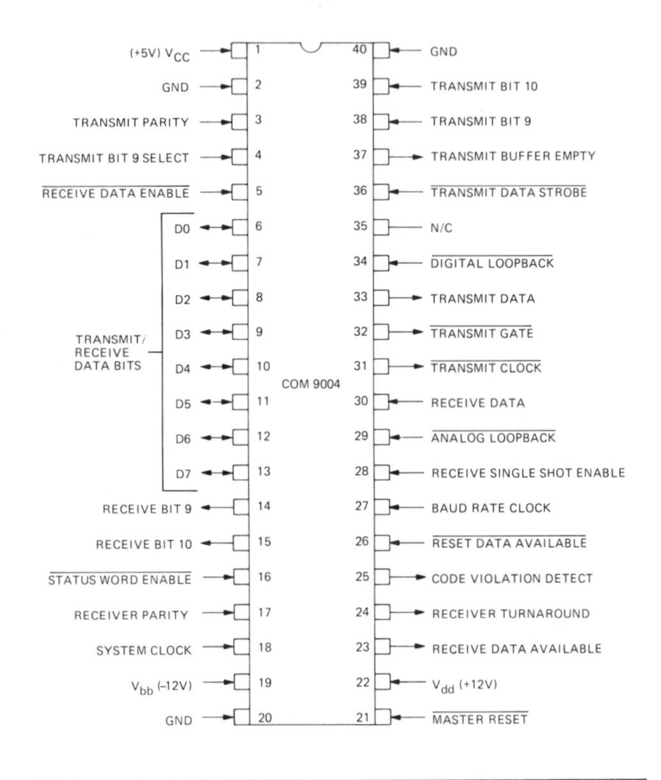

LOW it will respond to odd parity.

If everything is correct, the word is transferred to the *receiver buffer register* and the RECEIVER DATA AVAILABLE output (pin 23) set HIGH. The CPU can now read the word from the TRANSMIT/RECEIVE DATA BITS lines by engaging the output drivers with a LOW signal on the RECEIVE DATA ENABLE (pin 5) input. Bits 9 and 10, when available, are read from the RECEIVE BIT 9 (pin 14) and RECEIVE BIT 10 (pin 15) outputs. At the same time, the Receive Data Available flag is reset by strobing the RECEIVE DATA AVAILABLE input (pin 26) LOW.

This process continues until an Ending Sequence is detected, which immediately sets the RECEIVER TURNAROUND output (pin 24) HIGH, indicating the end of the message. This output can be reset only by loading the transmitter with an appropriate response or by holding pin 26 LOW for more than one clock cycle.

Clock

The chip is internally synchronized using a SYSTEM CLOCK (pin 18) input pulse. The BAUD RATE CLOCK (pin 27) sets the data rate and must run at 8 times the desired baud rate (typically 18.8696-MHz for 3274/3276 operation). This input is *not* TTL compatible and requires a minimum of 4.3-volts for the logic '1' level. The BAUD RATE CLOCK signal is divided in half and made available at pin 31, the TRANSMIT CLOCK output, to provide external pre-distortion timing.

Diagnostic Modes

The COM 9004 is capable of performing self-testing using two inputs: the ANALOG LOOPBACK (pin 29) and the DIGITAL LOOPBACK (pin 34). In the Normal Mode, both inputs are HIGH. The RECEIVE SINGLE SHOT ENABLE (pin 28) is used in conjunction with the Diagnostic Modes by limiting the HIGH level input to DATA RECEIVE to just 3 clock cycles.

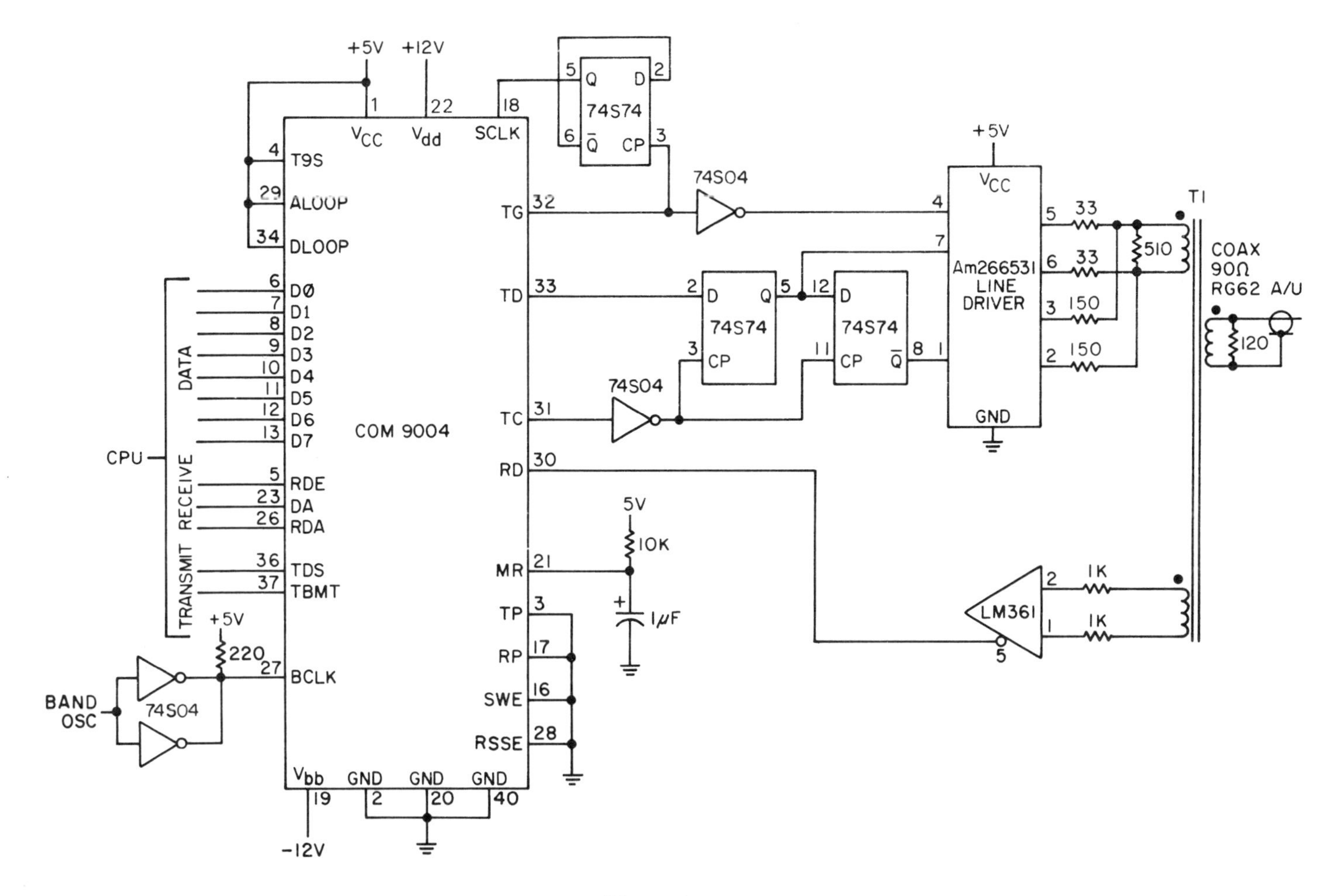
+5V
+12V
1
22
18
VCC
Vdd
SCLK
5
Q
D
2
74S74
6
Q̄
CP
3
4
T9S
29
ALOOP
34
DLOOP
74S04
TG
32
+5V
VCC
4
7
5
33
T1
6
33
510
COAX
90Ω
RG62 A/U
Am266531
LINE
DRIVER
3
150
2
150
120
6
D0
7
D1
8
D2
9
D3
10
D4
11
D5
12
D6
13
D7
DATA
TD
33
2
D
Q
5
12
D
74S74
74S74
3
CP
11
CP
Q̄
8
1
74S04
TC
31
GND
COM 9004
CPU
RECEIVE
5
RDE
23
DA
26
RDA
RD
30
5V
10K
MR
21
36
TDS
37
TBMT
TRANSMIT
1K
2
LM361
1
1K
5
+
1μF
TP
3
+5V
220
27
BCLK
RP
17
SWE
16
BAND
OSC
74S04
RSSE
28
Vbb
GND
GND
GND
19
2
20
40
-12V

Figure 1

Am7990

Local Area Network Controller for Ethernet

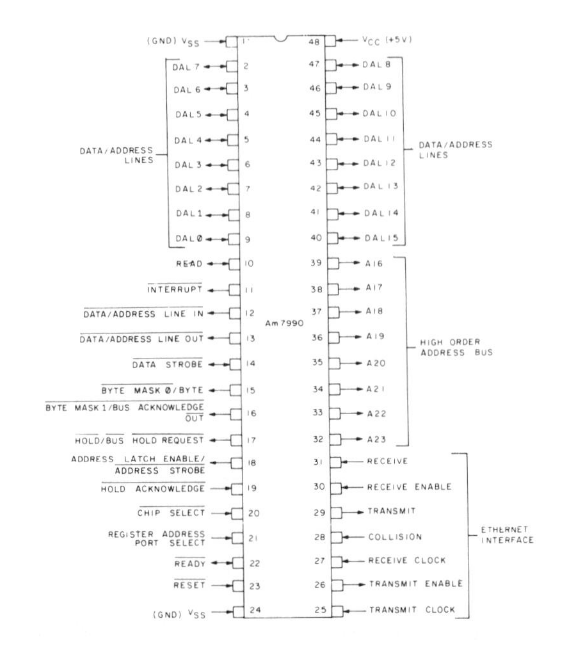

The AM7990 Local Area Network Controller for Ethernet, by Advanced Micro Devices, is a 48-pin VLSI device designed to greatly simplify interfacing a microcomputer or minicomputer to an Ethernet local area network. This chip is intended to operate in a local environment that includes a closely coupled memory and microprocessor.

Ethernet Protocol

Invented by Xerox at its Palo Alto Research Center, Ethernet is a 10-megabit-per-second packet-switched communications scheme for local networks. It can span distances up to 2.5-kilometers with a theoritical maximum of 1,000 nodes, or work stations. However, the maximum number of nodes one is likely to encounter in real situations will probably number in the hundreds.

Ethernet is a passive medium using a shared coaxial cable, with no computer or station performing centralized control. Instead, each node is a master, and all stations are given equal priority for use of the cable. Access to the Ethernet is set up by the nodes themselves using a statistical procedure in which each station listens to the cable to see if it is clear. If it is, the node proceeds to transmit the message.

Should two transmitters attempt to seize the cable at the same time, they will collide, and the data on the coax becomes garbled. When this occurs, the initiating transmitters stop transmission and jam the network by transmitting an Abort pattern. This jamming insures that all nodes will recognize the collision.

The node then attempts to transmit again using a truncated exponential binary back-off algorithm. This algorithm provides a defined time delay, which is a random number that varies from node to node, that prevents access to the cable while it is running. The first node to time out attempts control of the cable. If it is successful, communications proceed as before. Should another collision occur, however, the algorithm is incremented and another attempt made. After sixteen unsuccessful tries to avoid collision, an error flag is set and transmissions halted until new instructions are received.

Microprocessor Interface

The parallel processor interface of the 7990 has been designed to easily interface to a variety of popular 16-bit microprocessors, including: 68000, 8086, Z8000, and LSI-11. The controller itself uses multiplexed DATA/ADDRESS LINES (pin 9-2, 47-40) for chip control, but the CPU may be a mixed bag of bus configurations and still be acceptable to the 7990. The host processor communicates with the 7990 only during the initialization phase, for demand transmission, and periodically to read the contents of the *status registers*. All other transfers to and from the memory are handled under DMA control.

During initialization, the CPU loads the starting addressed of the initialization block into two *control registers*. Altogether, the 7990 utilizes four internal *control and status registers* to direct the various functions of the chip. Since many of the CPU buses (such as the Z-Bus and Q-Bus) are multiplexed, while others (like the Multibus, Versabus, and Unibus) are not multiplexed, a single

Next, the register programming is presented to the DATA/ADDRESS LINES. By changing the input byte to even parity, the REGISTER ADDRESS PORT SELECT goes LOW, thus transferring the data into the intended register. Once written, the address contained in the *register address port* will remain in effect until rewritten, and subsequent operations will occur within the last register specified.

All registers are a combination of read and write. The direction of data flow is determined by the READ (pin 10) input. A LOW input writes data into the chip, while a HIGH signal outputs data from the chip.

Please refer to the 68590 section for further instructions.

REGISTER ADDRESS PORT SELECT input (pin 21) selects the internal registers.

This input may be driven by the least significant address bit (A0) of a multiplexed bus (8086, etc.) or decoded from the address lines of a demultiplexed bus (68000, etc.). When this pin is HIGH, the *register address port* is engaged; a LOW input enables the *register data port*. Before any register may be accessed, however, the CHIP SELECT input (pin 20) must be valid.

By properly manipulating the DATA/ADDRESS LINES, the internal registers can be accessed in two steps. First, the CPU places the address of the desired register on the bus lines, while forcing the REGISTER ADDRESS PORT SELECT input HIGH. If the CPU specifies an odd numbered address location, the REGISTER ADDRESS PORT SELECT is automatically decoded to the proper level, regardless of the system bus (multiplexed or non-multiplexed) used. This loads the *register address port* with a pointer that will direct traffic to the selected register.

Next, the register programming is presented to the DATA/ADDRESS LINES. By changing the input byte to even parity, the REGISTER ADDRESS PORT SELECT goes LOW, thus transferring the data into the intended register. Once written, the address contained in the *register address port* will remain in effect until rewritten, and subsequent operations will occur within the last register specified.

All registers are a combination of read and write. The direction of data flow is determined by the READ (pin 10) input. A LOW input writes data into the chip, while a HIGH signal outputs data from the chip.

Please refer to the 68590 section for further instructions.

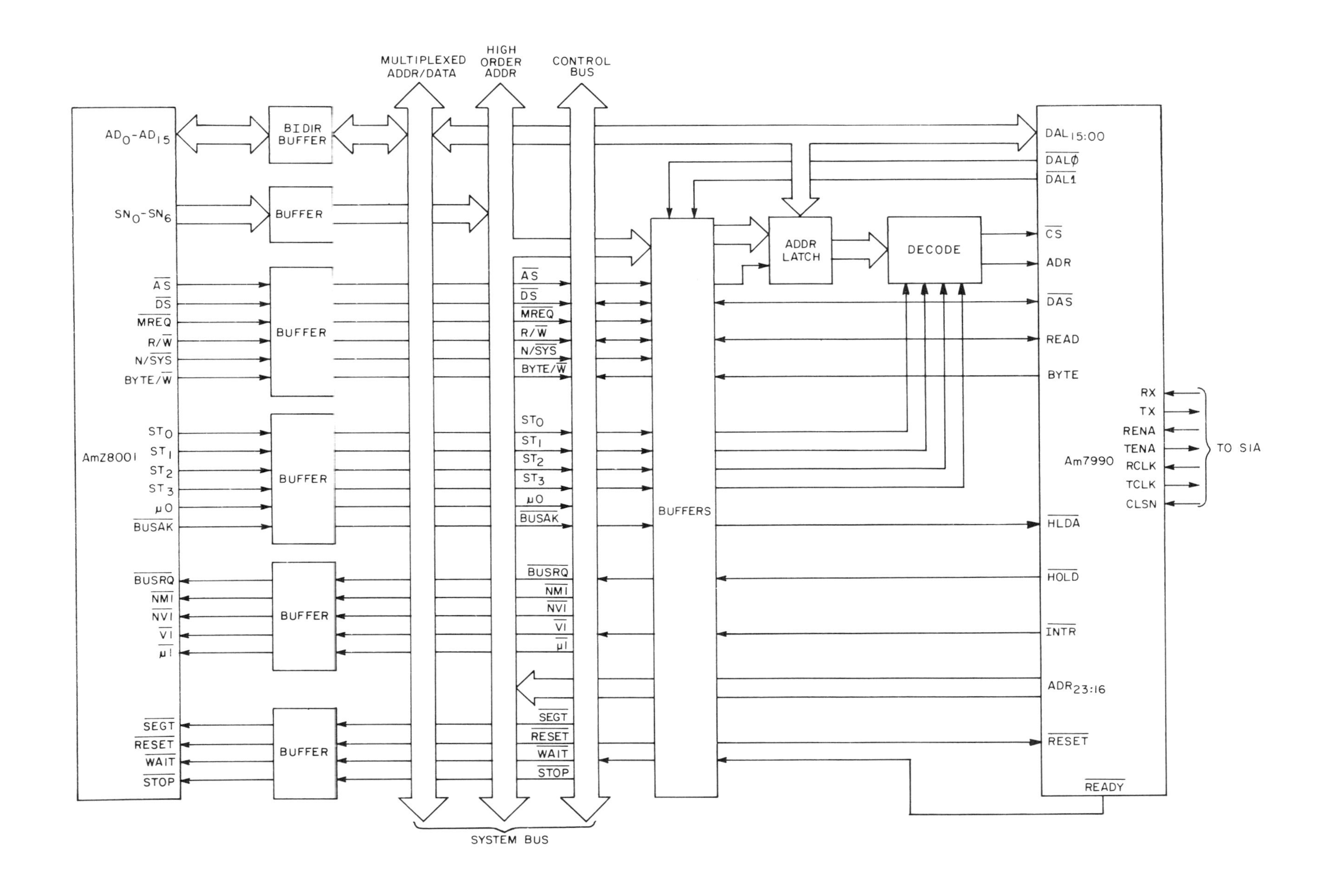

MULTIPLEXED ADDR/DATA
HIGH ORDER ADDR
CONTROL BUS
BIDIR BUFFER
BUFFER
AD0-AD15
SN0-SN6
AS
DS
MREQ
R/W
N/SYS
BYTE/W
ST0
ST1
ST2
ST3
μO
BUSAK
AmZ8001
BUSRQ
NMI
NVI
VI
μI
SEGT
RESET
WAIT
STOP
SYSTEM BUS
BUFFERS
ADDR LATCH
DECODE
DAL15:00
DALØ
DAL1
CS
ADR
DAS
READ
BYTE
RX
TX
RENA
TENA
RCLK
TCLK
CLSN
TO SIA
Am7990
HLDA
HOLD
INTR
ADR23:16
RESET
READY

MK68590

Local Area Network Controller for Ethernet

The MK68590 Local Area Network Controller for Ethernet, by Mostek, is a VLSI device designed to greatly simplify interfacing a microprocessor or computer to an Ethernet local area network. On-board DMA, advanced buffer management, and extensive error reporting coupled with diagnostics facilitate design and improve system performance.

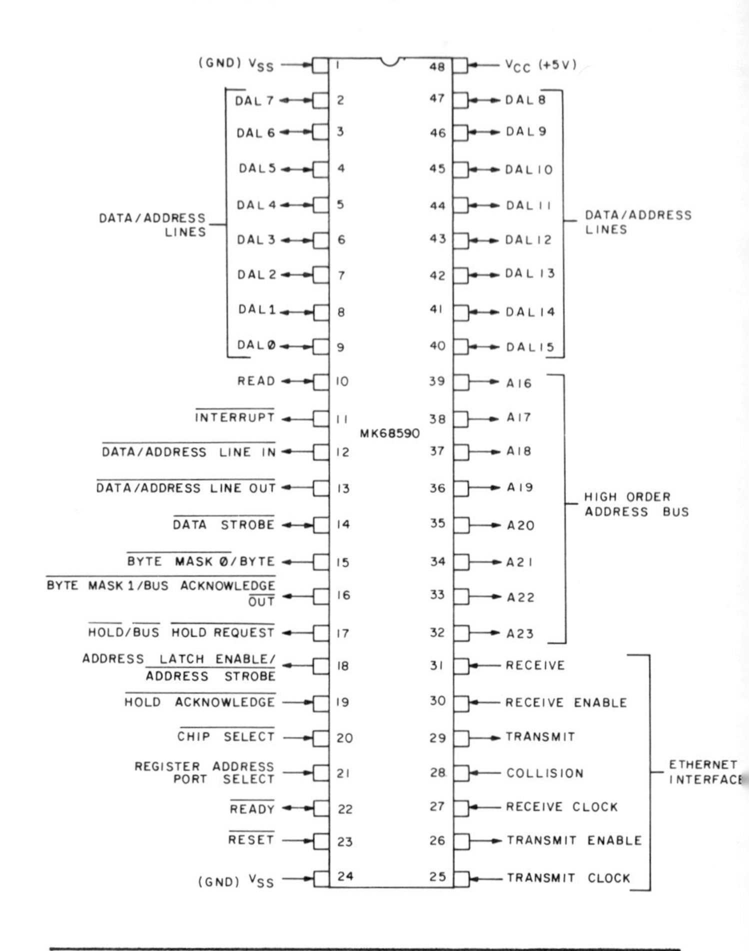

Buffer Management

The 68590 is intended to operate in a minimal configuration that requires close coupling between local memory and a processor. The local memory provides packet buffering for the controller chip and serves as a communication link between the chip and the CPU. It is only during the initial programming phase that the CPU talks directly to the 68950. All further communications are handled via a DMA controller, under microword control, contained within the Ethernet controller.

The basic organization of the buffer is a circular queue of tasks in memory called disruptor rings. There are two separate rings: One to describe transmit and the other for receiver operations. Up to 128 tasks may be queued up on a descriptor ring awaiting execution by the 68590. Each entry into the ring holds a location pointer to a data memory buffer and the buffer length, each in the sequence in which it is to be performed. Data buffers may be chained or cascaded together in order to handle a long data packet in multiplied buffer areas, thus increasing memory productivity.

Memory Access

The 68590 has a wide 24-bit linear address space which allows it to DMA directly into the memory address space of most 16-bit microprocessors. No segmentation or paging methods are used within the chip, and as such, the addressing closest resembles that of the 68000 CPU. However, programmable DMA structuring within the device makes it compatible with virtually all CPU memory contention schemes currently available.

Multiplexed processors such as the 8086 use a single signal in conjunction with the least significant address bit, A0, to establish whether the upper or lower byte is being accessed in memory. The non-multiplexed 68000, on the other hand, employs two separate strobe signals to identify the byte.

Conversion between the two schemes normally requires several external logic chips, and it's not advisable to mix the two within a system. But the 68590 eliminates these components and design compromises with a programmable BCON software bit, which permits a choice between the two DMA methods. In other words, a single command will place the 68590 controller into either DMA mode, with no loss in performance.

Multiplexed Memory Bus

DMA control over the memory is exerted only when the CHIP SELECT input (pin 20) is held HIGH. For transmit operations, data is read from memory into the 68590. At the beginning of the read cycle, the READ output (pin 10) is forced HIGH, and valid memory addresses are placed on the DATA/ADDRESS LINES (pins 9-2, 47-40) and HIGH ORDER ADDRESS BUS (pins 39-32).

These memory address outputs are then latched into an external address buffer using an address-strobe signal. To further enhance the controller's flexibility, the polarity of the address-strobe is programmable. For example, an active LOW ADDRESS STROBE is used with the Z8000, while an active HIGH ADDRESS LATCH ENABLE is needed for the 8086. Either polarity is available at pin 18.

Approximately 100-ns after strobing the address latches, the DATA/ADDRESS LINES go tristate. There is a 50-ns delay to allow for transceiver turnaround, then the DATA STROBE (pin 14) falls LOW to signal the beginning of the data portion of the cycle. At this point, the controller stalls, waiting for the memory to assert a READY (pin 22) signal. Once received, the DATA/ADDRESS LINES switch to their input state and the DATA STROBE makes a transition from a zero to a one, latching data into the chip. The automatic adjustment of the controller's bus cycle by the READY signal allows synchronization with variable cycle time memory due either to memory refresh or dual port access. Controller bus cycles are a minimum of 600-ns long, and can be increased through software in increments of 100-ns.

Upper and lower byte transfers associated with multiplexed bus timing is established by two control lines, as shown in fig. 2. The BYTE output (pin 15) could correspond to a BHE signal, while DAL0 represents the A0 address. These particular nomenclatures are equivalent, but not limited to an 8086 control bus.

Multiplexed bus operations for the MK68590 Ethernet controller are further described in the 68590 section of this book. The reader is urged to review that text.

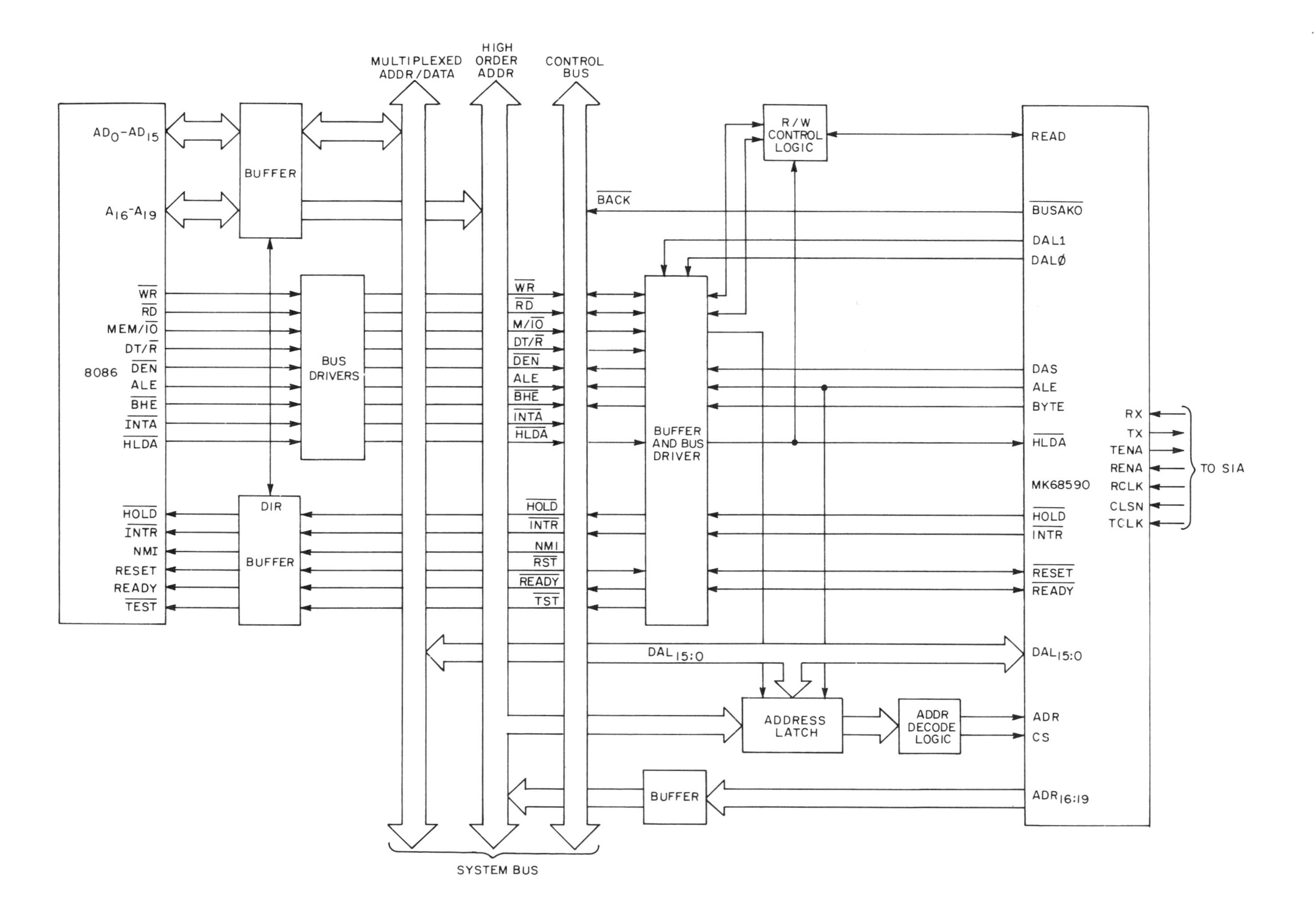
MULTIPLEXED ADDR/DATA
HIGH ORDER ADDR
CONTROL BUS
AD_0-AD_{15}
A_{16}-A_{19}
BUFFER
WR
RD
MEM/IO
DT/R
DEN
ALE
BHE
INTA
HLDA
8086
BUS DRIVERS
HOLD
INTR
NMI
RESET
READY
TEST
DIR
BUFFER
WR
RD
M/IO
DT/R
DEN
ALE
BHE
INTA
HLDA
HOLD
INTR
NMI
RST
READY
TST
BACK
R/W CONTROL LOGIC
BUFFER AND BUS DRIVER
READ
BUSAKO
DAL1
DALØ
DAS
ALE
BYTE
HLDA
MK68590
HOLD
INTR
RESET
READY
RX
TX
TENA
RENA
RCLK
CLSN
TCLK
TO SIA
$DAL_{15:0}$
$DAL_{15:0}$
ADDRESS LATCH
ADDR DECODE LOGIC
ADR
CS
BUFFER
$ADR_{16:19}$
SYSTEM BUS

68590

Local Area Network Controller for Ethernet

The 68590 Local Area Network Controller for Ethernet chip is designed to interface a processor to an Ethernet local area network. The on-chip DMA controller allows data packets to be transferred to and from a buffer memory without CPU intervention, while the internal buffer management scheme queues the packets during transmission and reception. The chip frees the processor from most network duties, thereby maximizing the bandwidth available for other CPU operations and reducing the probability of missing data packets on the network.

Multiplexed Memory Bus

Writing into memory is performed when data is received by the Ethernet controller. The write cycle begins exactly like a read cycle (see MK68590), with the memory address placed on the DATA/ADDRESS LINES (pin 9-2, 47-40) and HIGH ORDER ADDRESS BUS (pins 39-32) outputs, but with the READ line (pin 10) remaining LOW.

After the address-strobe (pin 18) pulse, the DATA/ADDRESS LINES change from memory addresses into data outputs. This data remains valid on the bus until the memory device asserts a READY (pin 22) signal. At this point, the DATA STROBE (pin 14) latches data into the memory. Data is held valid for 75-ns after the release of DATA STROBE to guarantee adequate transfer time. Here again, the chip uses the READY line to stretch the write cycle whenever necessary.

Two control lines are used to direct the flow of data through the bus transceivers. The DATA/ADDRESS LINE IN (pin 12) strobes data into the 68590, while the DATA/ADDRESS LINE OUT (pin 13) strobes data or memory addresses away from the controller. During a read cycle, pin 13 goes inactive before pin 12 to avoid spiking of the bus transceivers.

Non-Multiplexed Memory Bus

Demultiplexed memory designs require two separate strobe signals to identify the upper and lower data bytes. Setting the BCON software bit LOW allows the BYTE MASK 0 (pin 15) and BYTE MASK 1 (pin 16) outputs to represent these two lines. A chart of their interactions in respect to the word byte is presented in fig. 3.

With but this one exception, the read and write operations performed by the Ethernet controller follows the same sequence outlined in the multiplexed memory sections. In fact, the address architecture of the 68590 chip is very similar to the 68000 CPU, which also uses a non-multiplexed bus arrangement, and all timing sequences are straightforward and very natural to the controller.

DMA Bus Controller

If several devices with DMA capabilities must interface with a common bus, the 68590 provides either DMA controller or daisy-chain bus contention options to establish bus priority. For straight-out control, a bus HOLD request (pin 17) is made to secure control of the data bus. Access is granted with the receipt of a HOLD ACKNOWLEDGEment (pin 19).

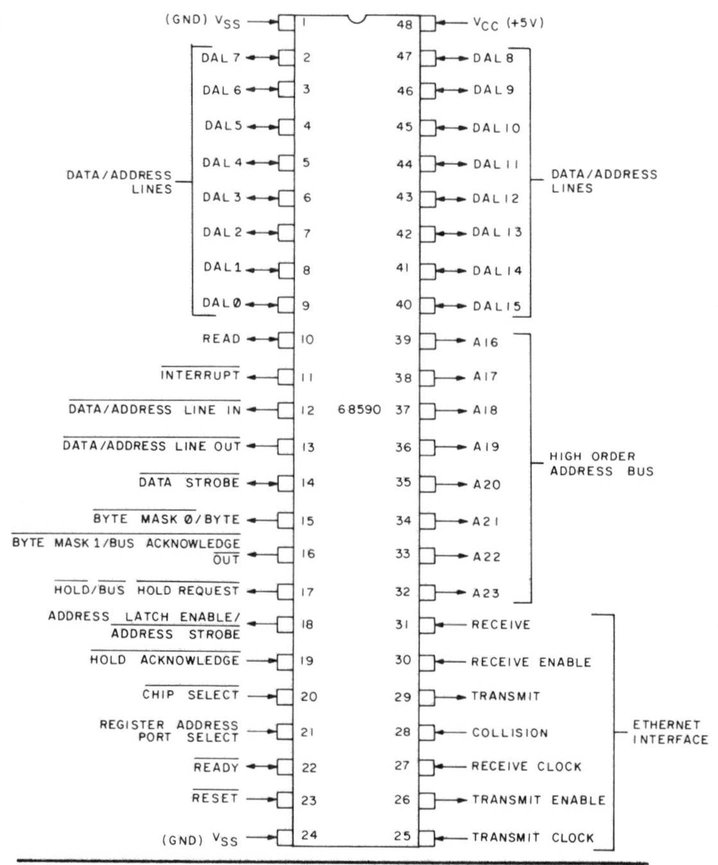

Daisy chaining priority adds a third signal to the bus contention scheme. When the controller requires access to the memory, it issues a BUS HOLD REQUEST (pin 17). A HOLD ACKNOWLEDGE will give access to the bus and drive the BUS ACKNOWLEDGE OUT output (pin 16) HIGH. This output is connected to the next lower priority device in the chain, suppressing it from DMA activity. Should the 68590 receive a HOLD ACKNOWLEDGE input without requesting acces to the bus, it forces the BUS ACKNOWLEDGE OUT CABLE LOW, giving bus control to the next device.

Cable Diagnostics

The time-domain reflectometry feature of the 68590 chip identifies and locates flaws on the Ethernet cable. A fault such as a short or an open acts as a discontinuity on the transmission line, causing reflections that a transmitting node sees as collisions. Therefore, the controller uses a *TDR counter* to measure the time between the initiation of a transmission and the detection of the collision. By comparing the counter values of several nodes, the defect can be located very quickly.

Additional Information

Due to the complexity of the 68590 Ethernet controller, three chips have been used to describe its operations: the Am7990, MK68590, and this 68590 section. Eventually, the controller will be second-sourced by several manufacturers, including Zilog and possibly Motorola.

You'll also notice that none of the Ethernet interface pinouts were described within this text. Their functions can be found by reading the information presented by their companion Serial Interface Adapter chips, the Am7991 and MK3891. Any one of the aforementioned controller chips in association with a Serial Interface Adapter will form a complete Ethernet link from CPU to cable transceiver.

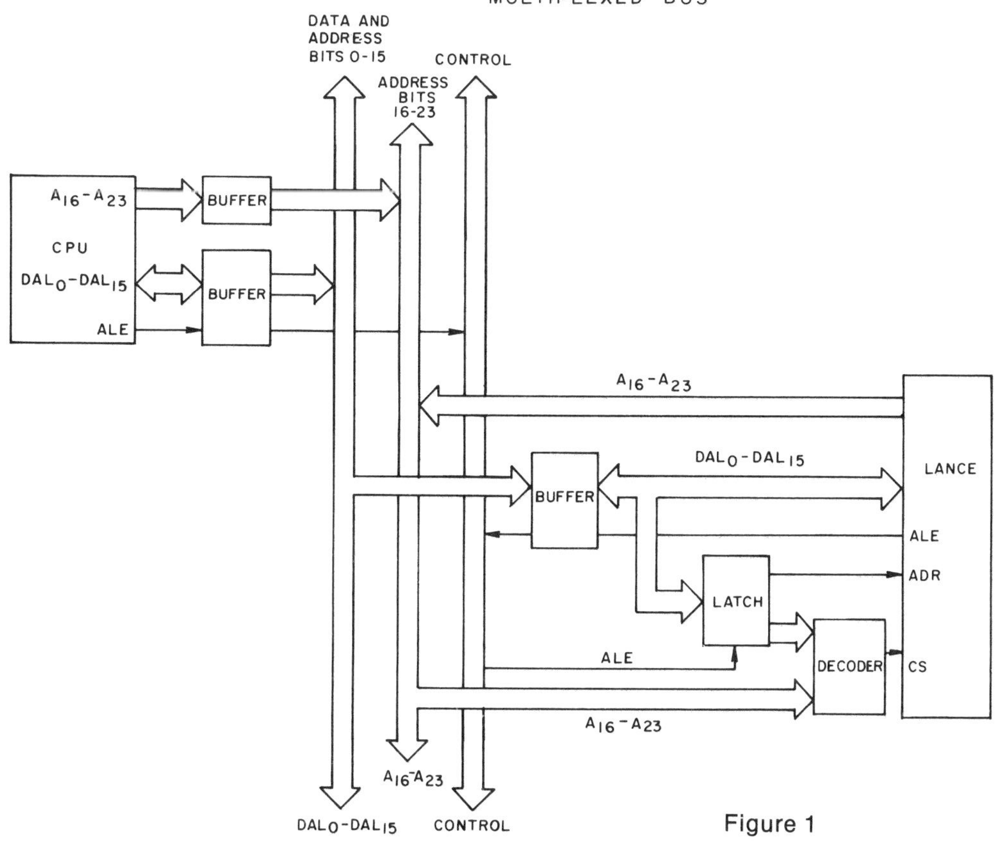

Figure 1

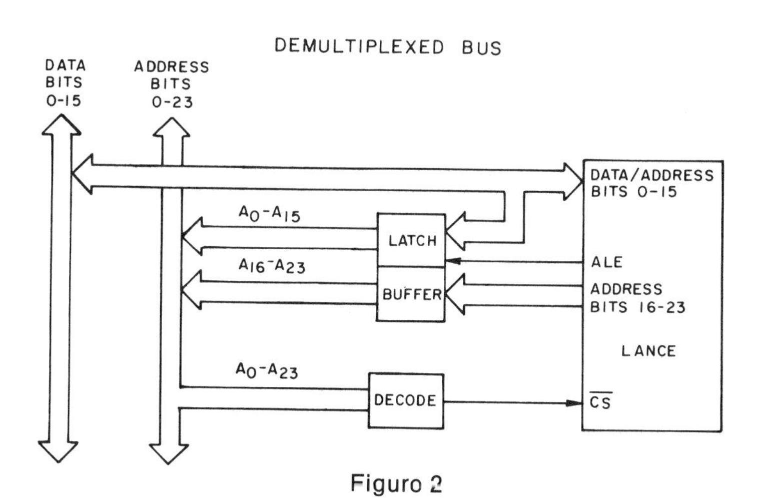

Figuro 2

DEMULTIPLEXED LOGIC

BM1	BMØ	
0	0	WHOLE WORD
0	1	UPPER BYTE
1	0	LOWER BYTE
1	1	NONE

MULTIPLEXED LOGIC

SIGNAL LINE \ MODE BITS	BSWP = Ø BCON = 1	BSWP = 1 BCON = 1
BYTE = 0 DALØ = 0	WORD	WORD
BYTE = 0 DALØ = 1	ILLEGAL	ILLEGAL
BYTE = 1 DALØ = 1	UPPER BYTE	LOWER BYTE
BYTE = 1 DALO = 0	LOWER BYTE	UPPER BYTE

Figure 3

Am7991

Ethernet Serial Interface Adapter

The Am7991 Serial Interface Adapter, by Advanced Micro Devices, interfaces the Am7990 Local Area Network Controller for Ethernet to a standard Ethernet cable transceiver. The chip provides node data and clock encoding/decoding functions using standard Manchester code to ensure clock and data synchronization throughout the network.

Functional Description

The 7991 has two sections to it: a transmitter and a receiver. The transmitter combines separate clock and NRZ data input signals and encodes them into a serial bit stream using standard Manchester II encoding. It also differentially drives up to 50-meters of twisted pair transmission line Ethernet Transceiver Cable.

The receiver performs three functions. For one, it detects the presence of data on the Ethernet cable and reduces the Manchester encoded data stream into separate clock and data outputs. It also receives collision information from the cable transceiver and provides a collision control output to the Am7990.

Crystal Controlled Oscillator

A 20-MHz fundamental mode crystal oscillator provides the basic timing reference for the 7991. The clock is divided by two to create the TRANSMIT CLOCK output. Both the 20-MHz and 10-MHz clocks are fed into the Manchester encoder to generate the transitions in the encoded data stream. The 10-MHz TRANSMIT CLOCK is used by the 7991 to internally synchronize the TRANSMIT DATA and TRANSMIT ENABLE inputs. The oscillator may use an external crystal across the XTAL inputs or an external TTL level 20-MHz input as a reference.

Receiver Signal Conditioning

The principle function of the receiver is the separation of the Manchester encoded data stream into clock signals and NRZ data. Before the data and clock can be separated, though, it must be determined whether there is actual data or just unwanted noise at the transceiver RECEIVE + and RECEIVE − inputs. The receiver provides a static noise margin of −175-mv to −300-mv for received carrier detection. These DC thresholds assure that no less than −175-mv is ever decoded, and that signals greater than −300-mv are always decoded.

The input signal conditioning section, see fig. 2, consists of two data paths. The receive data path is designed to be a zero-threshold, high-bandwidth receiver. The carrier detection receiver is similar, but with an additional bias generator to require that only data amplitudes greater than the bias level are interpreted as valid data. The noise rejection filter prevents noise transients from enabling the RECEIVE DATA output. Transient noise of less than 10-ns duration in the COLLISION path and 16-ns duration in the RECEIVE data path are rejected. This signal conditioning prevents unwanted idle sounds on the Ethernet cable from triggering a false start in the 7990 controller, while responding to actual data. A valid carrier detection is output to the 7991 controller on the CARRIER SENSE line.

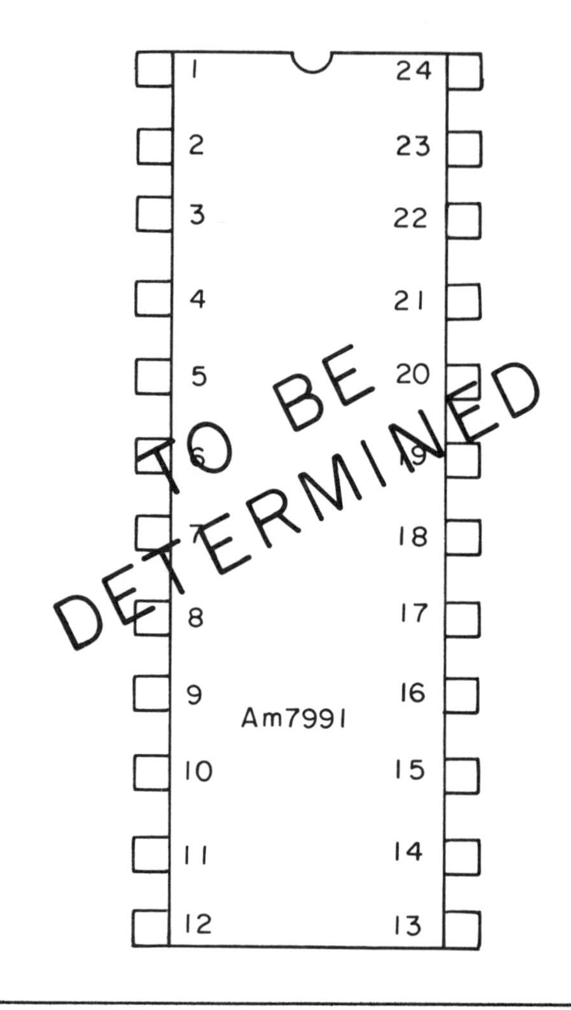

The clock pulse is separated from the data stream and output on the RECEIVE CLOCK line. Unfortunately, signals received from the Ethernet network will have finite rise and fall times. Consequently, a non-zero threshold is transformed into clock time distortion. To minimize the effect, the 7991's data receiver threshold is typically less than ±5-mv, thereby reducing its contribution to RECEIVE CLOCK jitters. In fact, the worst case error is less than 0.5-us.

Collision Detection

When two Ethernet nodes attempt to transmit through the cable at the same time, a collision occurs. Any collisions on the cable will be sensed by the Ethernet transceiver, subsequently generating a differential signal which is input to the COLLISION + and COLLISION − inputs. The 7991 processes the information and outputs it to the 7990 controller through the COLLISION output. The controller uses this signal to back off transmission and recycle itself. At the end of the time-out period, the controller will attempt another transmission of the data packet, provided the cable is clear at the time.

The Am7990 is identical in operation—and pinout—to the MK3891 Interface Adapter made by Mostek. Additional details concerning its operation are contained in that section. For a better overall view of the entire Ethernet interface system, the reader is urged to review the references presented by the 68590 text.

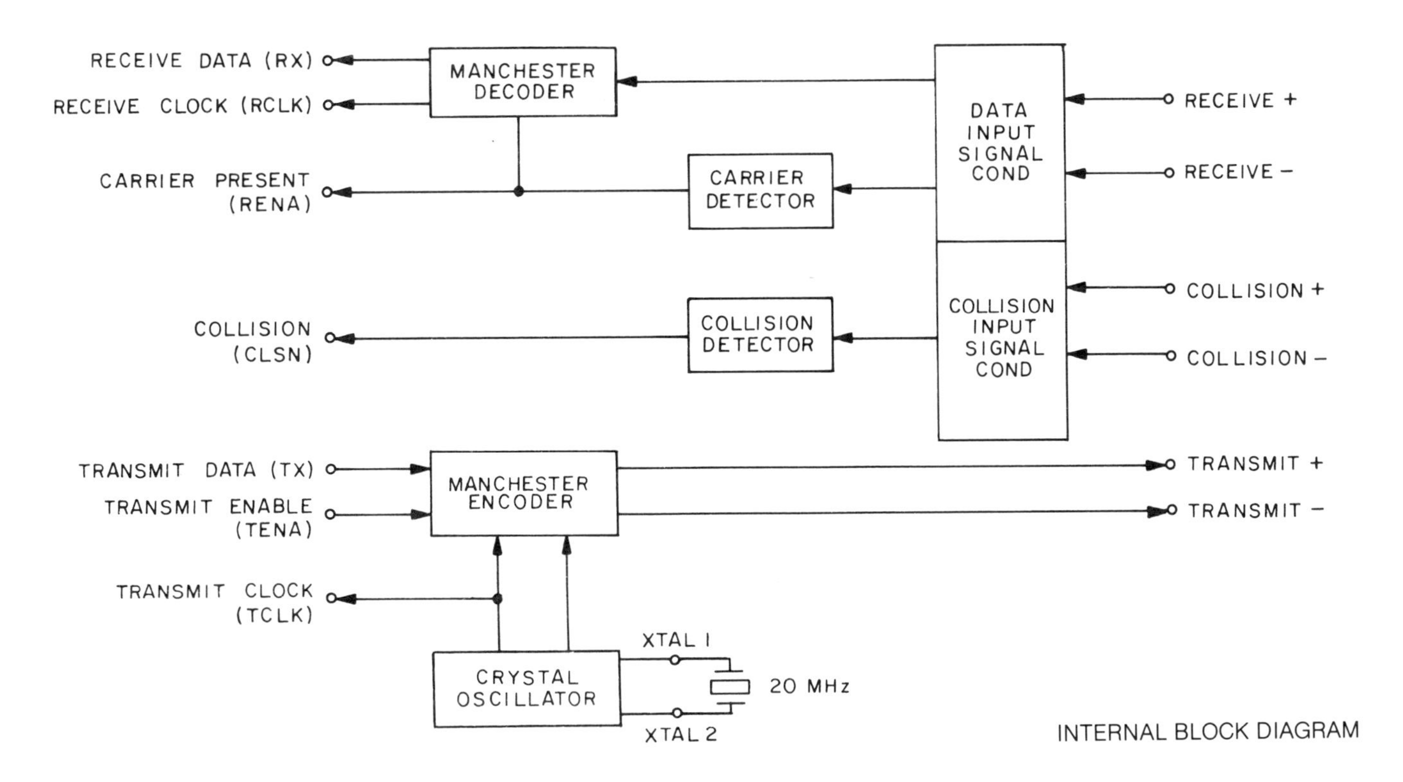

Figure 1

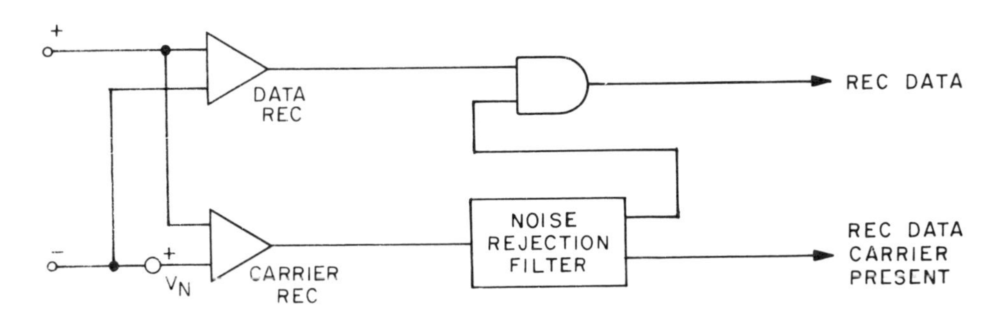

Figure 2

MK3891

Serial Interface Adapter

The MK3891 Serial Interface Adapter, by Mostek, is a VLSI device designed to greatly simplify the interfacing of a microcomputer or minicomputer to an Ethernet Local Area Network. The MK3891 is a companion device to the MK68590 Local Area Network Controller for Ethernet, and the two chips are intended to operate in an environment that includes a closely coupled memory and microprocessor.

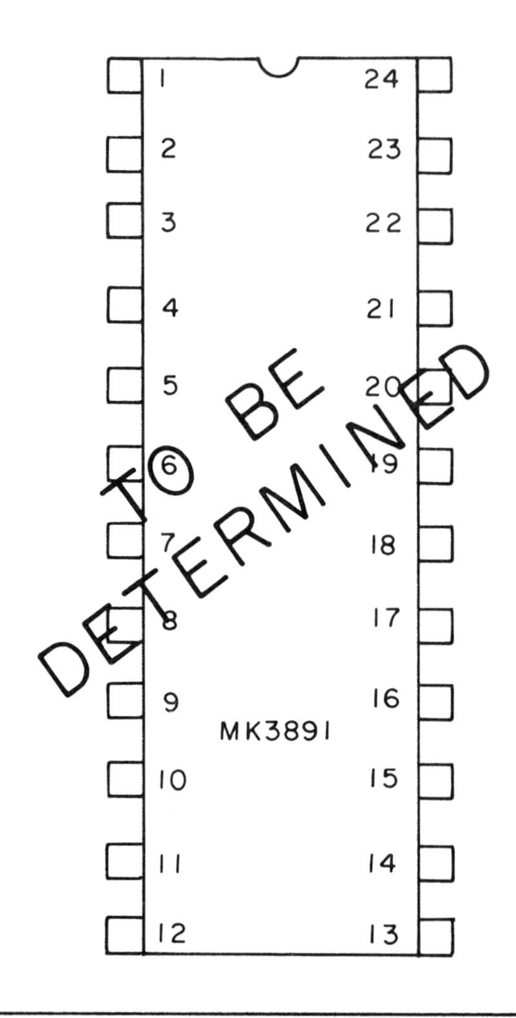

Receiver Section Timing

The principle function of the receiver is the separation of the Manchester encoded data stream into clock pulses and NRZ data. The CARRIER PRESENT output is the "receiver enabled" indicator established when a signal of sufficient amplitude and duration is present at the RECEIVE + and RECEIVE − inputs. Once the first frame transition is detected, the RECEIVE CLOCK and RECEIVE DATA outputs become available within six bit times, or approximately 600-ns from the time CARRIER PRESENT is asserted. At this point, the total phase error of RECEIVE CLOCK to the data stream is less than ±3-ns. The MK3891 is guaranteed to correctly decode a Manchester encoded serial bit stream with up to ±20-ns of jitter.

The receiver detects the end of a packet when the normal transitions on the differential RECEIVE inputs cease. Three half-bit periods after the last LOW-to-HIGH transition, the CARRIER PRESENT returns LOW. In that same three half-bit time, the RECEIVE CLOCK completes one last cycle, storing the last data bit. The RECEIVE CLOCK output then goes LOW, and remains LOW.

The collision detecting receiver is very similar in operation to the data receiver. The COLLISION output is asserted when a signal to the collision detector differential inputs, COLLISION + and COLLISION −, is of sufficient amplitude and duration. The COLLISION output will remain HIGH for a minimum of 80-ns and a maximum of 190-ns *after* the last HIGH-to-LOW transition on the COLLISION inputs.

Receive Clock Control

To insure quick capture of incoming data, the receiver phase-locked loop is frequency locked to the transmit oscillator whenever no receiver input is available. Once a frame pulse is detected, the PLL locks into its incoming edges.

Differential Pin Terminations

The differential RECEIVE + and RECEIVE − inputs to the Ethernet cable transceiver must be externally terminated at the chip by two 40.2-ohm resistors and one common-mode bypass capacitor. The chip's differential input impedence and common mode input impedence are engineered so that the Ethernet specification for cable termination impedence is met using standard 1% resistor terminators. The COLLISION differential inputs (+ and −) are terminated in exactly the same way as the RECEIVE inputs.

Transmit Section Timing

The TRANSMIT ENABLE and TRANSMIT DATA inputs are internally synchronized with the TRANSMIT CLOCK, and must meet setup and hold time requirements with respect to the TRANSMIT CLOCK rising edge. In the Transmit Mode, the MK68590 activates the TRANSMIT ENABLE input. At the same time, the first bit of data must be available on the TRANSMIT DATA INPUT.

As long as the TRANSMIT ENABLE remains HIGH, data is clocked into the 3891 interface chip by the TRANSMIT CLOCK and encoded for output to the cable transceiver. The Ethernet cable transceiver interfaces to the 3891 through the differential TRANSMIT + and TRANSMIT − outputs.

When TRANSMIT ENABLE goes LOW, the differential TRANSMIT outputs enter one of two idle states. The selection of Mode I causes the output to seek a HIGH level at the end of the last bit cell time. In Mode II, the output remains HIGH for approximately 50-ns, and then graduates to zero differential voltage across the two outputs. This prevents currents from flowing through the tranceiver's input transformer primary winding while in the idle state.

Power Supply

The 3891 chip is designed to operate from a single 5-volt power supply. The VC_{CC} pin represents the 5-volt input source, while the V_{SS} pin is the ground return.

The MK3891 Serial Interface Adapter is identical in operation—and pinout—to the Am7991 Ethernet Serial Interface Adapter made by Advanced Micro Devices. Additional details concerning its operation are contained in that section. For a better overall view of the entire Ethernet interface system, the reader is urged to review the references presented by the 68590 text.

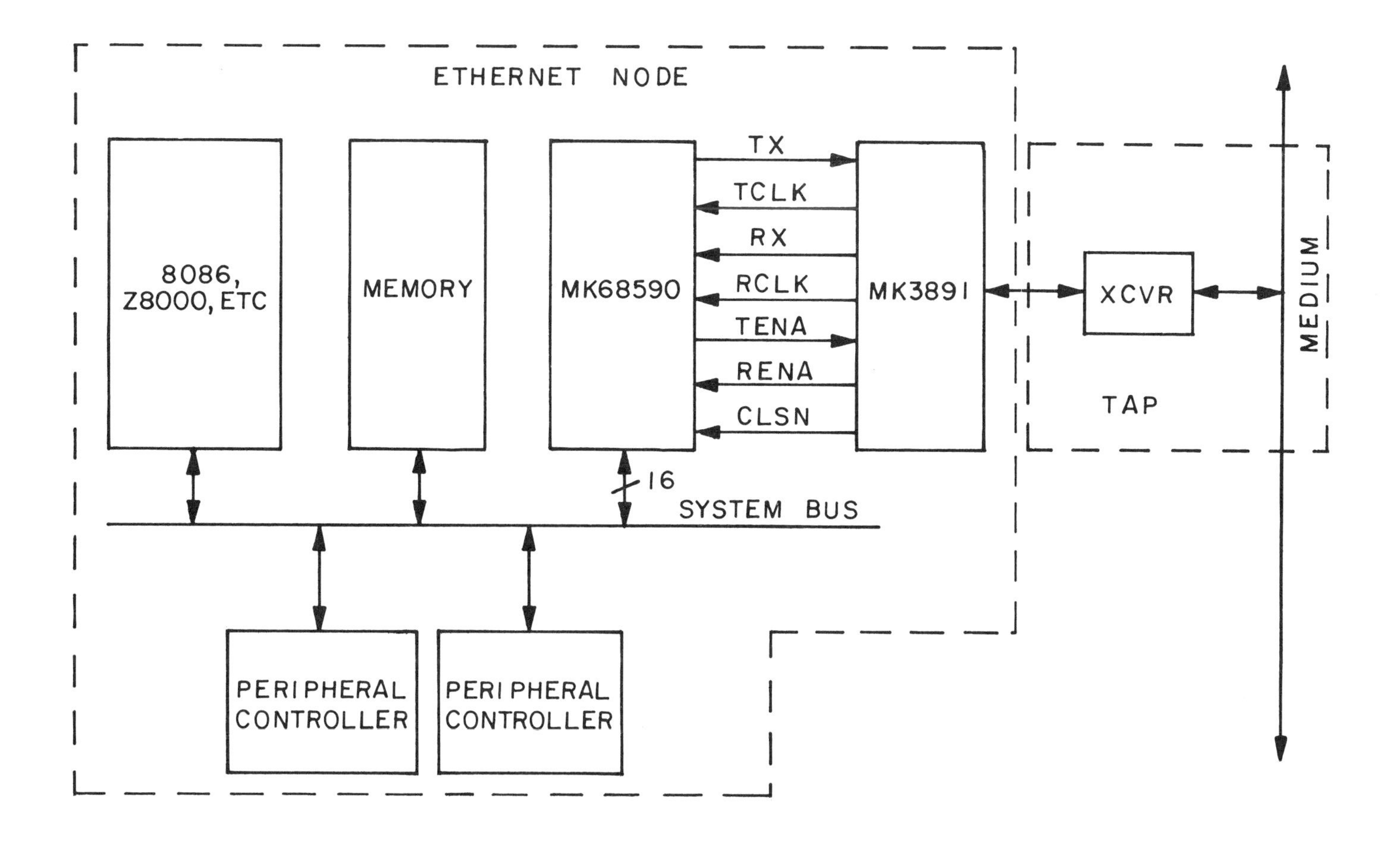

ETHERNET NODE
8086,
Z8000, ETC
MEMORY
MK68590
TX
TCLK
RX
RCLK
TENA
RENA
CLSN
MK3891
XCVR
TAP
MEDIUM
16
SYSTEM BUS
PERIPHERAL
CONTROLLER
PERIPHERAL
CONTROLLER

Figure 1

8001

Ethernet Data Link Controller

The 8001 Ethernet Data Link Controller, by SEEQ Technology, is designed to greatly simplify the development of Ethernet communication in computer-based systems. The chip contains all the logic necessary for interfacing an Ethernet network to a processor. Although Ethernet is a half-duplex system, the transmit and receive sections are independent and can operate simultaneously for diagnostic purposes.

Ethernet Framing

In an Ethernet communication network, information is transmitted and received in packets called frames. An Ethernet frame contains a Preamble, two address fields, a Type Field, a Data Field, and a Frame Check Sequence, in that order.

The Preamble is a 64-bit word consisting of 62 alternating "1"s and "0"s followed by an "11" ending indicator (10101...01011). The first of the address fields is a 6-byte Destination Address containing one of three specific address codes to which the packet is directed. The Destination Address may direct the message to an individual station, a group of stations, or be a general broadcast to all stations. The second address field is a 6-byte Source Address representing the address of the station from which the frame originated. The Type Field is a 16-bit (2-byte) directive used to identify the higher-level protocol associated with the packet.

The Data Field contains the actual message. The length of this field is not fixed, and may be adjusted to accommodate the data involved. However, it must not be shorter than 46-bytes nor longer than 1500-bytes. To ensure data fidelity, all fields, excluding the Preamble, are subjected to an error-detecting algorithm, the results of which are entered into a 32-bit Frame Check Sequence. This final field terminates the frame. The entire packet cannot be more than 1526-bytes long, and a minimum of 9.6-us must be allowed between adjacent packets, regardless of their length.

Data Transmission

The transmit section of the 8001 is responsible for the assembly and transmission of the data packet. The Destination Address, Source Address, Type Field, and Data Field for the frame are stored, in sequence, in an external memory buffer. The CPU simply loads the buffer at its convenience, and the 8001 does the rest.

When the controller gains access to the cable, it transfers these information bytes from the memory buffer into the chip, using an external DMA controller, through an 8-bit RECEIVE/TRANSMIT DATA BUS (pins 6-13). A TRANSMIT READY (pin 16) signal from the 8001 notifies the DMA controller that it is ready to accept buffer input. Data is subsequently transferred on each TRANSMIT WRITE (pin 15) input from the DMA. The transfer of the first byte initiates the transmit cycle, which begins with the 8001-generated Preamble.

Transmission begins with a TRANSMIT ENABLE (pin 3) output that activates the encoder circuit; data is then serially output through the TRANSMIT DATA (pin 4) port at the 10-MHz rate established by the TRANSMIT CLOCK (pin 14). If a collision occurs on the Ethernet cable, as identified by a HIGH level to the COLLISION (pin 24) input, the 8001 attempts a retransmission of the packet by strobing the TRANSMIT RETRANSMIT (pin 5) output. Pin 5 is also asserted if a data underrun occurs during transmission.

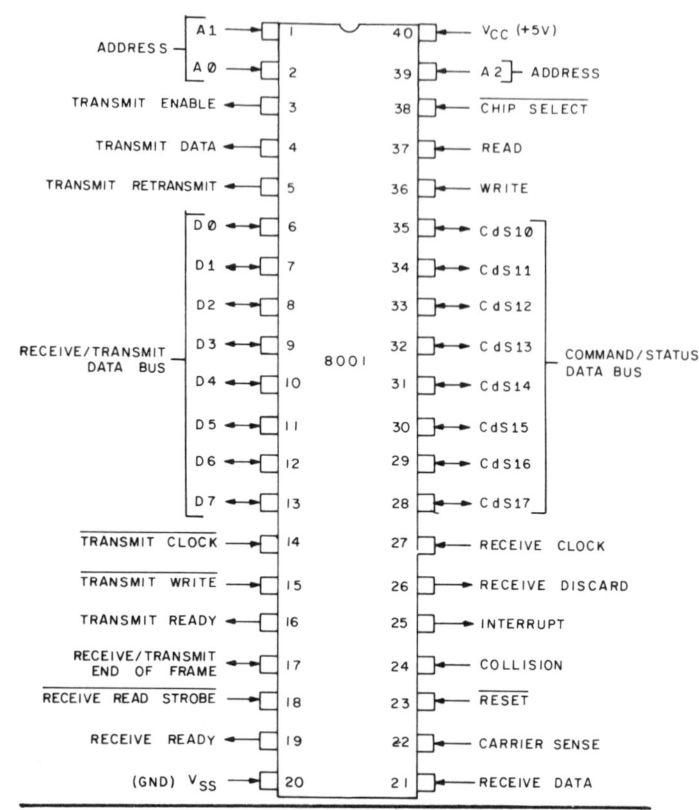

As the last byte of data is placed on the RECEIVE/TRANSMIT DATA BUS, the RECEIVE/TRANSMIT END OF FRAME input (pin 17) is forced HIGH, and the Frame Check Sequence automatically appended.

Data Reception

The 8001 receiver constantly monitors the Ethernet cable via the CARRIER SENSE (pin 22) input. When activity is noted (pin 22, HIGH), the receiver synchronizes the RECEIVE CLOCK (pin 27) with the incoming RECEIVE DATA (pin 21) during the Preamble, then examines the Destination Address.

If the incoming frame is addressed to the 8001 controller, the receiver will pass the frame—exclusive of the Preamble—to the memory buffer through the RECEIVE/TRANSMIT DATA BUS. As soon as the first byte is assembled, the RECEIVE READY output (pin 19) goes HIGH. Under DMA control, data is transferred on a RECEIVE READ STROBE (pin 18) input. A RECEIVE/TRANSMIT END OF FRAME (pin 17) output indicates the end of the message.

If, however, the Destination Address doesn't match, the 8001 will terminate reception as soon as the mismatch is recognized and issue a RECEIVE DISCARD (pin 26) command. This output is also set if the *receive FIFO* is not read before 16-bytes accumulate, creating an overflow condition.

CPU Interface

The CPU interfaces to the 8001 through eight bidirectional COMMAND/STATUS BUS (pins 35-28) lines. The ADDRESS inputs (pins 2, 1, 39) access the six registers which store the station's identification address used by the receiver for Destination Address matching. Furthermore, the ADDRESS inputs also provide access to the transmit and receive *command* and *status registers*.

The COMMAND STATUS DATA BUS, which is totally detached from the memory buffer RECEIVE/TRANSMIT DATA BUS, is tristated until the CHIP SELECT input (pin 38) is enabled. The READ (pin 37) and WRITE (pin 36) input controls the direction of data flow through the bus.

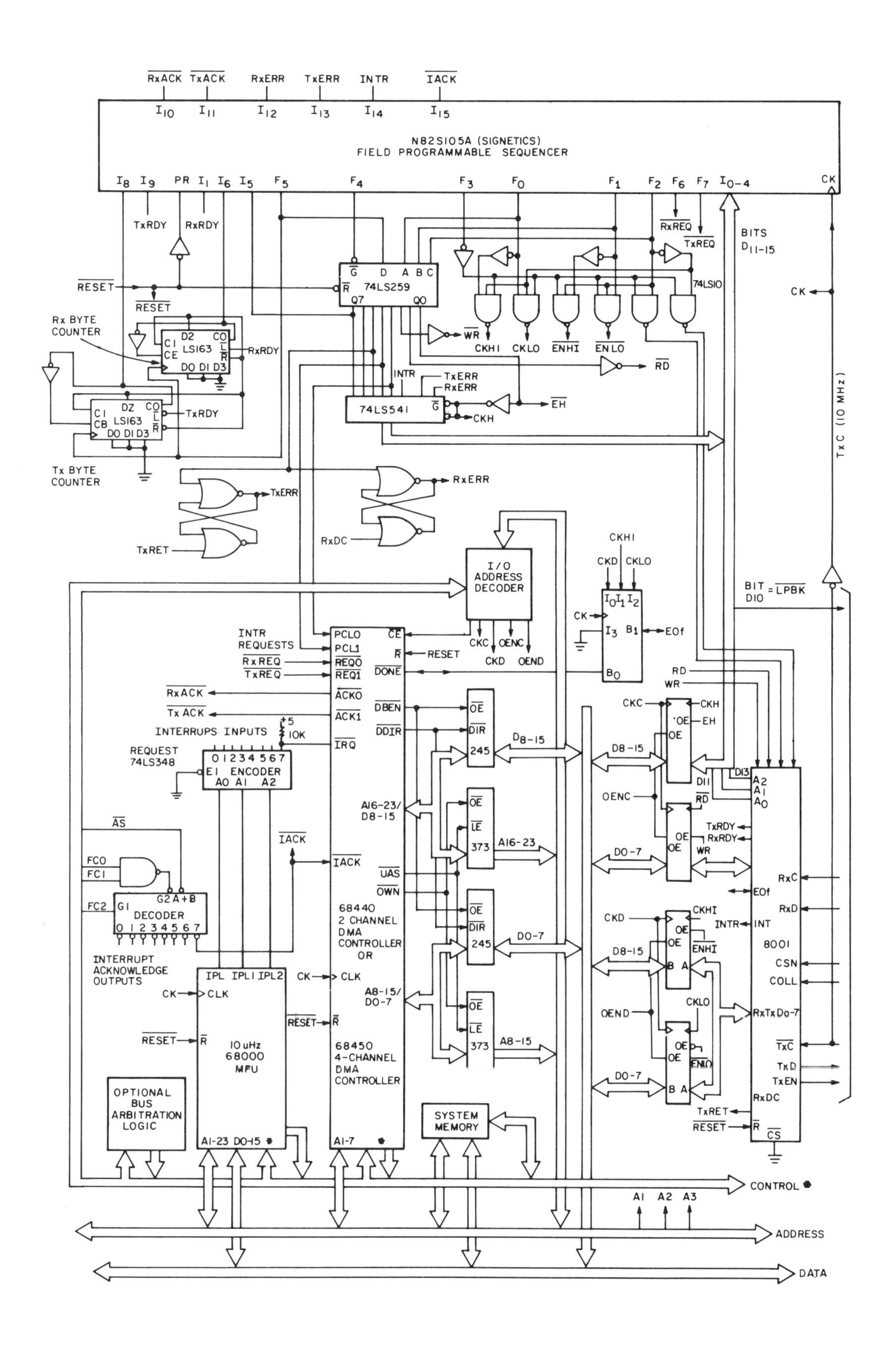

Figure 1

8002

Ethernet Encoder/Decoder

The 8002 Ethernet Encoder/Decoder chip, by SEEQ Technology, provides the Manchester data encoding and decoding functions specified by the cable transceiver, the 8002 provides a high-performance, minimum-cost interface for any system to Ethernet.

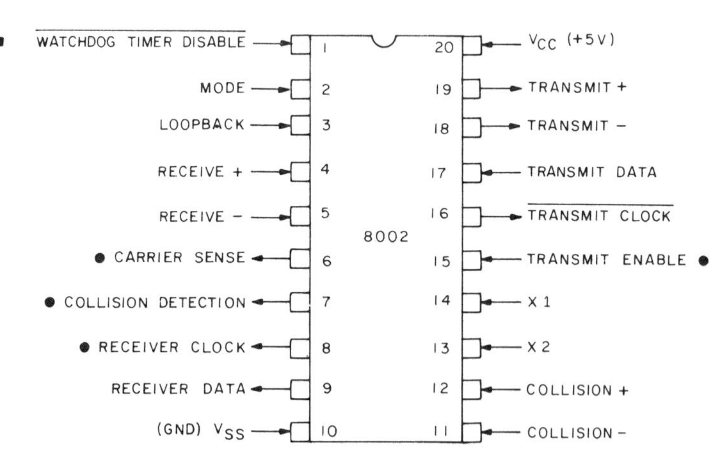

Manchester Encoding

The 8002 encoder combines TRANSMIT DATA (pin 17) information from the Ethernet controller with 10-MHz clock pulses to generate Manchester encoding for the cable transceiver. A 20-MHz crystal placed across the X1 (pin 14) and X2 (pin 13) inpots generates the 10-MHz TRANSMIT CLOCK (pin 16) reference signal. The encoder is enabled when the TRANSMIT ENABLE input (pin 15) is activated.

A pair of differential outputs, TRANSMIT + (pin 19) and TRANSMIT − (pin 18), drives a transformer-coupled Ethernet cable transceiver. When the node is at rest (not sending a message), the differential voltage between these two pins is reduced to less than 0.1-volts to limit the current flowing through the transformer's primary winding. The TRANSMIT drivers are source-followers which need external 240-ohm, 2-watt resistors for loading.

An internal 25-ms watchdog timer is built into the chip. It can be enabled or disabled by the WATCHDOG TIMER DISABLE (pin 1) line. The timer is started at the beginning of the frame transmission. Its purpose is to prevent "babbling" by the station in the event of a component failure. If the transmission ends before the timer expires, the timer is reset. If, however, the timer expires first, the frame is aborted and the TRANSMIT outputs clamped HIGH.

Manchester Decoder

The chip receives Manchester encoded data from the Ethernet through its RECEIVE + (pin 4) and RECEIVE − (pin 5) pins. The receiver input includes a noise filter that provides a static noise margin of −200-mv to −300-mv. The DC threshold and noise rejection filter properties assure that differential RECEIVE data signals less than −160-mv or narrower than 20-ns are always rejected, while signals greater than −240-mv and wider than 30-ns are always accepted.

The filtered data is processed by the data and clock recovery circuit using phase-locked loop technology. The PLL is designed to lock onto the incoming Preamble, obliging a transition width asymmetry of +12-ns to −12-ns, within 12-bit cell times. Whenever there is activity detected on the RECEIVE inputs, the CARRIER SENSE (pin 6) is asserted. This signal activates the 8001 controller receiver circuits.

The PLL separates the clock pulses from the data and transports them out the RECEIVER CLOCK (pin 8) output. The NRZ decoded data is available at the RECEIVER DATA (pin 9) output.

The encoder/decoder chip also includes a collision detection circuit that interfaces with the cable transceiver at the COLLISION (pins 12, 11) inputs. These are differentail inputs that contain a filtering circuit similar to the RECEIVE input, but with modified timing and amplitude parameters. When a collision is detected on the cable, the COLLISION DETECTION output (pin 7) is asserted HIGH.

Differential Input Termination

Both differential inputs, the RECEIVE and COLLISION, must be terminated by a pair of 39-ohm series resistors for proper impedence matching. The center tap of the termination is tied to an external voltage divider, which develops a voltage (in the range of V_{CC} − 0.5V to V_{CC} − 2.5V) that sets the common mode voltage requirements.

MODE CONTROL

Although specifically designed to directly interface to the SEEQ 8001 Ethernet controller, the 8002 will also work with other standard Ethernet controller chips. For instance, Intel's 82586 Local Communications Controller.

However, the Intel controller uses negative logic for four of its control signals that link the encoder/decoder and controller. To satisfy these conditions, the 8002 employs a MODE input (pin 2) that effectively inverts these four functions when asserted.

For on-chip testing of the signal path, the 8002 can be placed into a Loopback Mode by enabling the LOOPBACK (pin 3) input. In this mode, the transmitter output is internally tied to the receiver input, and all internal functions—including the noise rejection filter—are tested. No signals, however, ever exit the chip; the differential output and input lines are clamped. At the end of a test frame transmission, a 600-ns COLLISION DETECTED signal is generated to evaluate the collision circuit. The watchdog timer remains enabled in this mode.

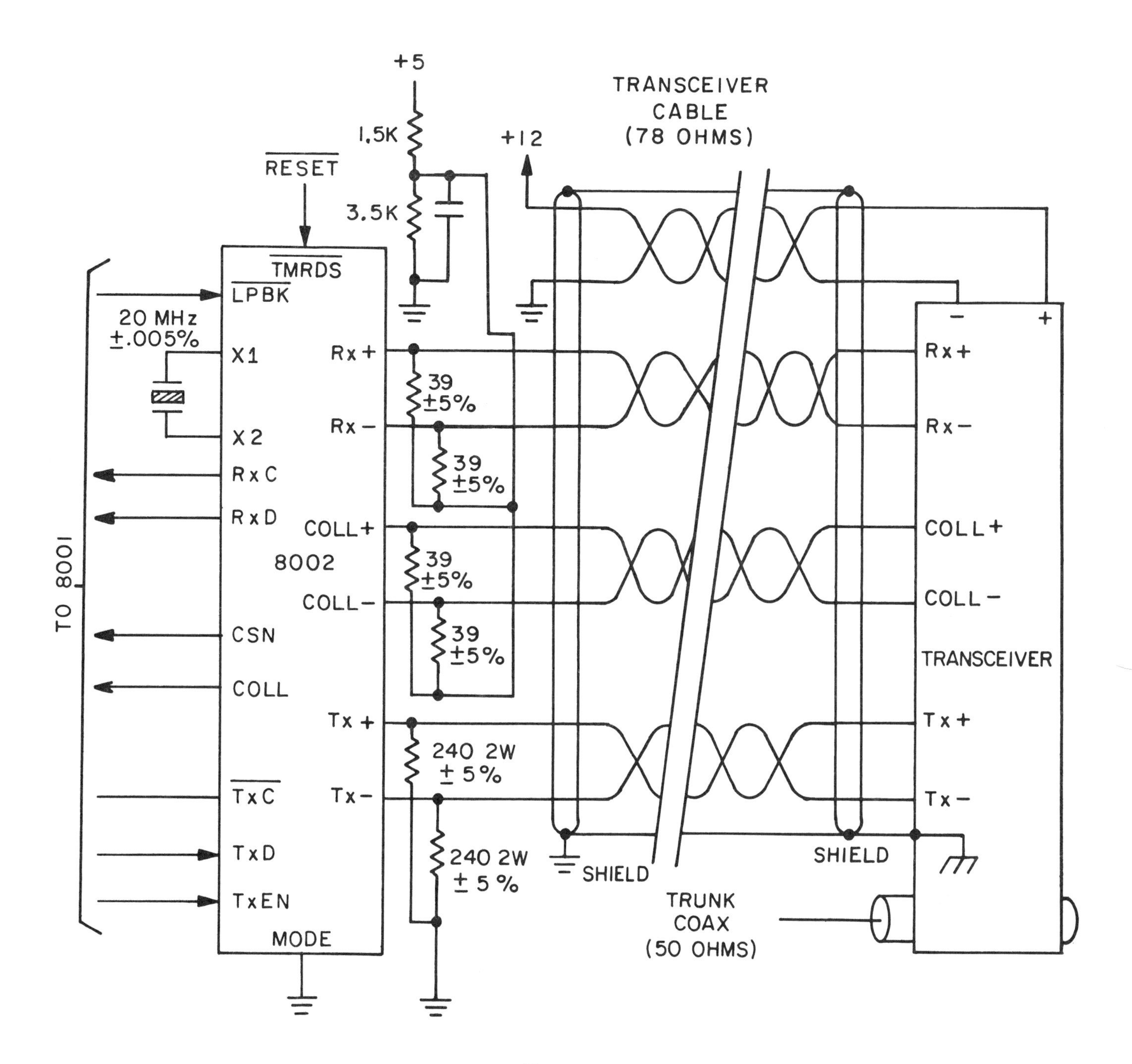
+5
1.5K
3.5K
+12
TRANSCEIVER
CABLE
(78 OHMS)
RESET
TMRDS
LPBK
20 MHz
±.005%
X1
X2
RxC
RxD
8002
CSN
COLL
TxC
TxD
TxEN
MODE
Rx+
Rx−
COLL+
COLL−
Tx+
Tx−
39
±5%
240 2W
±5%
TO 8001
SHIELD
TRUNK
COAX
(50 OHMS)
TRANSCEIVER
−
+

Figure 1

82586

Local Communications Controller

The 82586, by Intel, is an intelligent, high-performance Local Communications Controller designed to relieve the host CPU of many of the tasks associated with controlling local network communications. The device is basically a CSMA/CD communications controller with very flexible framing that permits a wide variety of different techniques to be specified, including Ethernet. Couplid with the 82501 Ethernet serial interface, this chip-set represents a complete solution for Ethernet and other emerging CSMA/CD network applications.

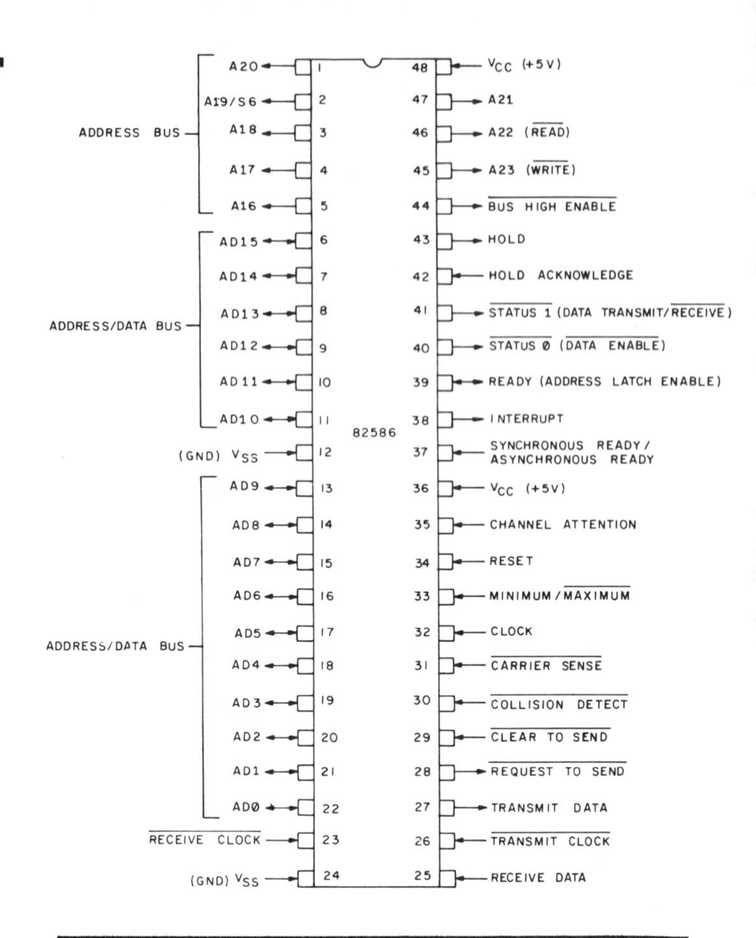

System Architecture

The 82586 implements the entire frame transmission and reception process, including transmit and receive buffer memory management, without CPU intervention. In fact, the controller never communicates with the CPU! The 82586 is essentially a parallel processor that shares a common memory with the CPU, and the only control the CPU exercises over the controller is indirect.

The memory is divided into three major blocks: System Control, Command Block List, and Receive Frame Area. Each block represents a part of the whole network operation.

Memory Structure

Memory is accessed using 24-bit wide addresses. There are two types of addresses: real addresses and segmented addresses. A real address is a single 24-bit entry primarily used to address the transmit and receive data buffers. The segmented address, on the other hand, consists of a 24-bit base with a 16-bit offset code used for acquisition of all Command Blocks, Buffer Descriptors, Frame Descriptors, and System Control Blocks. In general, only the 16-bit portion of the addressed entity is ever specified when accessing one of these memory blocks.

Based on an architecture similar to that of Intel's iAPX 86 processor, the memory is accessed through sixteen bidirectional ADDRESS/DATA BUS (pins 22-13, 11-6) lines. For full 24-bit addresses, the ADDRESS BUS (pins 5-1, 47-45) represents the upper eight bits. Physically, the memory is organized as a high bank (D15-D8) and a low bank (D7-D0) of 512K 8-bit bytes addressed in parallel.

Byte data with even addresses is transferred on the D7 to D0 bus lines, while odd addressed data is transferred on the D15 to D8 bus lines. The controller provides two enable signals, BUS HIGH ENABLE (pin 44) and A0 (pin 22), to selectively allow reading/writing of either or both locations. Whole word transfers (16-bits wide) are restricted to even addressing; therefore, operands are constrained to be arranged with the LSByte located at an even address for maximum performance.

Minimum/Maximum Systems

However, the 48 unique pins surrounding the chip are insufficient to handle the diversity of system and CPU configurations that the 82586 is likely to encounter. Therefore, the user is given an option. When all 24 address bits are required, bus control and management is turned over to an 8288 bus controller chip. This chip decodes the STATUS outputs (pins 40 and 41) to generate, read, write, and enable signals. The READY (pin 29) input synchronizes operations between the 82586 and CPU when in this mode. Pin 37 is a programmable SYNCHRONOUS or ASYNCHRONOUS version of the READY line.

For smaller systems, or stand alone applications, the 82586 can operate in a Minimum Mode. The two modes are determined by the MINIMUM/MAXIMUM (pin 33) input; when HIGH, the chip operates in the Minimum Mode. Displacing the upper 8-bits of ADDRESS, the 82586 chip can now generate its own control signals. Data is transferred to and from memory using the READ (pin 46) and WRITE (pin 45) outputs in conjunction with the ADDRESS LATCH ENABLE (pin 39). DATA ENABLE (pin 40) provides an enable output for the 8286/8287 bus transceivers in stand-alone applications, while the DATA TRANSMIT/RECEIVE (pin 41) controls the direction of data flow through the transceiver.

Controller Commands

The 82586 executes commands from the Command Block List portion of the memory. The controller has a repertoire of eight commands, all of which are fetched and executed in parallel with the host CPU's operations. When the CPU issues a command, placing it into memory, it notifies the controller by forcing the CHANNEL ATTENTION input (pin 35) HIGH—which happens to be one of only two lines that directly links the 82586 and CPU. Access to memory is requested by raising the HOLD output (pin 43) HIGH, and is granted when a HOLD ACKNOWLEDGE (pin 42) is received. After the command has been executed, the 82586 responds with an INTERRUPT (pin 38), which remains in effect until another command is ordered.

One such command is the Transmit command. For a full description of Ethernet network functions, refer to the 82501 chip.

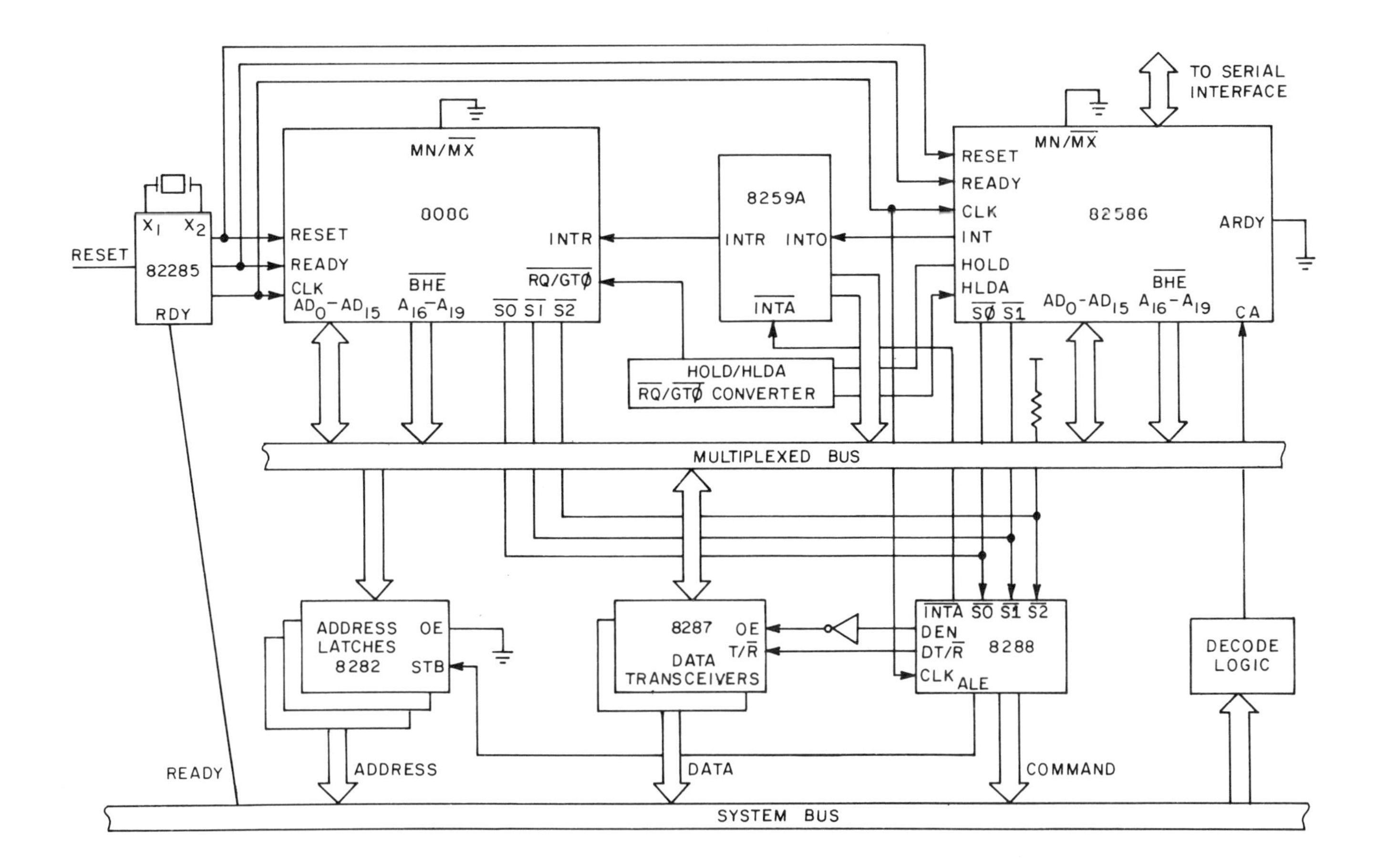
TO SERIAL
INTERFACE
MN/MX
RESET
READY
CLK
INT
HOLD
HLDA
SØ S1
82586
ARDY
AD0-AD15
BHE
A16-A19
CA
8259A
INTR INTO
INTA
X1 X2
RESET
82285
RDY
8086
RESET
READY
CLK
AD0-AD15
BHE
A16-A19
INTR
RQ/GTØ
SO S1 S2
HOLD/HLDA
RQ/GTØ CONVERTER
MULTIPLEXED BUS
ADDRESS
LATCHES
8282
OE
STB
8287
DATA
TRANSCEIVERS
OE
T/R
INTA SO S1 S2
DEN
DT/R
8288
CLK
ALE
DECODE
LOGIC
READY
ADDRESS
DATA
COMMAND
SYSTEM BUS

Figure 1

82501

Ethernet Serial Interface

The 82501 Ethernet Serial Interface and encoder chip, by Intel, is designed to work directly with the Intel 82586 controller in Ethernet and non-Ethernet 10Mbps local area network applications. When combined with the 82586 controller chip, the Ethernet interface is complete from the CPU to the cable transceiver.

Transmit

Outgoing data is fed from the 82586 controller into the 82501 TRANSMIT DATA (pin 17) input. From here the message is Manchester encoded and presented to the network when the TRANSMIT ENABLE (pin 15) pin is LOW. The TRANSMIT ENABLE input is activated by a REQUEST TO SEND signal from the 82586.

The interface chip supplies encoded data to the Ethernet cable via a cable receiver tap. An output driver, composed of a complementary pair of TRANSMIT (pins 18, 19) outputs, generates a differential signal for the transceiver. Since the first bit of any frame is always a "1," the first transition of the TRANSMIT (pin 19) output is always negative, while the last transition is always positive. Following the last transition, the $\overline{\text{TRANSMIT}}$ (pin 18) output is slowly brought to its HIGH state so that zero differential voltage exists across the outputs, thus eliminating DC currents in the primary of the transceiver's coupling transformer between messages. Please note that the cable driver is a differential gate requiring either external pull-down resistors or a 20-mA current sink on both outputs.

An on-chip fail-safe watchdog timer circuit prevents the station from locking up in a continuous transmit mode.

Clock

The 82501 also generates the 10-MHz clock signal necessary to implement Ethernet communications. A 20-MHz crystal placed across the X1 (pin 14) and X2 (pin 13) clock inputs provides the basic clock source. This 20-MHz signal is then divided by two to supply the TRANSMIT CLOCK (pin 16) signal for the 82586 TRANSMIT CLOCK input.

Receive

The cable transceiver transforms the single-ended coax signal into a balanced differential voltage, which is input to the 82501's RECEIVE (pins 4, 5) inputs. This balanced line must be terminated with a pair of 39-ohm resistors across the inputs. The center tap of the terminating resistance (refer to fig. 1) is tied to an external voltage reference (source impedence of 18.5-ohms minimum) to establish a common mode bias voltage that is needed for the 82501 receive circuitry.

A noise filter is provided at the RECEIVE pair inputs to prevent spurious signals from improperly triggering the receiver. As soon as the first valid negative pulse is recognized by the noise filter, the CARRIER SENSE (pin 6) is asserted, alerting the 82586 controller to the beginning of a frame.

An internal analog phase-locked loop is used to extract the receive clock signal from the data stream. The PLL will acquire a lock within 12-bit periods, during which time the RECEIVE CLOCK (pin 8) output is held LOW. Some bit cell timing distortion can be tolerated, as the PLL corrects its frequency to match the incoming signal transitions. An external capacitor across C1

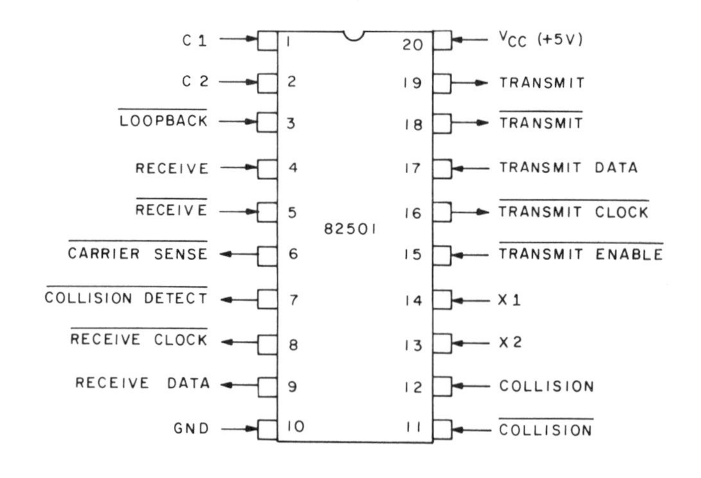

(pin 1) and C2 (pin 2) establishes the oscillator's center frequency. The receiver decodes the cable input from Manchester to NRZ, then serially transfers it, in step with the RECEIVE CLOCK, to the 82586 controller through the RECEIVE DATA (pin 9) output.

Immediately after the end of a message, the filter blocks all incoming signals for 5-us minimum and 7-us maximum, thus removing unwanted pulses that may occur on the coaxial cable following a transmission. The receiver clock detects the end of a frame, and reports it to the 82586 by disengaging the CARRIER SENSE output while holding RECEIVE DATA HIGH.

Collision Detection

The cable transceiver monitors the Ethernet cable for signal collisions. It reports its findings to the interface chip through a pair of COLLISION (pins 12, 11) inputs. A valid collision-presence signal will activate the COLLISION DETECT output (pin 7), which is tied to the COLLISION DETECT input of the 82586 controller. The common mode voltage and line termination requirements for the COLLISION inputs are identical to the RECEIVE inputs.

Loopback

In the normal mode, the 82501 operates as a full-duplex device, capable of transmitting and receiving simultaneously. When the LOOPBACK input (pin 3) is asserted, the 82501 internally routes the serial data from its TRANSMIT DATA input through the transmitter logic, retiming and Manchester encoding it, then returns it through its receiver logic, where it is decoded and the RECEIVE CLOCK generated. This loopback function provides a complete check of the chip's operations, as if the signal were processed from the cable, without actually passing the signal through the cable.

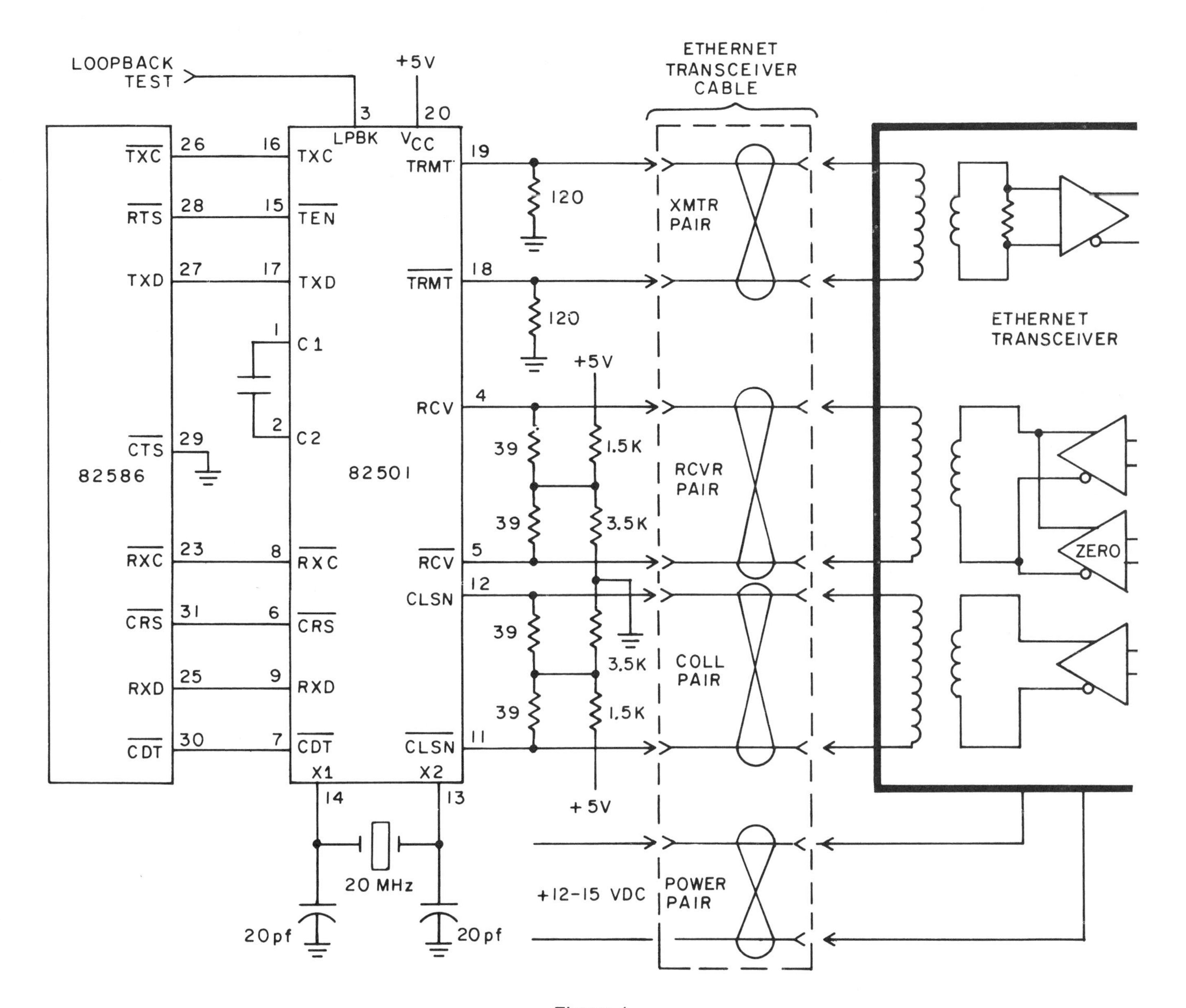

LOOPBACK TEST
+5V
ETHERNET TRANSCEIVER CABLE
3
20
LPBK
VCC
TXC
TEN
TXD
C1
C2
CTS
82586
82501
RXC
CRS
RXD
CDT
TRMT
RCV
CLSN
X1
X2
20 MHz
20pf
120
39
1.5K
3.5K
XMTR PAIR
RCVR PAIR
COLL PAIR
POWER PAIR
+12-15 VDC
ETHERNET TRANSCEIVER
ZERO

Figure 1

CHAPTER TWO
POWER SUPPLY AND SPECIAL PURPOSE

UC1524

Regulating Pulse Width Modulator

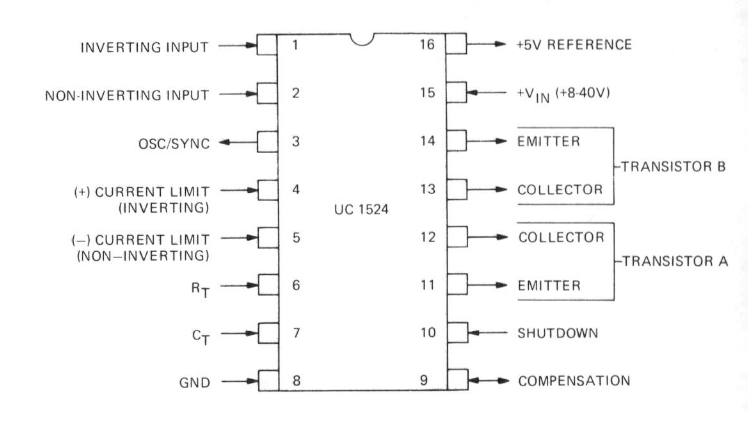

The UC1524 is a fixed-frequency Pulse Width Modulated voltage regulator chip made by Unitrode. This chip contains all the functions necessary for the construction of a switching power supply. It includes a programmable sawtooth oscillator, error amplifiers, a stable 5-volt reference source, pulse steering flip-flop, current limiting and shutdown circuitry, and a pair of uncommitted output transistors capable of controlling 200-mA at 40-volts.

Oscillator

The regulator operates by sampling the output voltage and comparing it to a fixed reference. The sampling rate is determined by an internal sawtooth oscillator whose frequency is programmed with one timing resistor, R_T, and one timing capacitor, C_T. Programmable from -0.03-mA to -2-mA, resistor R_T (pin 6) establishes a constant current through pin 7, the timing capacitor C_T, resulting in a linear voltage ramp that the comparator uses to determine the pulse width. Practical values for R_T fall between 1.8k and 100k; the range for C_T is 0.001-uf to 0.1-uf.

An external synchronous clock signal of 3-volts can be applied directly to the OSC/SYNC (pin 3) output to force synchronization of the switching frequency, provided R_T and C_T are selected to run slightly higher than the clocking frequency. If two or more regulator chips are used in a single design, it is advantageous to have them synchronized. When this is the case, all OSC/SYNC outputs must be tied together, and all C_T pins are wired to a common capacitor. In this configuration, only one R_T connection is required; all other R_T pins are left unconnected.

Error Amplifiers

The chip contains two error sensing amplifiers; one for voltage, the other for current. The chip also contains an on-board 5-volt regulator that serves as a REFERENCE source (pin 16) for the error amps, as well as powering the UC1524's internal circuitry. Since the error amps are powered by the reference source, however, the REFERENCE voltage falls outside the common mode input limits, and it is necessary to use a resistive voltage divider to bring it within the common mode range.

The current limiting amplifier is designed to sense the voltage drop across a small resistor placed in the ground line. Gain of both amplifiers can be controlled with a feedback resistor between the COMPENSATION input (pin 9) and the INVERTING INPUTs (pins 1 & 4, voltage and current, respectively), but are most often used in the open loop mode. In either case, a frequency compensating network to ground is recommended at the COMPENSATION pin.

Output Circuit

The amplifiers' output voltages are fed to the voltage comparator and compared to the voltage ramp generated by C_T. The result is a pulse with modulated width (variable duty cycle) that is steered to the output transistors by the flip-flop. Each transistor is alternately driven, and may be used in either single-ended or push-pull circuits.

When used in the single-ended mode, both transistors are paralleled, resulting in a switching frequency equal to the oscillator frequency. In push-pull designs, on the other hand, the output pulses are divided between the transistors and the operating frequency is only half the switching frequency.

Blanking

In addition to providing a linear ramp for the comparator, timing capacitor C_T also generates a blanking pulse to the output transistors, assuring that the two are never on simultaneously.

If small values of C_T are required for frequency control, the blanking time can be increased by shunting a 100-pf capacitor from pin 3 to ground. Still greater dead times can be realized by clamping the output of the error amplifier, as shown in fig. 3.

Shutdown

The outputs of the error amplifiers are active HIGH, and share a common input with the comparator. Also tied to the comparator is a shutdown transistoir. The base of this transistor is accessible through pin 10, the SHUTDOWN CONTROL, and a HIGH input to this pin will cause the transistor to conduct, thus shorting the comparator to ground. Until the SHUTDOWN signal is removed, further operations of the modulator are halted.

Improving the UC1524

A family of UC1524A ICs have been developed to retain the same desirable versatility of the UC1524, while offering some substantial improvements. Most noteworthy is the fact that the common mode range of the error amplifiers has been extended to include the reference voltage, thus eliminating the REFERENCE voltage divider, and the current sensing amp has been modified to monitor currents in the power supply output line as well as the ground lines. The "A versions" are totally pin compatible with the UC1524.

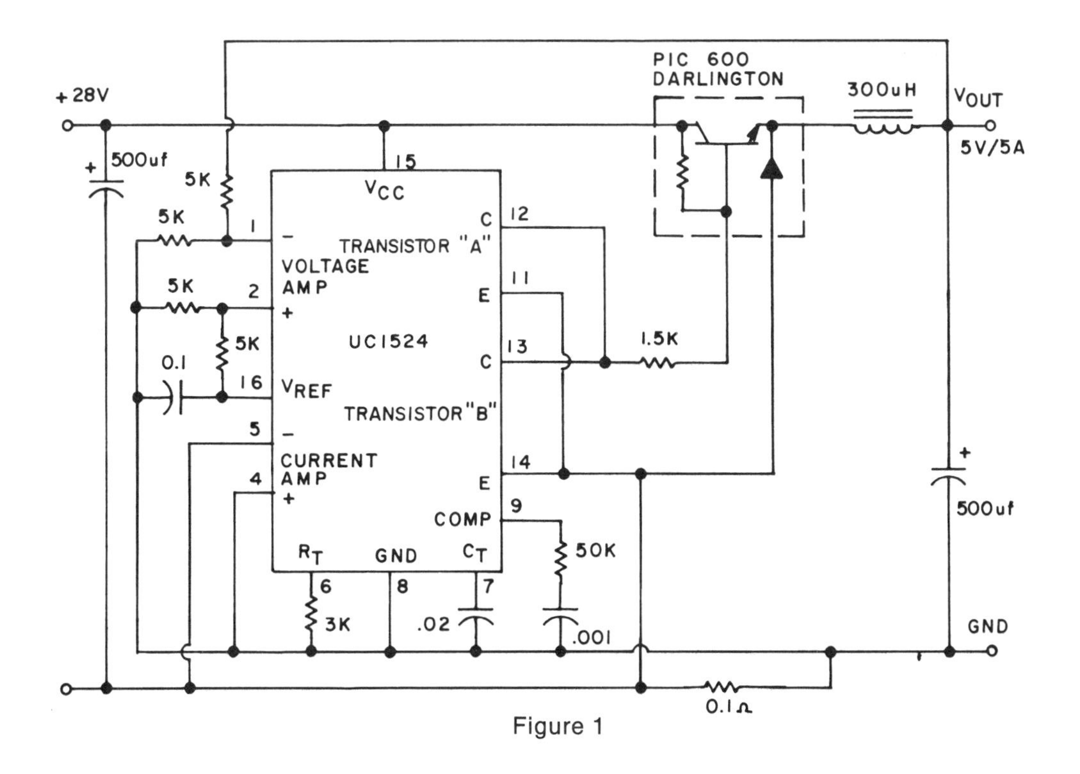

Figure 1

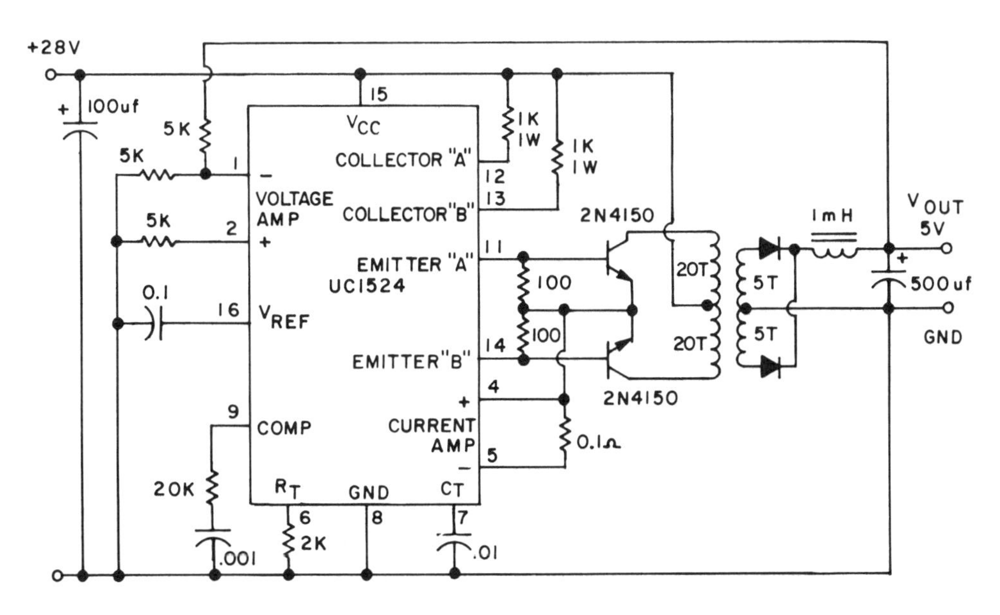

Figure 2

Figure 3

$$f = \frac{1.18}{R_T C_T}$$

MC34060

Switchmode Pulse Width Modulation Control Circuit

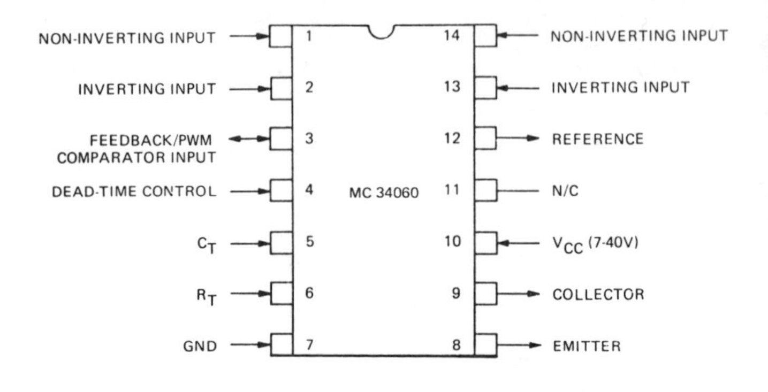

The MC34060, introduced by Motorola, is a fixed-frequency Pulse Width Modulation Control Circuit designed primarily for use in switching power supplies. The chip contains a programmable sawtooth oscillator, two error sensing amplifiers, a stable 5-volt reference source, and an uncommitted output transistor capable of sinking or sourcing 200-mA.

Operating Frequency

Switching power supplies regulate voltage by outputting a constant-amplitude fixed-frequency pulse with a variable duty cycle. As the duty cycle decreases, less average power is available to the load, and the output voltage—when integrated by an output filtering circuit—decreases. As more power is demanded, the pulse width is increased to maintain the proper voltage level.

The operating frequency determines the number of voltage samples taken every second. The recommended oscillator frequency spans the spectrum from 1-kHz to 200-kHz, with all specifications measured at 25-kHz. The switching frequency is controlled by an internal sawtooth oscillator that is programmed with two external components; a resistor (R_T) and a capacitor (C_T), which connect between ground and the R_T (pin 6) and C_T (pin 5) control pins, respectively.

Ranging in value from 1.8k to 500k, the timing resistor is configured to supply a constant current to the charging capacitor. Voltage charge across the capacitor develops a linear ramp, which is monitored by an internal voltage comparator. Pulse width modulation is accomplished by comparing the sawtooth voltage across C_T to control signals from two error amplifiers.

Error Amplifiers

The chip contains two operational amplifiers, each with a common mode input range of − 0.3-V to V_{CC} − 2-V, which monitor the output performance of the regulator. Amplifier 1 is intended for voltage sensing, and uses the internal 5-volt source for reference. Amplifier 2 is used as a current limiter, sensing current flow by measuring the voltage drop across a small resistor in the negative line.

The output of the amplifiers are actively HIGH, and both outputs are ORed together and fed to the comparator, which in turn adjusts the duty cycle. With this configuration, the amplifier that demands the minimum ON time dominates control of the loop.

The ORed outputs are also made available to the user through pin 3, the FEEDBACK/PWM COMPARATOR INPUT. The gain of the error amplifiers is set by looping a feedback resistor from the FEEDBACK pin to the INVERTING inputs (pins 2 & 13 for amps 1 and 2, respectively). The NON-INVERTING inputs are pinned out at pin 1 for amp 1 and pin 14 for amp 2. The controller is capable of regulating both positive and negative voltages, as demonstrated by the two sensing arrangements shown on the schematic.

Duty Cycle

Although the duty cycle is primarily the responsibility of the MC 34060, the user can exercise some control over it externally. The maximum duty cycle is set by the DEAD-TIME CONTROL (pin 4) and is 96% maximum. With no input to pin 4, an offset of approximately 120-mV exists, limiting the duty cycle to 96%. However, as the control voltage increases from 0-V to 3.3-V, the duty cycle decreases until it reaches zero.

The inclusion of a DEAD-TIME CONTROL allows the designer to build circuits which have soft turn-on times. Of course, the final decision for duty cycle is made by the controller; voltage on the DEAD-TIME CONTROL pin simply limits the *maximum* duty time the pulse may have.

Internal Reference

A highly stable internal REFERENCE (pin 12) voltage has been incorporated for accurate voltage control to the error amplifiers. This source is capable of supplying up to 10-mA of current and can be used to power external circuits when the need arises. The source is temperature stabilized to less than 50-mV over its entire operating temperature range, and has a nominal value of ±5%.

Output Circuits

The output circuit contains an on-chip transistor capable of handling 200-mA of current, which is adequate for many applications. However, the transistor can be used as a current controller for external pass transistors, thus multiplying the current controlling capabilities.

The transistor is totally uncommitted, with the exception of the base input, and is available at pin 9 for the collector and pin 8 for the emitter. Operating voltages across this transistor must not exceed 42-volts; the chip itself is limited to 40-volts.

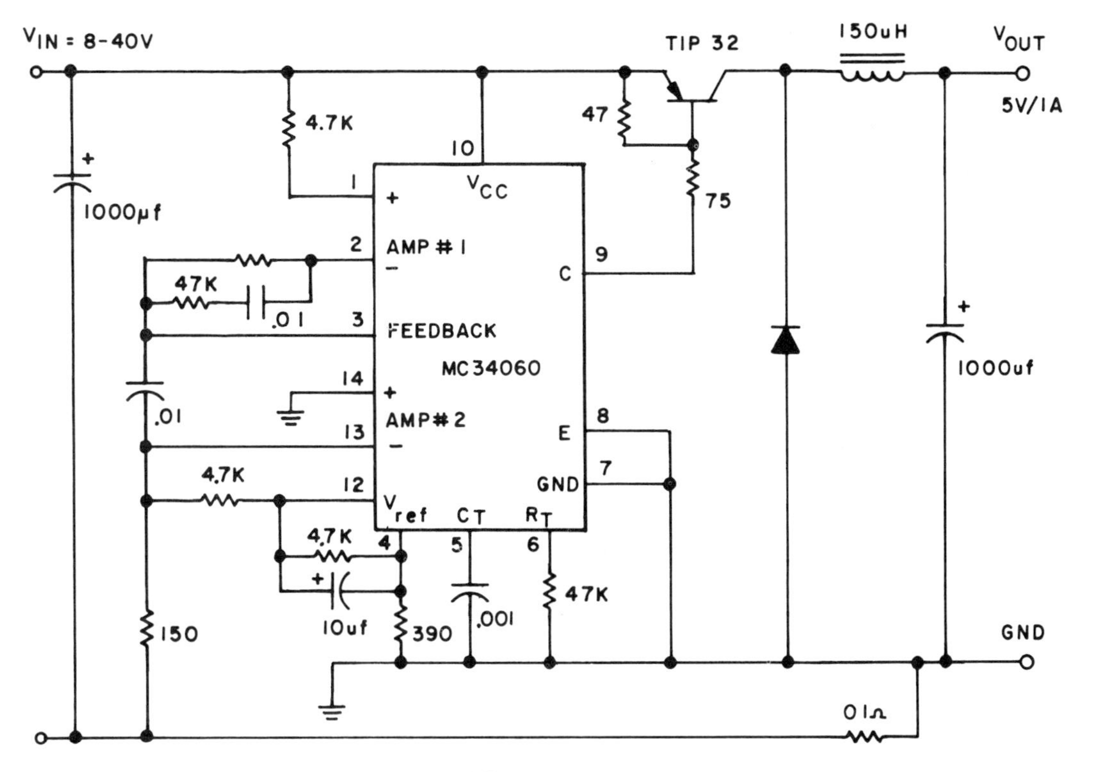

Figure 1

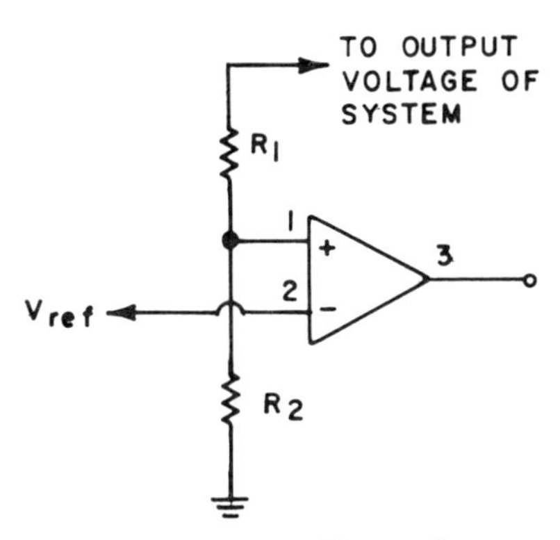

Figure 2

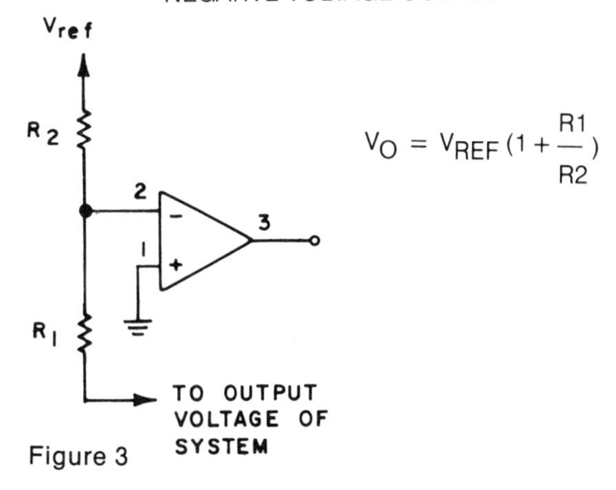

$$V_O = V_{REF}\left(1 + \frac{R1}{R2}\right)$$

Figure 3

DEAD-TIME CONTROL

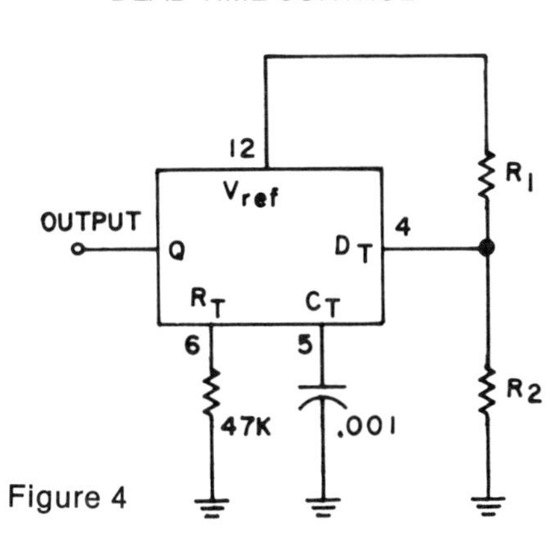

Figure 4

SOFT START

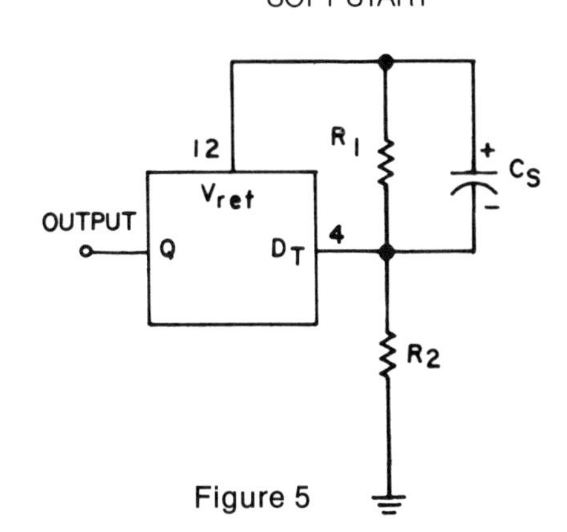

Figure 5

CDP 1871

Keyboard Encoder

The CDP 1871, made by RCA, is a Keyboard Encoder designed to directly interface between the RCA CDP1802 or CDP1804 microprocessor and a mechanical SPST keyboard array. With the proper controls, however, it will easily interface to any 8-bit bus line. The encoder chip scans and generates code for 53 ASCII coded keys plus 32 HEX coded keys.

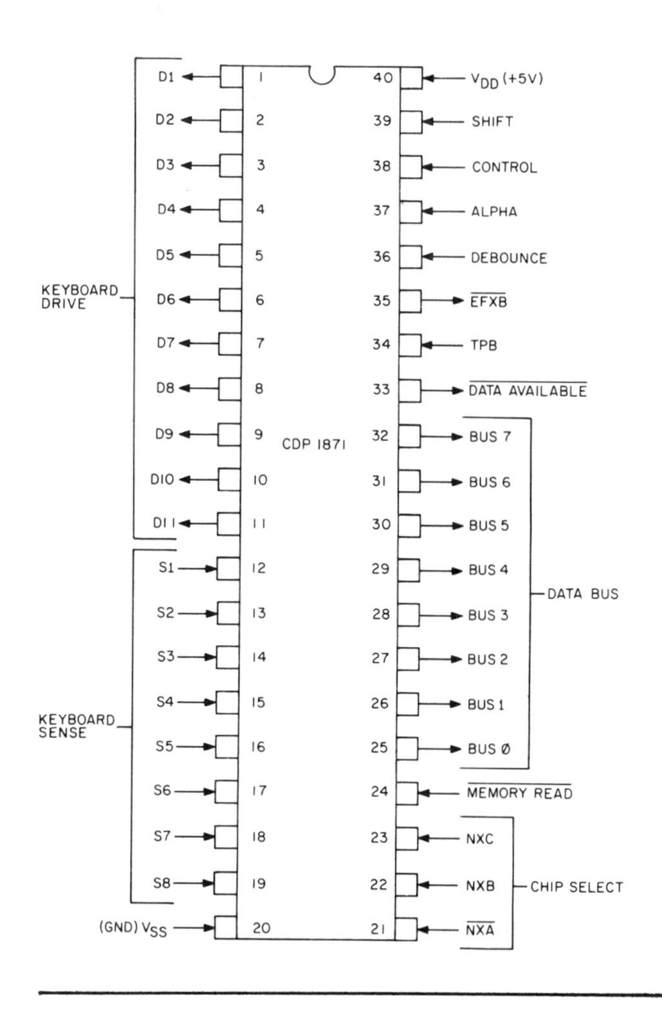

Keyboard Interface

The 1871 is made up of two major sections: the counter/scan logic and the control logic. The counter/scan logic drives the D1 through D11 KEYBOARD DRIVE lines (pins 1-11) HIGH one at a time, which scan the input lines to the keyboard. As the pulse ripples across the lines, activated keys are matrix sampled by the KEYBOARD SENSE inputs (pins 12-19) one keyboard output at a time. The drive lines have open-drain outputs which source the internal pull-down resistors of the KEYBOARD SENSE inputs, causing them to go HIGH. Each output line is capable of driving up to eight keys. Once a keystroke is detected, line scanning ceases and the DATA AVAILABLE output (pin 33) goes LOW.

Control Logic

The drive lines are, in turn, scanned by an internal 5-stage counter. The scan counters are clocked by the TPB input (pin 34), a clock pulse that is normally derived from the CPU. When a keydown condition is detected on a sense line, the control logic inhibits the clock pulse to the scan counters and sets the DATA AVAILABLE output LOW. This signal is interpreted by the microprocessor as an Interrupt Request.

The information stored inside the scan counters represents the ASCII or HEX key code of the depressed key, which is presented to the DATA BUS outputs (pins 25-32) when requested by the CPU. The DATA BUS outputs, which are normally in the tri-state, are enabled by decoding the NXA (pin 21), NXB (pin 22), and NXC (pin 23) chip select inputs. Inputs NXB and NXC must be HIGH; NXA is LOW. A strobe pulse to the MEMORY READ input (pin 24, LOW) is also required before the data is transferred.

A TPB clock pulse at the end of the read cycle resets the DATA AVAILABLE output HIGH and resumes scanning procedures.

Alpha, Shift, and Control

The outputs of the scan counter are conditionally encoded by the ALPHA (pin 37), SHIFT (pin 39), and CONTROL (pin 38) inputs. When any of these inputs is set HIGH, drive and sense decoding is modified. Although all three inputs are connected to the keyboard, they produce no code by themselves.

The SHIFT and CONTROL inputs have internal pull-down resistors to simplify use with momentary-contact switches. The ALPHA input, on the other hand, is designed for a standard SPDT switch to provide an alpha-lockout function. In all cases, the CONTROL function takes precedence over both the SHIFT and ALPHA.

Debounce

After the depressed key is released, a debounce delay is enabled. The DEBOUNCE input (pin 36) provides a pinout for an external circuit to eliminate false triggering caused by the key's release. This pin is connected to the junction of a series RC timer circuit. If the debounce feature is not required, the pull-down resistor must still be included.

Once the debounce delay has occurred, the clock inhibit is removed, allowing the scan count to continue. This provides an N-key lockout, which prevents the entry of erroneous codes when two or more keys are engaged at the same time. Only the first key is recognized, while all other key depressions are ignored until the first key is read by the CPU.

Auto Repeat

If the first key remains depressed after the DATA AVAILABLE output is reset, an auxiliary data available signal (EFXB, pin 35) is generated and made available to the CPU to indicate a possible auto-repeat condition. The auto repeat mode will depend upon the decision by the microprocessor to perform another I/O read cycle, or ignore the condition until the key is returned to normal. This output signal is usually connected to a flag input of the CPU.

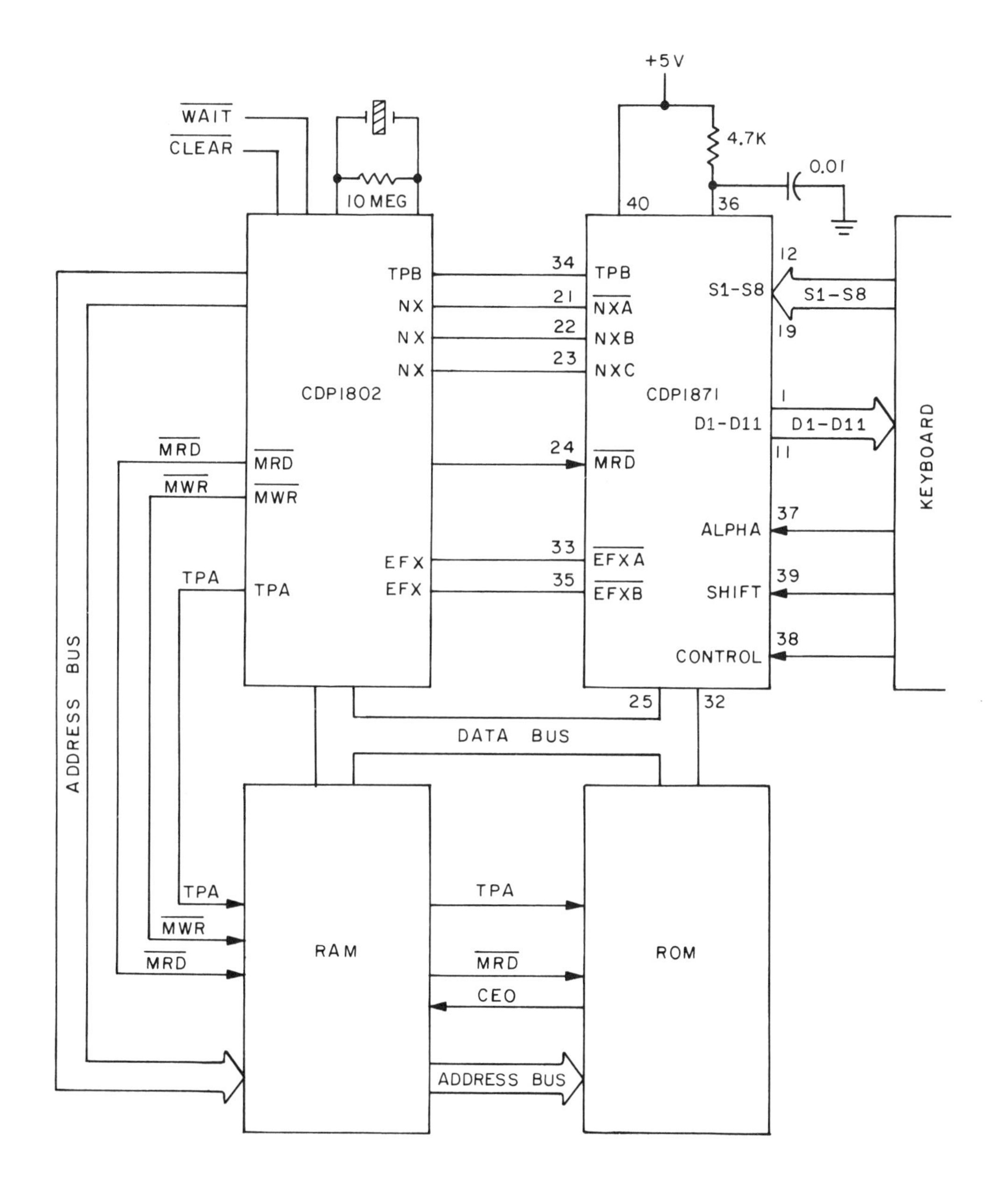
+5V
WAIT
CLEAR
10 MEG
4.7K
0.01
40
36
CDP1802
CDP1871
TPB
NX
NX
NX
34
21
22
23
TPB
NXA
NXB
NXC
12
S1-S8
S1-S8
19
1
D1-D11
D1-D11
11
KEYBOARD
MRD
MRD
MWR
MWR
24
MRD
37
ALPHA
33
EFX
EFXA
TPA
TPA
EFX
35
EFXB
39
SHIFT
38
CONTROL
ADDRESS BUS
25
32
DATA BUS
TPA
MWR
MRD
RAM
TPA
MRD
CEO
ADDRESS BUS
ROM

Figure 1

8279

Programmable Keyboard/ Display Interface

The 8279, by Intel, is a general purpose Programmable Keyboard and Display Interface device. Since data input and display are an integral part of most microprocessor systems, the 8279 has been designed to provide these functions without placing a large burden on the CPU.

The 8279 has two sections: keyboard and display. The keyboard section can interface to any typewriter-style keyboard—or random toggle/thumb switch array—while the display section drives a matrix-scanned numerical or alphanumeric display up to 16 characters in length.

Keyboard Interface

The Keyboard portion can provide an ASCII interface to any 64-contact key matrix with 2-key lockout and N-key rollover functions. The 8279 can also interface to an array of sensors or a strobed keyboard, such as the hall effect and ferrite variety, when operated in the Sensor Matrix and Strobed modes.

As each row is activated, the controller scans for a key closure using eight RETURN LINE (pins 38, 39, 1, 2, 5-8) inputs. If a down key condition is detected, a debounce circuit stops the scanning for about 10-msec then samples the keyswitch again. If it is still valid, the address of the key plus the status of the CONTROL (pin 37) and SHIFT (pin 36) inputs are transferred into the *FIFO RAM register*. This register can accumulate up to eight characters, and any time there is data in the *FIFO RAM register*, the INTERRUPT REQUEST (pin 4) flag is set.

The Sensor Matrix Mode allows the 8279 to detect keystrokes other than those generated by mechanical switches. In this mode, any logic that can be triggered by the SCAN LINEs may be used to enter data on the RETURN LINE inputs. The input maps directly to an *FIFO RAM* position—except the CONTROL and SHIFT inputs, which are ignored.

In the Strobed Mode, the data is also entered directly into the *FIFO RAM* through the RETURN LINEs on the rising edge of a STROBE (pin 37) input pulse. The RETURN LINE inputs can also be used as a general purpose strobed input when in this mode.

CPU Interface

The CPU reads the contents of the *FIFO RAM register* through eight DATA BUS (pins 12-19) lines. Data is read with the READ input (pin 10) set LOW and a strobe to the CHIP SELECT (pin 22) input. When the chip is not selected, the DATA BUS assumes a high-impedence state.

To set the operating mode, the control registers are loaded with directives via the same DATA BUS. The BUFFER ADDRESS (pin 21) is used to identify the DATA BUS word. When the BUFFER ADDRESS input is HIGH, the data on the DATA BUS is a command or status. A logic LOW indicates information data is present.

Information data is input to the 8279 for display purposes. With the WRITE input (pin 11) held LOW, data enters the 8279 on the receipt of a CHIP SELECT strobe.

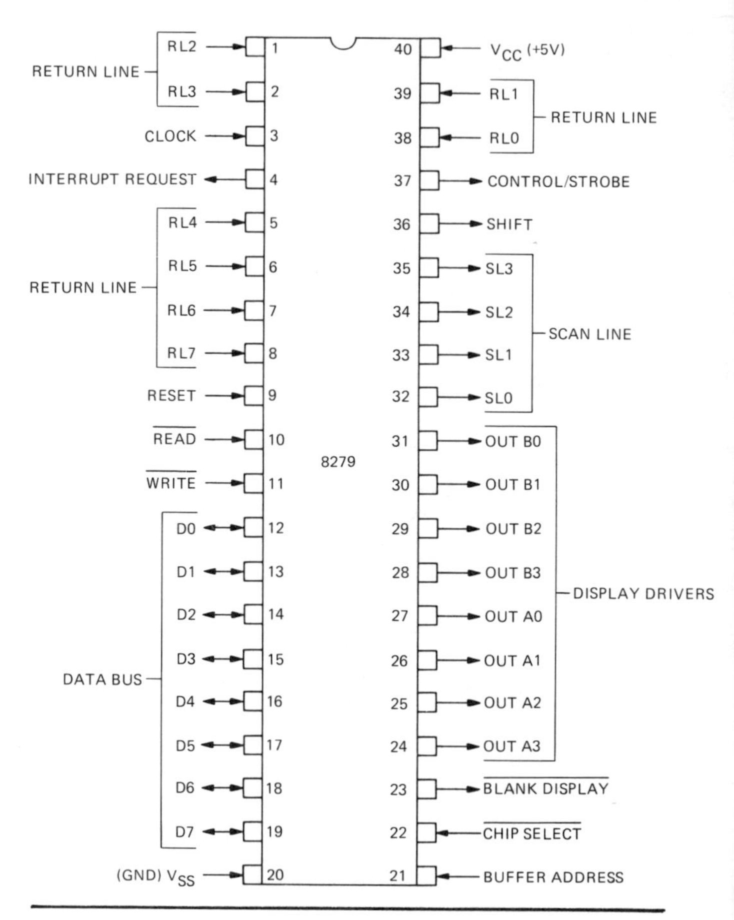

Display Interface

The display section has a 16×8 *display RAM* which stores the data for display and provides a matrix scanned interface for LED, incandescent, and other popular display technologies.

The *display RAM*, which can be split into dual 16×4 RAMs, is loaded with display data by the CPU. The *display RAM* has no formal organization itself, it merely reflects the input of the CPU, and can even represent ASCII or other popular codes. Each bit of the *RAM* is buffered through the *display registers* and output to the DISPLAY DRIVERS.

The DISPLAY DRIVERS are grouped into two 4-bit nibbles to facilitate display interfacing. The first half of the *display RAM* is accessible at OUT A0 to OUT A3 (pins 27-24) and the second half at OUT B0 to OUT B3 (pins 31-28). When these two ports are concantenated, they form a single 8-bit output.

The DISPLAY DRIVERS can be used to directly drive the segments of a numerical display, or they can be encoded to drive alphanumeric and dot matrix readouts, depending upon the configuration of the *display RAM*. When used in conjunction with the SCAN LINE drivers, the 8279 will manage a display of up to 16 characters, and provide automatic refresh with no further CPU intervention.

The CLOCK input (pin 3) generates all the internal timing, and has a programmable prescaler which divides the input frequency to yield the 100-kHz necessary for proper operation.

In the traditional Scanned Keyboard Mode, each keyboard row is successively driven by the SCAN LINE (pins 32-35) outputs. The scan line counter can operate in two modes: encoded and decoded. When encoded, the outputs are a binary representation of the row to be driven, and a decoder is required to select the row addressed. Some keyboards include this feature, others do not. For direct row addressing, the 8279 can decode the least two significant scan counter bits to provide a 1 of 4 linear scan at the SCAN LINE outputs.

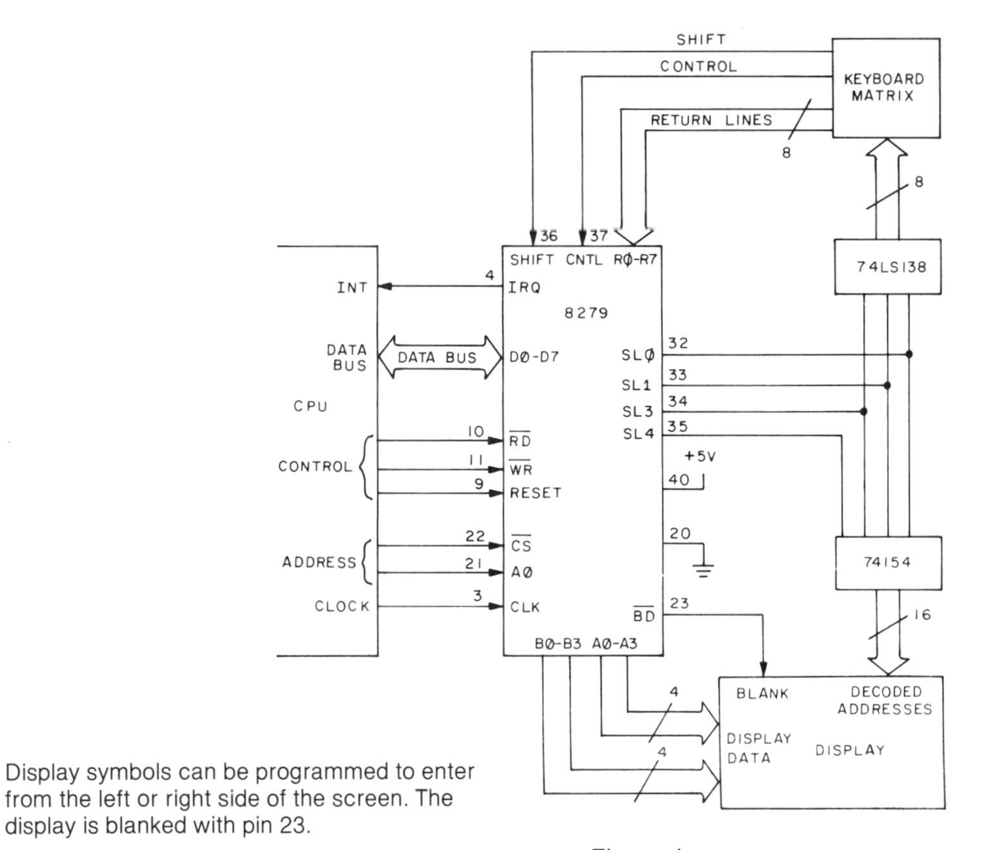

Display symbols can be programmed to enter from the left or right side of the screen. The display is blanked with pin 23.

Figure 1

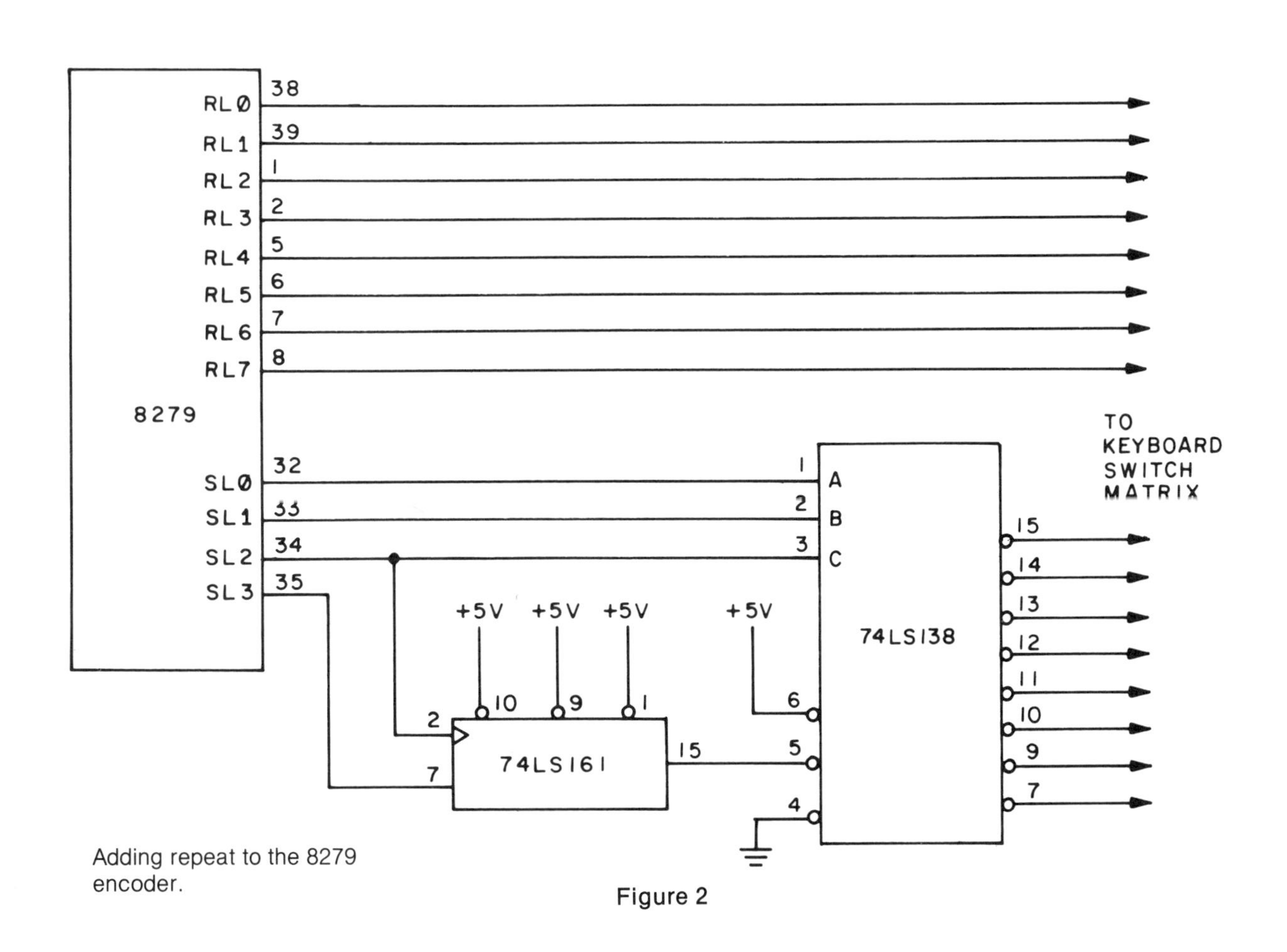

Adding repeat to the 8279 encoder.

Figure 2

CY300

Alphanumeric LCD Controller

The CY300, by Cybernetic Micro Systems, is a TTL compatible CMOS device configured to control the alphanumeric display of a 16-character LCD dot matrix display. The characters are internally generated and displayed in standard ASCII format. The chip is primarily designed to operate as a display console for microcomputers and can even be used to replace a CRT in many systems.

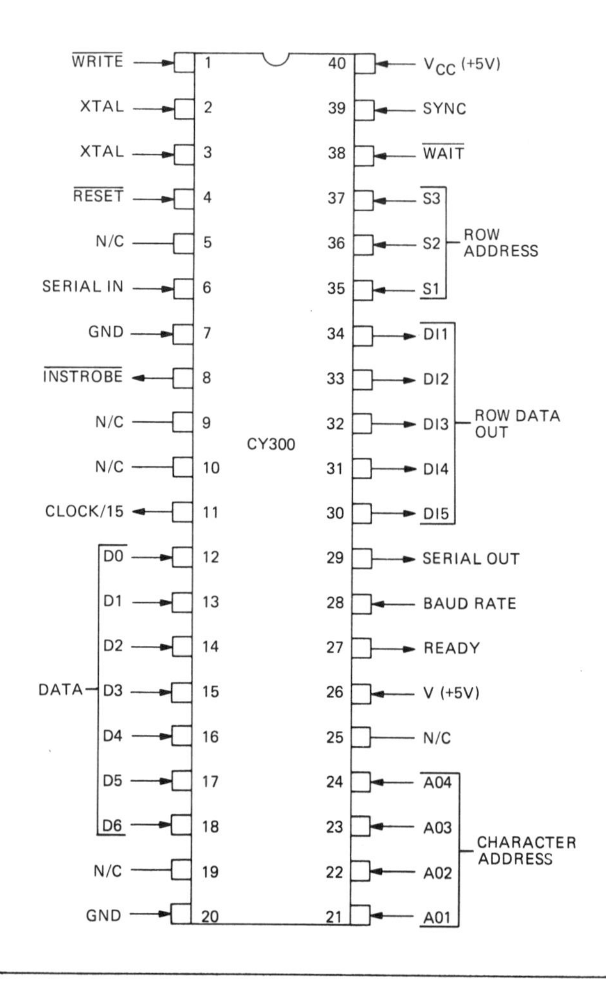

Operation

The CY300 has been designed to drive a 5×7 dot matrix LCD display of 16 characters in length using the Toshiba T3891 matrix driver module. Although only 16 characters can be reviewed at any one time, the internal register is capable of storing up to 32 individual characters.

An internal pattern generator converts the ASCII inputs into the dot pattern which makes up the displayed symbols. The LCD display is scanned by the T3891 module in conjunction with the CHARACTER ADDRESS (pins 21-24), ROW ADDRESS (pins 35-37) and ROW DATA OUT (pins 34-30) CY300 interface. The display is synchronized using the SYNC (pin 39) input. As the vertical column of dots parade across the screen, the controller selects the dots to be excited for proper reproduction of the character.

Parallel Operation

The CY300 can operate in either a parallel or serial mode, depending upon your needs. In the parallel mode, ASCII data is input through the seven DATA lines (pins 12-18), which are generally connected to a keyboard. The data must be held on line until the controller's READY output (pin 27) goes HIGH, which is subsequently followed by a WRITE strobe (pin 1) input from the keyboard. The transfer is acknowledged by an INSTROBE (pin 8) pulse. On the trailing edge of the INSTROBE, data enters the registers and the symbol is displayed.

This process is repeated for each stroke of the input until the registers are full or a return key is actuated. The return command causes the registers to dump their holdings and resets them for the next line of information.

Serial Operation

Data can also be fed into the display controller through the SERIAL IN (pin 6) input. Data on this line is accumulated in the registers and displayed on the LCD readout. As before, a return command clears the registers.

The data stored inside the registers is output in serial form when the return command is initiated. If the display console is to be used as an intermediary between the keyboard and a microprocessor, the SERIAL OUT (pin 29) is fed into the processor in serial ASCII format, otherwise it is lost.

The baud rate for serial transmission is governed by the BAUD RATE (pin 28) input. With three rates to choose from, a HIGH input will set the rate at 1200-baud; a LOW input will drop it to 300-baud. If left unconnected, a 600-baud rate is assumed. Although the characters are sent as a group, the chip tests the WAIT line (pin 38) before sending each character. If the line is LOW, the CY300 holds off transmission until it goes HIGH, thus enabling the output to synchronize with modems and other interfacing devices.

Displays

The display controller can be used strictly for display purposes. When in the "dumb" mode, data received in either parallel or serial form will simply be displayed on the screen.

Used as a keyboard controller, however, the CY300 is able to take the input from the keyboard, display it, and *then* transfer it to a microprocessor via the SERIAL OUT port.

Editing

The use of the controller as an interface between a keyboard and the CPU has a distinct advantage over a hardwired interface. Besides displaying the information as it is typed in, relieving the processor of the duty, the CY300 contains the necessary functions to edit the material before it is dumped into the processor.

The editor displays itself in the form of a cursor which precedes the typed character by one position. If a mistake occurred in the last character, a delete command from the keyboard will back up the cursor one space and wipe out the character. If, however, you wish to alter a character in midline, it is possible to move the character to the desired position without erasing the line. Cursor movements are controlled by the keyboard, and its position is identified by highlighting the symbol. Once in the proper position, a simple keystroke will erase the former character and replace it with the new input data.

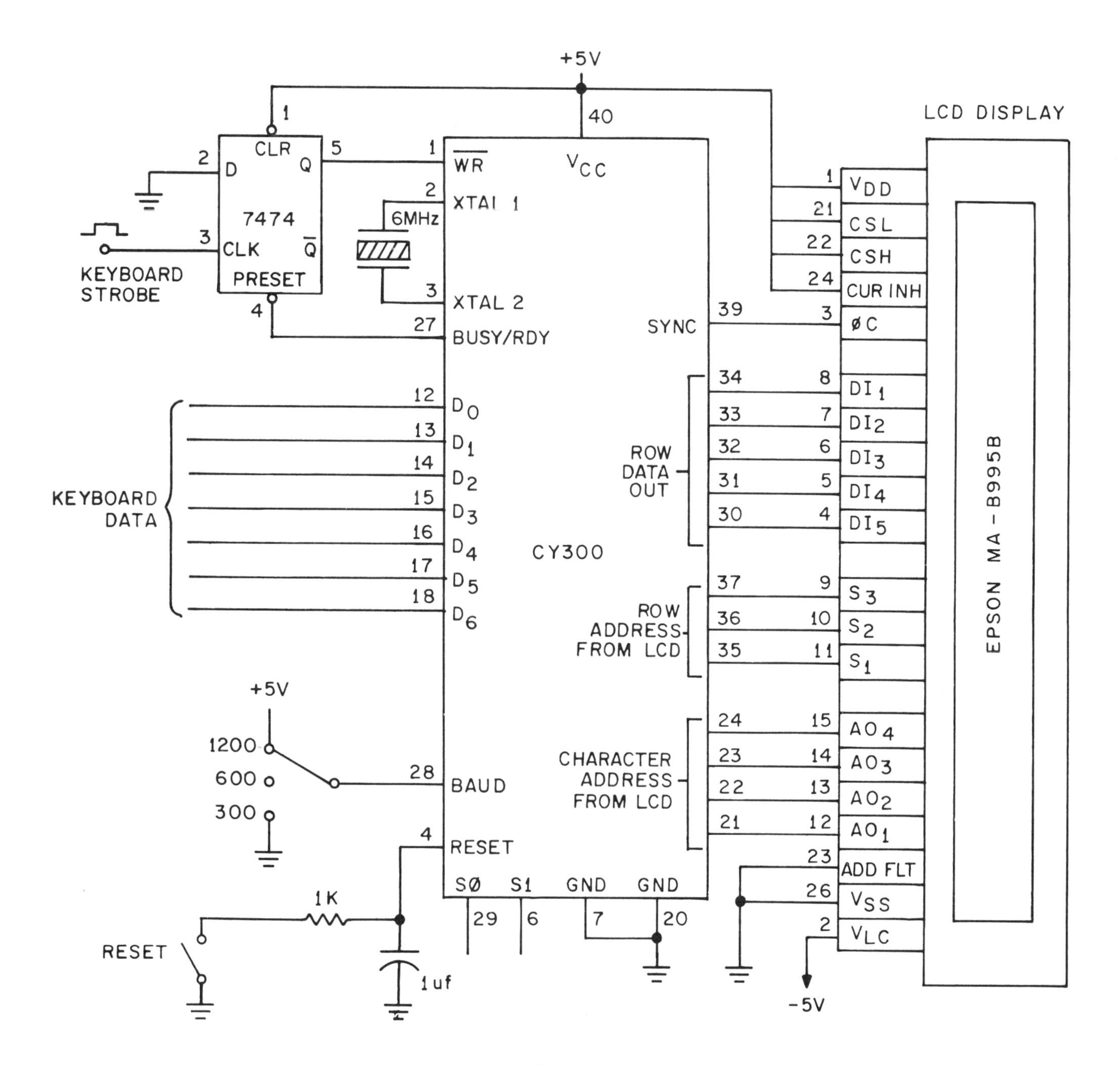
+5V
40
LCD DISPLAY
1
CLR
D
Q
7474
CLK
Q
PRESET
2
3
4
5
KEYBOARD
STROBE
WR
VCC
XTAL 1
6MHz
XTAL 2
BUSY/RDY
SYNC
VDD
CSL
CSH
CUR INH
ØC
DI1
DI2
DI3
DI4
DI5
ROW
DATA
OUT
KEYBOARD
DATA
D0
D1
D2
D3
D4
D5
D6
CY300
ROW
ADDRESS
FROM LCD
S3
S2
S1
EPSON MA-B995B
+5V
1200
600
300
BAUD
CHARACTER
ADDRESS
FROM LCD
AO4
AO3
AO2
AO1
RESET
ADD FLT
VSS
VLC
SØ
S1
GND
GND
1K
RESET
1uf
-5V

Figure 1

10937

Alphanumeric Display Controller

The 10937, introduced by Rockwell International, is an Alphanumeric Display Controller designed to interface to segmented displays, including gas discharge, vacuum fluorescent, and LED. The controller will display up to 16 characters of 14 or 16 segments plus a decimal point and comma.

All functions required to store and map upper case ASCII characters and control the display are generated in the 10937 device, without any refresh input from the host processor.

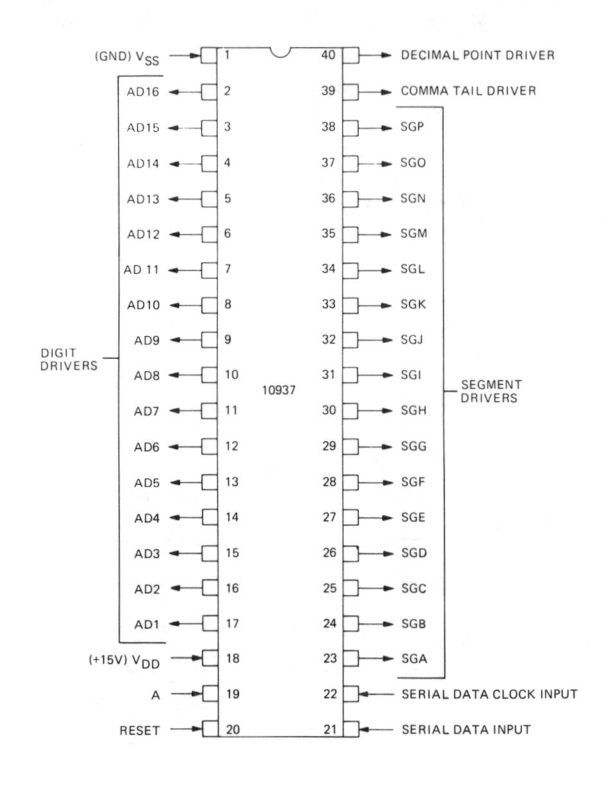

Operation

Input data is loaded into the *display data buffer* through the SERIAL DATA INPUT (pin 21) port. The input rate is asynchronous and determined by the SERIAL DATA CLOCK INPUT (pin 22) frequency, which is normally supplied by the transmitting device.

The input consists of an 8-bit word; the first input bit of the word is a Control Command which instructs the controller whether the following data is display or control information.

Display Drivers

The 10937 drives the display using 32 output lines. The 16 DIGIT DRIVERS (pins 17 to 2) select each one of the 16 display digits sequentially. The SEGMENT DRIVERS (pins 23-38) drive the individual display segments to produce the character. The display is multiplex scanned, and a readout is illuminated when both the DIGIT DRIVERS and SEGMENT DRIVERS for a particular character are energized simultaneously.

There is also a separate COMMA TAIL DRIVER (pin 39) and DECIMAL POINT DRIVER (pin 40) which drive the comma and decimal character of the display, respectively. They, too, are multiplexed and excite when the DIGIT DRIVER output coincides.

The display controller will manage both 14- and 16-segment displays. When a 14-segment display is used, the SGA (pin 23) output is the top segment and SGF (pin 28) becomes the bottom segment. SGB (pin 24) and SGE (pin 27) are floated.

Control Data Words

When the Control bit (first bit of an incoming word) is logic 1 (HIGH), the incoming data is a Control Data Word. There are three functional control codes for the 10937 chip. They are as follows:

The Load Buffer Pointer command positions the *display data buffer* to the desired character position. The *load buffer pointer register* is programmed with the decimal equivalent of the desired location *less* 2. For example, to address character 6 of the display, a value of 4 is loaded into the register. When this command is followed by a display data word entry, the previous character is erased and replaced.

The Load Duty Cycle instruction sets the duty cycle of the display. The time allocated for each character during the scanning process is 32 clock cycles. Thirty-one of these cycles are dedicated to the display and one is reserved for inter-digit off time. The Load Duty Cycle code contains a 5-bit word that determines the amount of ON time the display will receive: a 0 indicates no ON time, while 31 gives maximum time. Anything in between can be used to adjust the brightness of the display, or to modify the display timing for gas discharge displays. The entire display can be made to blink by alternating ON and OFF instructions.

The Load Digit Counter code determines the length of the line to be displayed. When all 16 characters are required, a 0 is entered, otherwise the decimal value is specified. This code maximizes the duty cycle for the display.

There is the possibility of addressing a fourth command: the Test Mode. This code is not intended as a user function and if executed will lock the device into a test mode, from which the only recovery is a power-on reset.

Display Data Words

If the control bit of any word is a logic 0 (LOW), the incoming word is a display data word. Display data words are 6-bit ASCII words that the 10937 decodes into 64 characters. The *buffer pointer register* is automatically incremented before the word is stored in the *display data buffer*. Therefore, the next input will occupy the next space. At the end of the line, as determined by the *digit counter register* count, the pointer will return to position 1. Any new entries will replace the number 1 character and advance the pointer to position 2.

However, the entry of either a comma or a decimal point (period) will *not* increment the *buffer pointer register* count. The decimal point or comma is always associated with the previous character, and if you wish to leave a space following the punctuation, you must specify it.

Reset

The controller must be reset after each power-up. This is done automatically by holding the RESET input (pin 20) LOW for at least 10-milliseconds after power has been applied. A RESET during normal operation will clear all registers and the display.

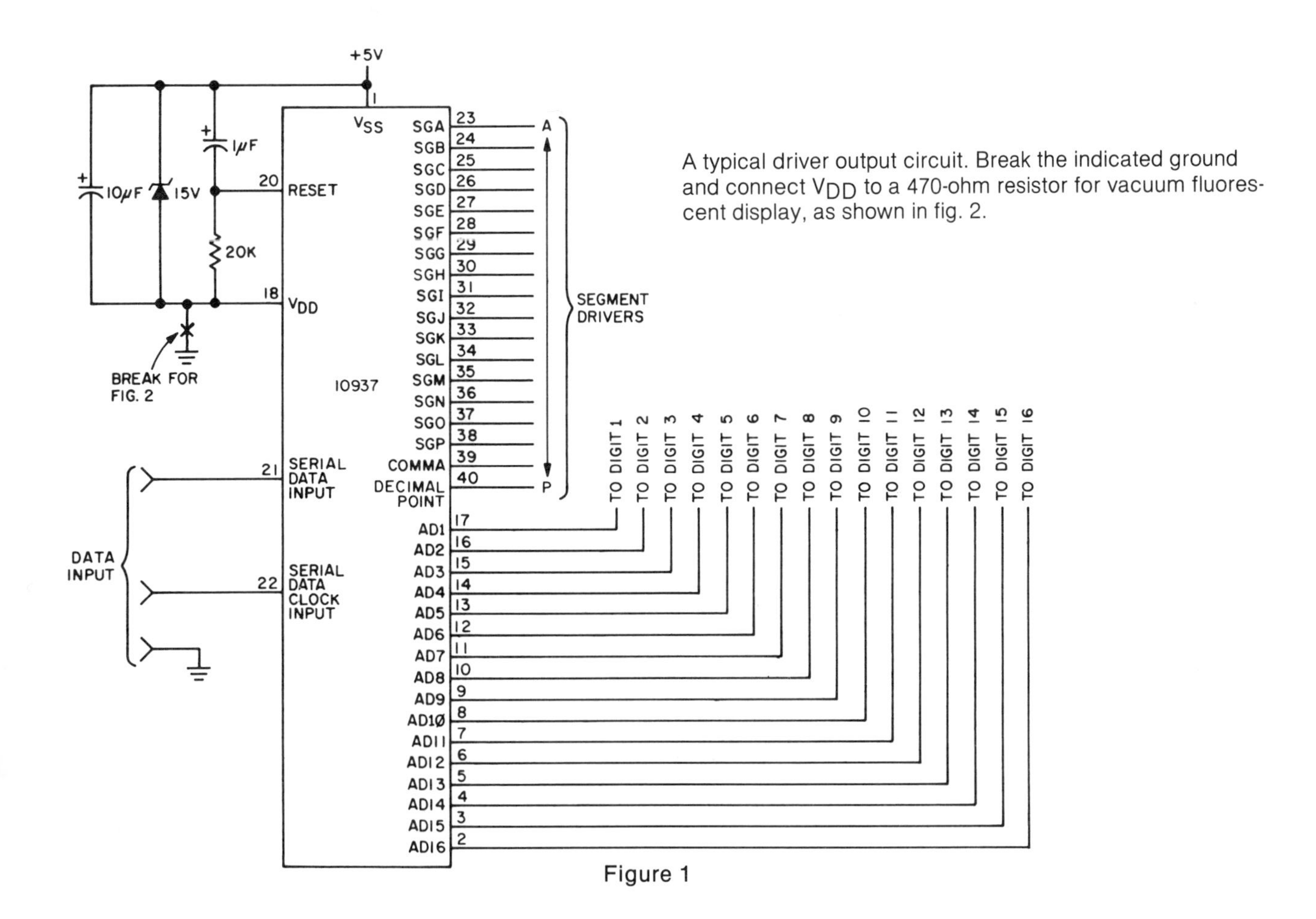

A typical driver output circuit. Break the indicated ground and connect V_{DD} to a 470-ohm resistor for vacuum fluorescent display, as shown in fig. 2.

Figure 1

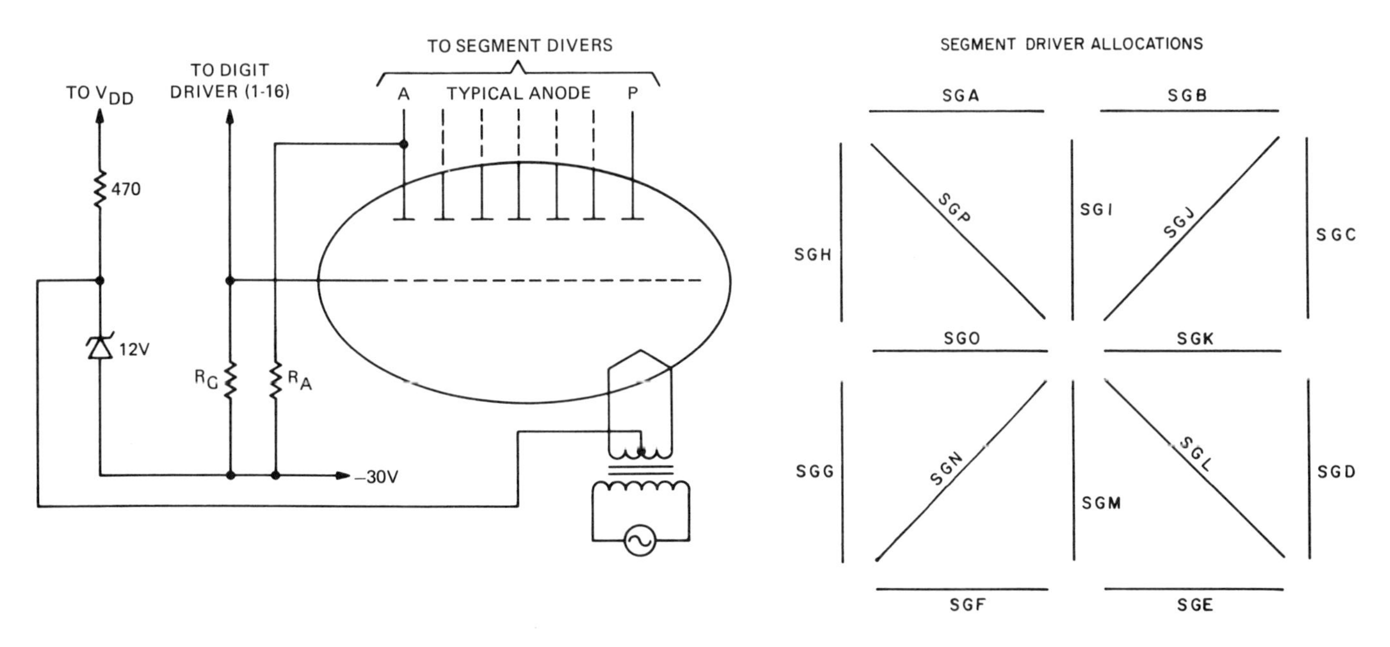

Figure 2

Figure 3

10938 & 10939

Dot Matrix Display Controller

The 10938 and 10939 Dot Matrix Display Controller, by Rockwell International, is a two-chip display controller set designed to interface to gas discharge, vacuum fluorescent, or LED dot matrix displays. The chips will display up to 20 characters in 5 × 7 character font, and can be concantenated to provide 80 or more characters.

Data is input to the controller in either serial or parallel form and translated into dot patterns for the full 96-character ASCII set. The display is programmed by specifying internal software routines using Control Data words.

Control Data Words

Control Data words can either set the display parameters prior to entering display data, or can modify existing display, and are distinguished from Data Display words by a Control Prefix code equal to 0000 0001. Whenever this 8-bit code is entered into the controller chip, it recognizes the next input byte as a control command. The controller will attempt to display all input data not preceded by a Control Prefix.

Control Data commands can position a cursor or limit the length of the message line. The Load Buffer Pointer Command determines the position of the line cursor, and consequently the placement of the character. The length of the message is limited by entering the number of characters to be displayed into the *digit counter*, which can be as few as one character.

The display is refreshed using a multiplexed timing cycle, and the time slot for each character refresh is 16, 32, or 64 cycles apiece, as determined by the *load digit time register*. For displays with 40 or more characters, or where the display is subject to sudden movements during viewing, it may be necessary to increase the refresh rate by limiting the refresh period to 16 or 32 scans per character. The actual number of refresh lines, however, is controlled by the Load Duty Cycle control, and can be any number less—down to zero. This controls the ON-time to OFF-time ratio for the display, and adjusts the brightness accordingly. The Load Duty Cycle code can also be used to modify the display timing, as is required for gas discharge displays.

Data Input

Display data can be input to the *display data buffer* in either serial or parallel form. DATA BUS inputs D0 through D7 (pins 6-13) on the 10939 chip make up the parallel input ports. Words on the DATA BUS lines are transferred into the *display data buffer* when the INPUT DATA STROBE (pin 5) is pulsed HIGH.

For serial input, the DATA BUS lines D2 through D7 (pins 9-13) must be tied to V_{DD} (pin 37). Raising any one of these lines to the V_{SS} (pin 36) level automatically shifts the controller into the Parallel Mode. Serial data is input to D1 (pin 7), Most Significant Bit first. The baud rate is controlled by D0 (pin 6) and is used in conjunction with the INPUT DATA STROBE to clock in the data.

The Data Display words are loaded as 8-bit codes. Each ASCII character is represented by the lower seven bits of the display word. When the controller is in the Normal Display Mode, the eighth bit is ignored—but it becomes a control bit for the Blank and Inverse Modes. In the Blank Mode, any character with an MSB equal to "1" will be blanked; in the Inverse Mode, it will be displayed with all SEGMENT DRIVER outputs inverted, or the equivalent of an "inverted video" CRT format. These controls allow individual characters—or groups of characters—to be blinked or blanked by simply changing the mode (using Data Control words) without disturbing the contents of the *display data buffer*.

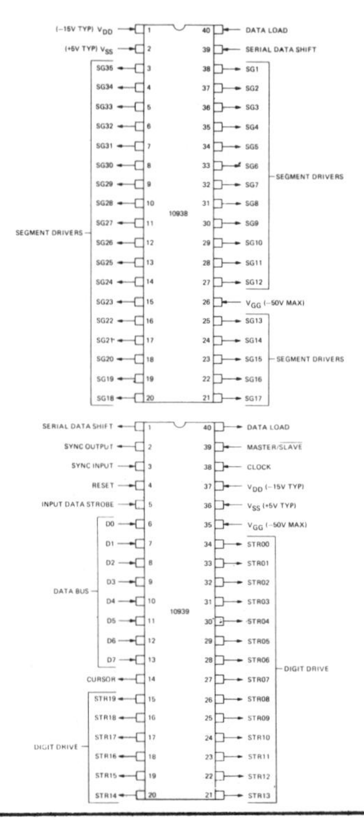

Twenty display words are required to completely load the *display data buffer*. The *buffer pointer register* automatically increments after each data word is stored in the buffer, and will reset the pointer to character position 0 when its value is equal to the *digit counter*.

Display Drivers

The 10938 contains a 96 × 35 PLA that drives the cathode dots of the display. These drivers are output through the SEGMENT DRIVER outputs (pins 38-27, 25-3, representing SG1-SG35, respectively). the character anode (grid) drives are supplied by the 10939 chip through the DIGIT DRIVE outputs, pins 34 to 15. the display segments are illuminated when the SEGMENT DRIVERS and DIGIT DRIVEs for a particular character coincide. A CURSOR output drive (pin 14) is located on the 10939 chip.

Chip Interconnections

Since it requires two chips to operate the display, the two must communicate between each other and synchronize their operations. The SERIAL DATA SHIFT clock (pin 1 on 10939, pin 39 on 10938) and DATA LOAD (pin 40 for both) lines are joined together to perform this function.

When more than one 10939 is linked together to expand the display line, they are synchronized with the SYNC OUTPUT (pin 2) and SYNC INPUT (pin 3) signals in association with the CLOCK (pin 38) control. The MASTER/SLAVE (pin 39) input determines the related chips mode. At power on, the 10939s are held in the Halt Mode. Normal display refresh sequencing starts upon receipt of a Start Refresh control code.

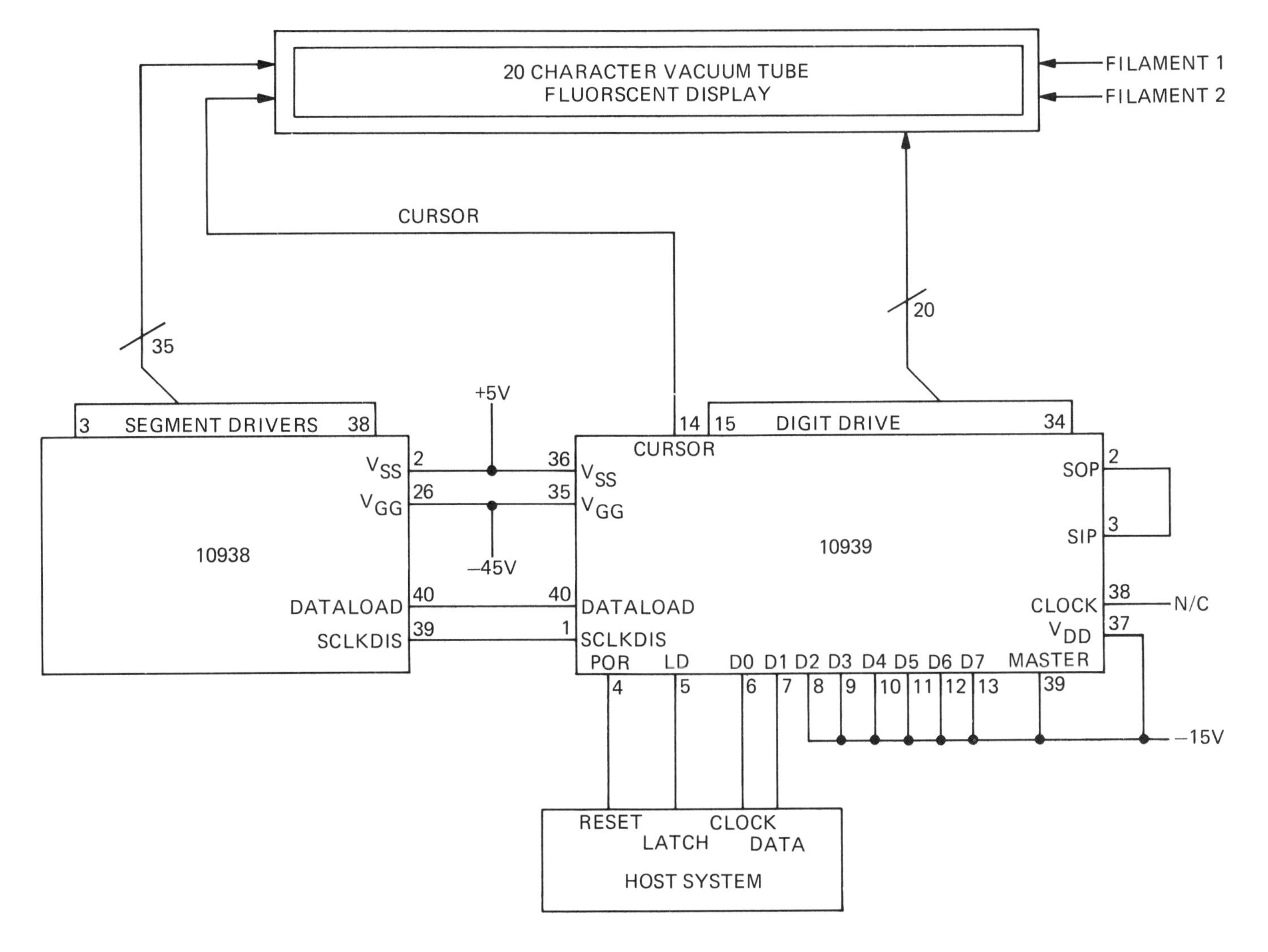

Figure 1

SGØ1	SGØ2	SGØ3	SGØ4	SGØ5
SGØ6	SGØ7	SGØ8	SGØ9	SG10
SG11	SG12	SG13	SG14	SG15
SG16	SG17	SG18	SG19	SG20
SG21	SG22	SG23	SG24	SG25
SG26	SG27	SG28	SG29	SG30
SG31	SG32	SG33	SG34	SG35

DOT MATRIX PATTERN ASSIGNMENT FOR A 5X7 ASCII CHARACTER SET.

Figure 2

8254

Programmable Interval Timer

The 8254, made by Intel, is a counter/timer device designed to solve the common timing control problems associated with microprocessor system design. The chip contains three independent 16-bit counters, each capable of handling clock inputs up to 10-MHz. All modes of operation are software programmable.

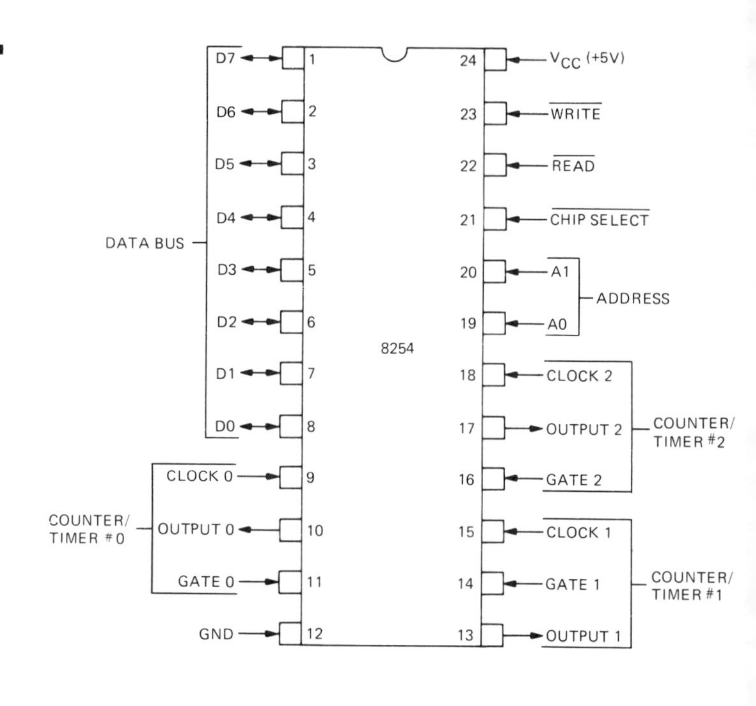

General

The 8254 eliminates the often time-consuming chore of generating accurate time delays by replacing software programs with hardware timers. Instead of operating timing loops in software, the desired delay is simply plugged into the timer chip. The counters are then decremented by the CLOCK signal until the count is clocked out. After the specified delay, the 8254 will output an interrupt, thus leaving the CPU free to carry on other duties during the timing sequence, rather than expend its efforts on a software loop.

Counters

The three internal counters are identical in function, yet individual in operation. Each counter can be pre-loaded with any count up to 64K binary or 10K BCD, and are decremented on the falling edge of the CLOCK input. CLOCK inputs are pins 9, 15, and 18 for their respective counters, 0, 1, and 2.

When the count is completed, an output pulse occurs at the OUTPUT pin for each counter (pins 10, 13, 17 for counters 0, 1, and 2, respectively). However, the counter does not halt when it reaches zero. Except for the periodic modes, it continues counting by wrapping around to the highest input number and down counting it. In the periodic modes, the counter reloads itself with the initial count and proceeds counting from there.

Each counter also has a GATE input (pins 11, 14, 16 for 0, 1, and 2) which can control the countdown process. In fact, certain modes require a GATE signal before the counter is activated. Detection of a GATE input is a mixture of level and transition sensing, depending upon the mode selected. Regardless of mode, though, the GATE is always sampled on the rising edge of the CLOCK pulse, and its status stored in a flip-flop memory until the next CLOCK cycle.

Counter Programming

When the timers are first powered up, their state is undefined and they must be programmed before use. The programming procedure for the 8254 is very flexible, and only two things need be remembered.

First, for each counter a Control Word must be written before the initial count is entered. The Control Word sets the operating mode of that particular counter. Second, the initial count must adhere to the format specified by the operating mode.

All Control Words are written into the *control word registers* by selecting both ADDRESS inputs (pins 19 & 20) HIGH. Control commands are transferred through eight bidirectional DATA BUS lines (pins 8 through 1), with the intended counter addressed by the D6 and D7 bits of the input byte. After the registers are initialized, the timer count is entered into the *count register*. Since the counters are 16-bits wide, and the input bus is only 8-bits wide, the count is entered using a double fetch operation, least significant byte first. Data is written into the chip by enabling the WRITE input (pin 23) LOW.

It is often desirable to read the value of a counter without disturbing the count process. This is entirely possible with the 8254 chip by pulling the READ input (pin 22) LOW. Both READ and WRITE are qualified by the CHIP SELECT input (pin 21), which must be held LOW during the operation.

Operating Modes

Altogether, there are six modes of operation. They can be divided into two categories: conditional timing, of which there are four modes, and periodic. The period modes include Mode 2 and Mode 3. Mode 2 is a divide-by-N counter, and is typically used to generate a Real Time Clock interrupt. Mode 3 is a square wave generator, and is very similar in operation to Mode 2, except that the initial count is divided exactly in half to give a 50 percent duty cycle. Mode 2 is often used for baud rate generators.

The conditional modes contain two different strobe functions. Mode 4 provides a software triggered strobe input, while Mode 5 uses the GATE input for a hardware triggered strobe.

Mode 0 is a terminal count interrupt function. The counter is loaded with the desired time delay, and when it expires, an OUTPUT interrupt is performed.

Finally, there is Mode 1, which is also a conditional mode. This mode is defined to perform a hardware triggerable one-shot OUTPUT when the GATE is activated. If a new count is written into the register during the one-shot pulse, the current one-shot is not affected, but the new number is entered upon completion for use with the next GATE input signal.

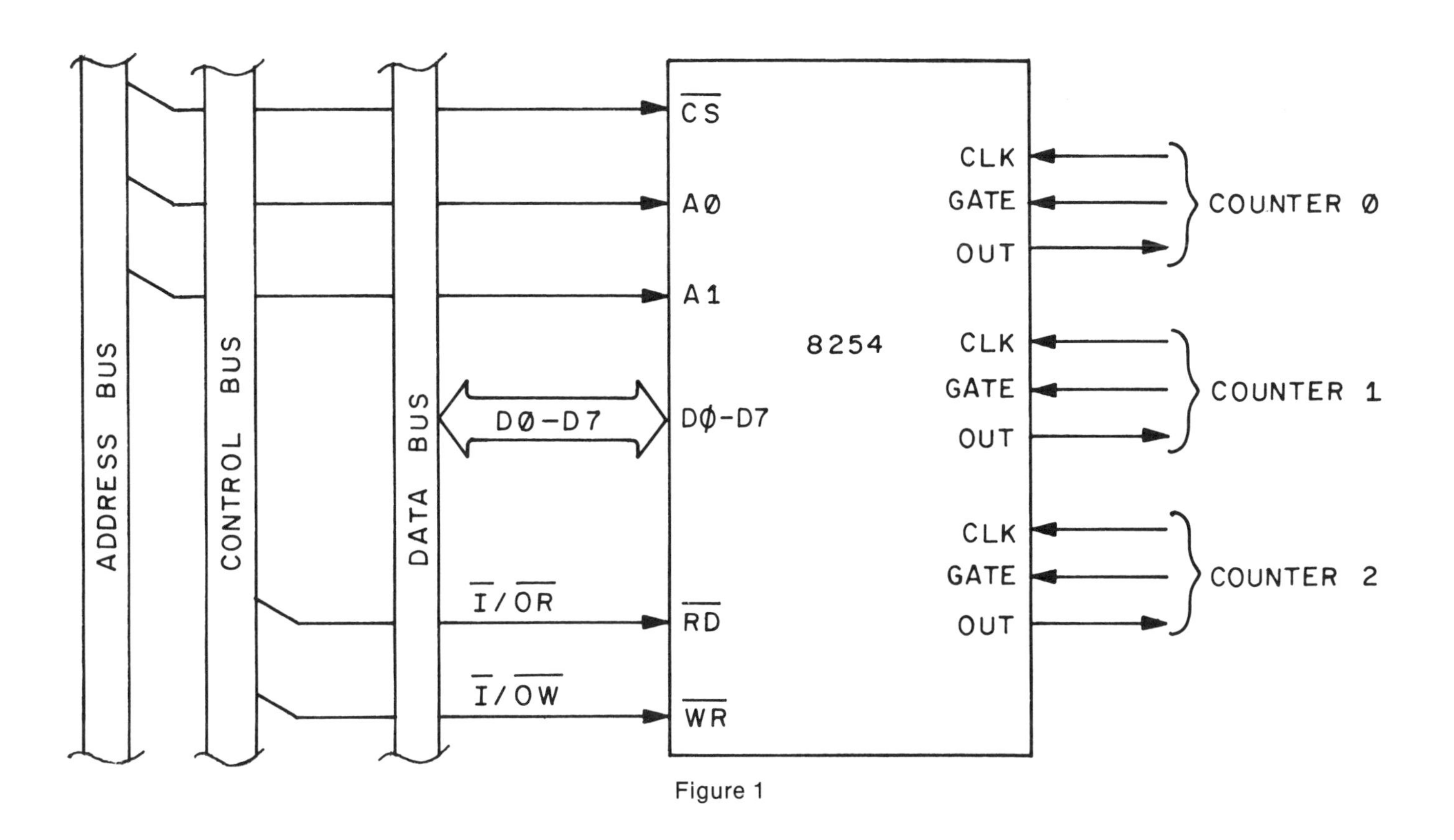
ADDRESS BUS
CONTROL BUS
DATA BUS
CS
A0
A1
D0-D7
D0-D7
I/OR
RD
I/OW
WR
8254
CLK
GATE
OUT
COUNTER 0
CLK
GATE
OUT
COUNTER 1
CLK
GATE
OUT
COUNTER 2

Figure 1

MC6840

Programmable Timer

The MC6840, from Motorola, is a Programmable Timer chip which relieves the microprocessor of the tedious routine of initiating software loops to measure timing events. It contains three 16-bit binary counters, three corresponding Control Registers, and a Status Register. The three counters are under software control, and may be used to create system interrupts and/or generate output signals.

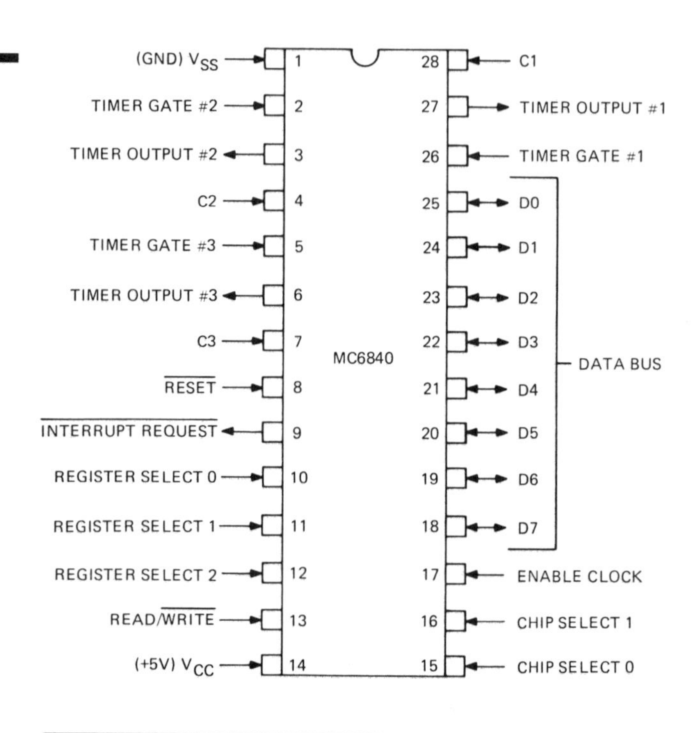

Operation

The MC6840 contains three 16-bit countdown counters, each of which may be clocked separately. In a typical operation, a timer will be loaded with the desired count by first storing two bytes of data into an associated *counter latch*. This data is then transferred into the counter during the initialize cycle. When the counter is enabled, it decrements the contents on each subsequent clock period until one of several predetermined conditions causes it to halt or recycle.

Counter Loading

The timer chip interfaces to the CPU through eight bidirectional DATA BUS lines (pins 25-18). Since the DATA BUS is only 8-bits wide and the counters are 16-bits wide, a *temporary storage register* is provided to accumulate the binary inputs, with the Most Significant Byte loaded first.

Writing into the registers begins by first enabling the CHIP SELECT inputs, CS0 (pin 15) and CS1 (pin 16). With CS0 LOW and CS1 HIGH, the device is selected and data transfer can occur. A register is addressed by configuring the REGISTER SELECT lines (pins 10, 11, 12), followed by a LOW input to the READ/WRITE pin (pin 13) which enables the input buffers and transfers data from the CPU into the register on the trailing edge of the ENABLE CLOCK (pin 17).

Alternately, a HIGH input to pin 13 will transfer data from the timer to the CPU, as is done when reading the contents of the registers. The DATA BUS output drivers are three-state devices which remain in the high-impedence mode until a read or write operation is performed.

Initialization

Once the bits are in the *counter latches,* they are loaded into the counters on an initialization cycle. Counter initialization always occurs when a reset condition, either through the RESET input (pin 8, LOW) or an internal software bit, is recognized. In certain modes, initialization can also happen when a *write timer latch* command or a negative transition of the TIMER GATE input occurs.

Counter re-initialization automatically occurs when a negative transition of the ENABLE CLOCK is received after the counter has reached an all-zero state, thereby reloading the counter with the number contained in the *counter latches.* It is important to note that a software reset (internal) has no effect on the contents of the *counter latches,* while a hardware RESET (pin 8) presets them to their highest value (64K).

Clock

The MC6840 is driven by a master ENABLE CLOCK (pin 17). The clock synchronizes data transfer between the CPU and the timer chip, and should be derived from the processor itself.

The clock inputs to the counters, however, can be independent of the ENABLE CLOCK. Input pins C1 (pin 28), C2 (pin 4), and C3 (pin 7) will accept asynchronous TTL signals to decrement timers 1, 2, and 3, respectively. Clock input C3 is unique in that it has a divide-by-eight prescale option available in software.

The external clock pulses are clocked in by ENABLE CLOCK pulses. Therefore, it is necessary that the external clocks have a lower frequency than the ENABLE CLOCK, as determined by the ENABLE CLOCK pulse width plus the input hold time (with the exception of the divide-by-eight mode). Although the ENABLE CLOCK performs a synchronization function on the external clock inputs, it in no way affects the input frequencies.

Timer Inputs/Outputs

The timers can be utilized in several ways. First, an all-zero condition can initiate an INTERRUPT REQUEST (pin 9) to the CPU. The INTERRUPT REQUEST line is an "open drain" output which permits similar request lines to be tied together in a wire-OR fashion.

In addition to the interrupt function, three output pins—pins 27, 3, and 6—supply a TTL signal from each of the counters, 1, 2, and 3, respectively. These outputs can be configured to provide single-shot pulses, continuous pulses, or square waves to other devices.

The MC6840 timer also incorporates TIMER GATE inputs for each counter: pin 26 for #1, pin 2 for #2, and pin 5 for #3. These inputs are used for triggers or clock gating operations, and directly affect the internal 16-bit counter, even in the divide-by-eight mode.

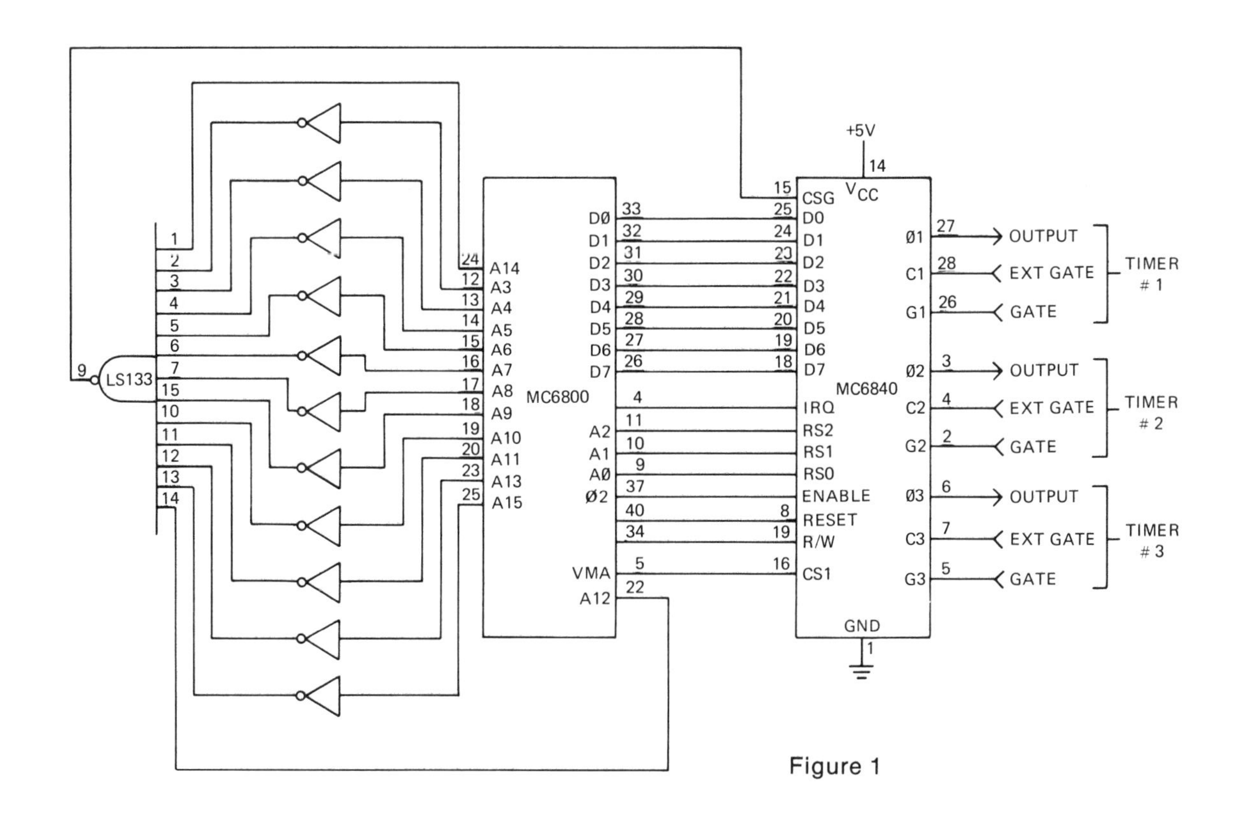

Figure 1

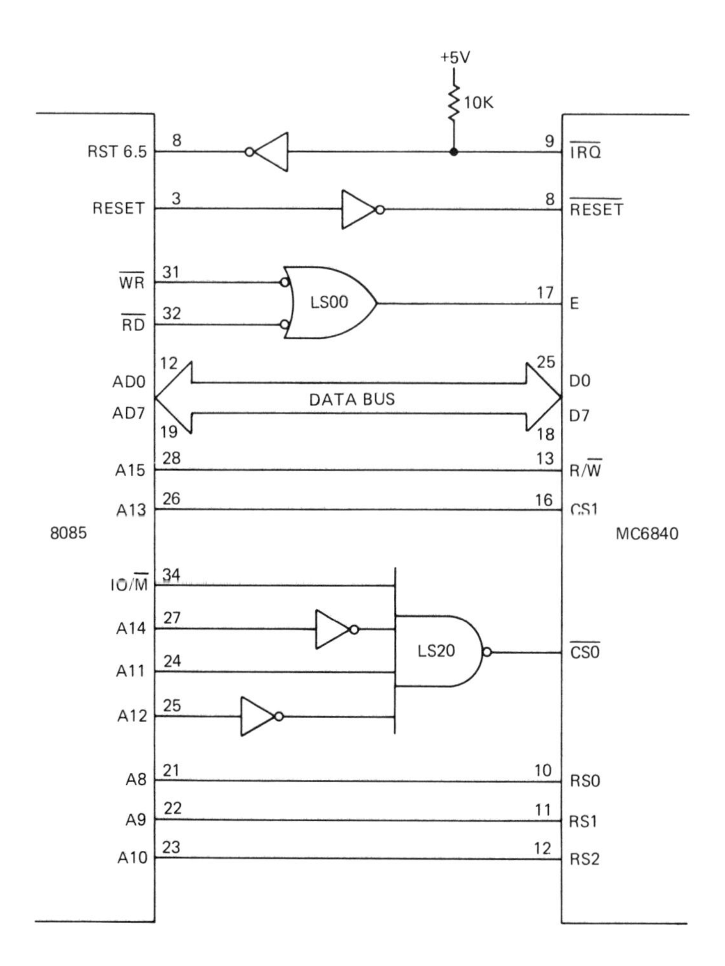

Figure 2

Z8430

Counter/Timer Circuit

The Z8430, made by Zilog, is a four channel Counter/Timer designed to replace the software timer loops so common with CPU time delay requirements. Each channel is independently software programmable to function as either a timer or counter. The chip connects directly to the Z-80 microprocessor, and needs no additional logic in most cases.

Programming

Programming the Z8430 is straightforward; each channel is programmed prior to operation using only two bytes. Program instructions and data are entered into the device through eight bidirectional DATA BUS lines (pins 25-28, 1-4).

The chip does not receive a specific write signal; rather, it internally generates its own from the inverse of an active READ request (pin 6, LOW) signal. During the write cycle, the INPUT/OUTPUT REQUEST input (pin 10) and CHIP ENABLE (pin 16) are LOW, while the READ input is HIGH. In a read cycle, on the other hand, all three inputs (pins 6, 10, and 16) are pulled LOW.

The channels are selected by configuring the CHANNEL SELECT inputs (pins 18 & 19). The first program word is a control word that defines the operating mode. There are two modes: Timer and Counter.

Timer

Once the mode is set, a time constant is entered into the counter. The time constant is a binary word with a value of 1 to 256. The timer has two trigger modes, as determined by the D3 input. When D3 is reset to 0, the timer is automatically triggered on the rising edge of the second CLOCK (pin 15) pulse following the write operation. When D3 is set to 1, though, the timer is triggered externally through the CLOCK/TRIGGER inputs (pins 23, 22, 21, 20, representing counter 0 through 3, respectively).

Once started, the timer decrements from the master clock and runs continuously, with no interruption or delay, unless stopped by a reset. The chip has both software and hardware RESET (pin 17) capabilities.

Each counter also includes a prescaler which divides the system clock by a factor of either 16 or 256. The prescaler is available only in the Timer Mode.

Counter

When used in the Counter Mode, the value set into the down counter is decremented by every active edge (rising) of a clock pulse presented on the CLOCK/TRIGGER input. The trigger is asynchronous, but the count is synchronized with the CLOCK pulse; therefore, the trigger period must always be at least twice as long as the clock period.

Outputs

A ZERO COUNT/TIMEOUT output occurs immediately after the down counter reaches zero. These outputs—pins 7, 8, and 9 —are driven by counters 0, 1, and 2, respectively. The fourth channel (counter #3), however, does not have a ZERO COUNT/TIMEOUT output; an INTERRUPT REQUEST (pin 12) is the only output available from this counter/timer.

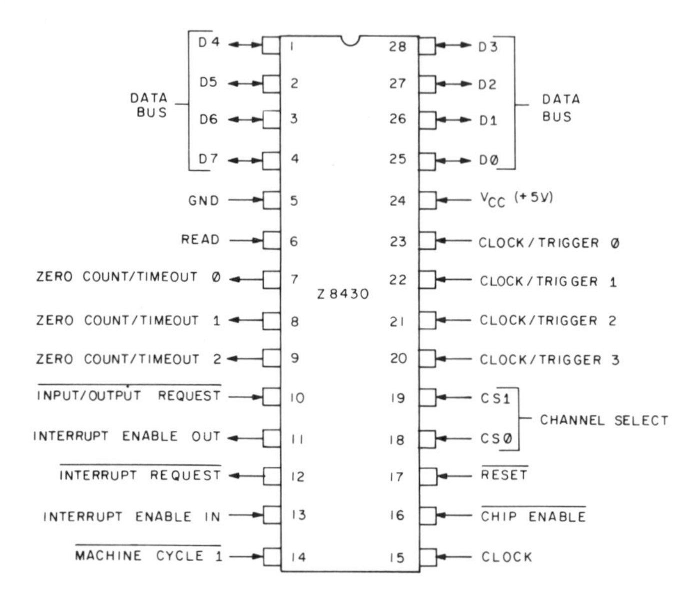

When an interrupt occurs, which can be programmed by any of the four channels accordingly, the CPU acknowledges its existence by driving the INPUT/OUTPUT REQUEST (pin 10) and MACHINE CYCLE 1 (pin 14) inputs LOW. The Z8430 then places an interrupt vector on the DATA BUS.

Interrupt Priority

This device, like most in the Zilog family, has a daisy chain priority interrupt system. Each of the counter/timers is assigned an interrupt priority, with channel 0 having the highest priority and channel 3 the lowest.

If a channel is being serviced with an interrupt routine, it cannot be interrupted by a channel with lower priority until service is complete. Higher priority channels, however, may interrupt the servicing of lower priority channels.

This same priority sequencing can also be carried out to supporting chips in the system through two pins: the INTERRUPT ENABLE IN input (pin 13) and the INTERRUPT ENABLE OUT output (pin 11). When an INTERRUPT REQUEST is made by the Z8430, the INTERRUPT ENABLE OUT is forced LOW. This output remains LOW until the interrupt servicing is completed. The INTERRUPT ENABLE IN, on the other hand, is an input that, when pulled LOW, effectively halts all interrupt servicing to the chip until it is returned HIGH.

By daisy chaining the desired chips in the system, it is possible to assign each one its own priority status. The chip with the highest priority outputs its INTERRUPT ENABLE OUT to the INTERRUPT ENABLE IN of the next chip in the chain. As long as the first chip has no active interrupts in progress, the CPU will service any interrupt generated by the second device. If, however, the first chip requests an interrupt, the INTERRUPT ENABLE OUT will pull the INTERRUPT ENABLE IN of the second chip LOW, cutting off its interrupt service.

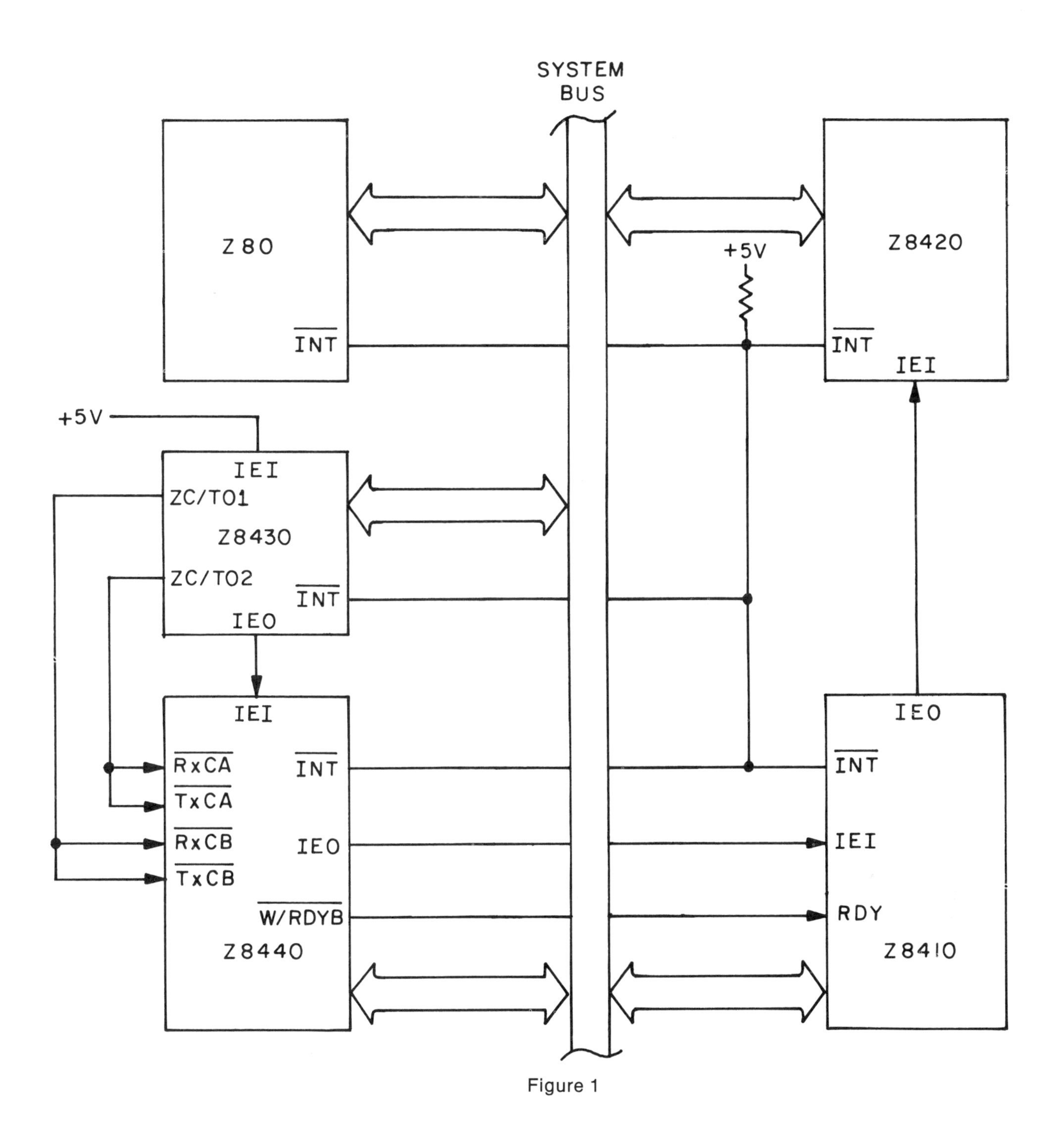

Figure 1

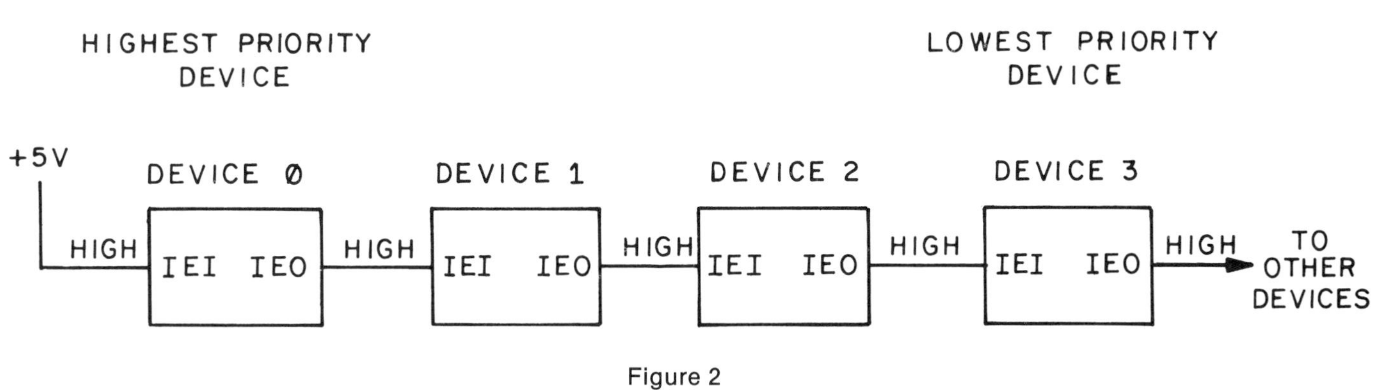

Figure 2

8294

Data Encryption Unit

The 8294 Data Encryption Unit, by Intel, is a microprocessor peripheral device designed to encrypt and decrypt data using the algorithm specified in the Federal Information Processing Data Encryption Standard. Because the 8294 uses this NBS approved algorithm, it can be used in a variety of Electronic Funds Transfer and banking applications where data encryption is a must.

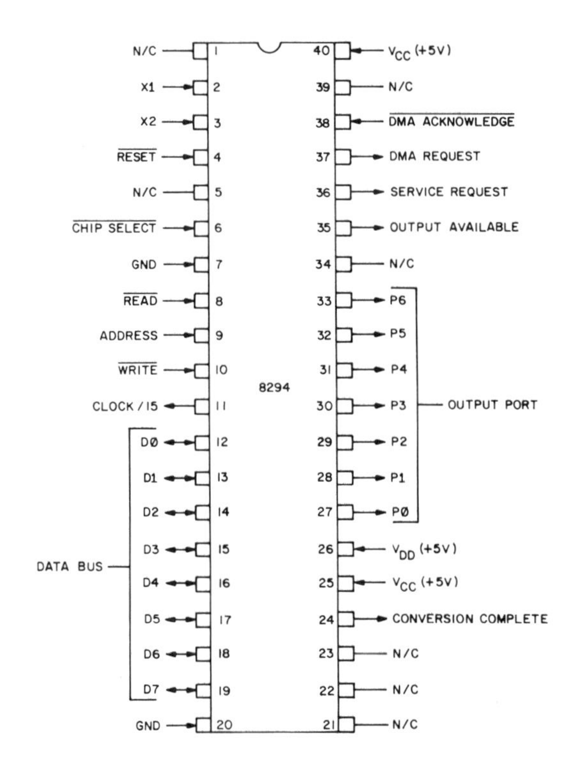

Operation

Data is enciphered according to a programmable 56-bit encoding key. This key determines what operations are to be performed on the data, and can be changed on a daily or minute-by-minute basis. However, the key only needs to be entered one time for all enciphering or deciphering purposes. New keys are entered only when the code is changed.

A 64-bit message block is placed into the 8294 after the key is established, eight bytes at a time. Each byte must have odd parity in order to be accepted. The message block is then processed according to the mode of operation set in the *command register* (encode or decode). After the algorithm is completed, the converted data is output and replaced with the next data block.

CPU Interface

All data and software commands are shuttled between the CPU and the 8294 through eight bidirectional DATA BUS (pins 12-19) pins. Data is input to the chip when the WRITE (pin 10) line is LOW, and removed when the READ (pin 8) control is LOW. The READ and WRITE commands are enabled when the CHIP SELECT input (pin 6) is LOW.

Only four internal registers are needed to manage the chip's entire operation. Each register is addressed by a combination of the READ, WRITE, and ADDRESS (pin 9) pins, and accessed through the DATA BUS.

Interrupts

Interrupts are internally software generated in the 8294, and may be inspected by polling the contents of the *status output buffer.* To minimize software overhead, though, three hardware interrupt outputs are also available to the user.

The SERVICE REQUEST output (pin 36) signals the CPU when the *input buffer* is empty, and awaiting data input or a command. Once data has been loaded into the *input buffer,* it is processed according to the prescribed code and the results placed in the *output buffer.* Whenever the *output buffer* is full, the OUTPUT AVAILABLE (pin 35) interrupt goes HIGH, indicating a read operation is necessary. Another indication that a conversion has been completed is displayed by the CONVERSION COMPLETE (pin 24) output signal. Even though all three hardware interrupts are dedicated and individually pinned out, they are not effective unless specifically requested by a Set Mode command.

DMA Operation

To reduce CPU throughput, data can be transferred directly to or from memory using two DMA control signals. To initiate a DMA transfer, the CPU must first initialize the two DMA channels and issue a DMA Enable command. Following this command, the number of 8-byte message blocks to be processed is entered (up to 256 blocks). The 8294 then generates a DMA REQUEST signal (pin 37) whenever one data block has been converted and is ready for transfer to memory. Transfer takes place when the DMA controller grants access to the memory by pulling the DMA ACKNOWLEDGE input (pin 38) LOW. The next block of data is then loaded and the cycle continues, with no further intervention by the CPU until the specified number of data blocks are processed. Once the DMA operational number is depleted, the DMA mode is terminated and control reverts back to the CPU.

Oscillator

All internal timing is controlled by a self-contained oscillator and timing circuit. A choice of crystal, L-C, or external clock can be used to generate the basic oscillator frequency of 1- to 6-MHz.

A crystal or L-C network connected between X1 (pin 2) and X2 (pin 3) provides a feedback path for the oscillator's high-gain amplifier stage, giving the proper phase shift for oscillation. An external clock signal can also be used as a reference frequency for the 8294; however, the input voltage levels are *not* TTL compatible. The external input must be in the range of 1- to 6-MHz and must be driven by a buffer with a suitable pull-up resistor to guarantee that a logic "1" is above 3.8-volts. A recommended connection is shown in figure 3.

Output Port

Totally detached from the encryption/decryption function is an OUTPUT PORT (pins 27-33). These output lines, which have absolutely nothing to do with the encryption, simply reflect any 7-bit word which is asserted by the CPU using a Write To Output Port command.

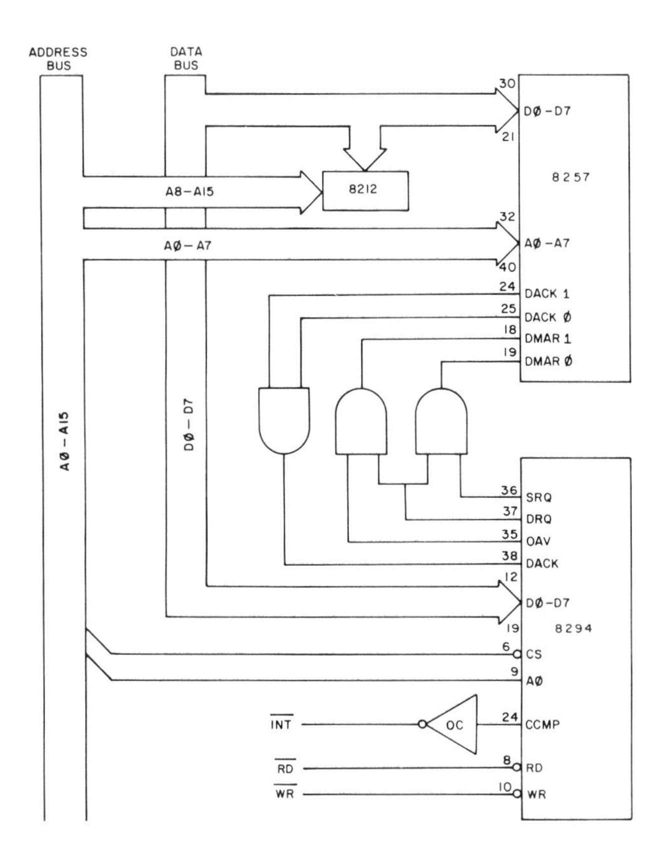

Figure 1

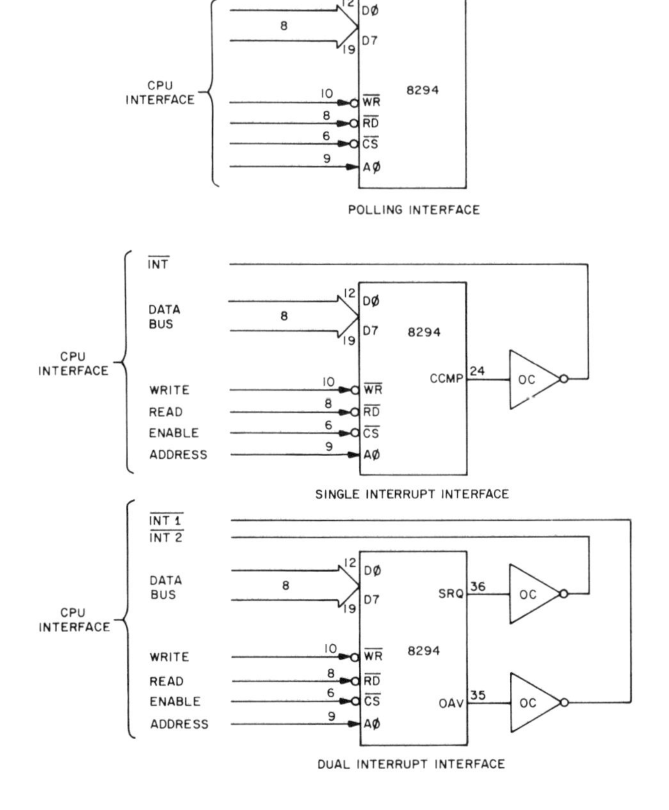

Figure 2

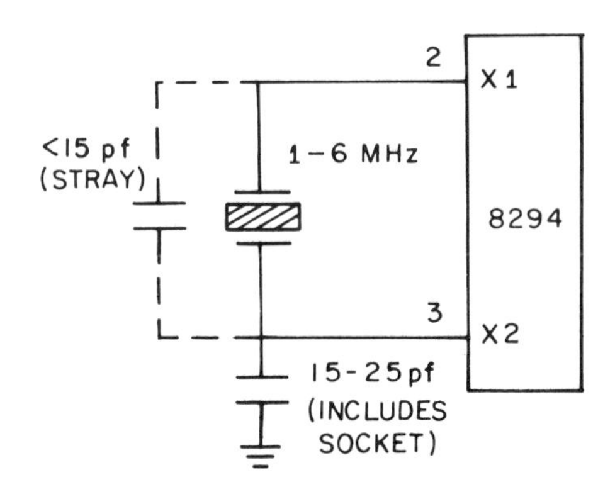

Figure 3a

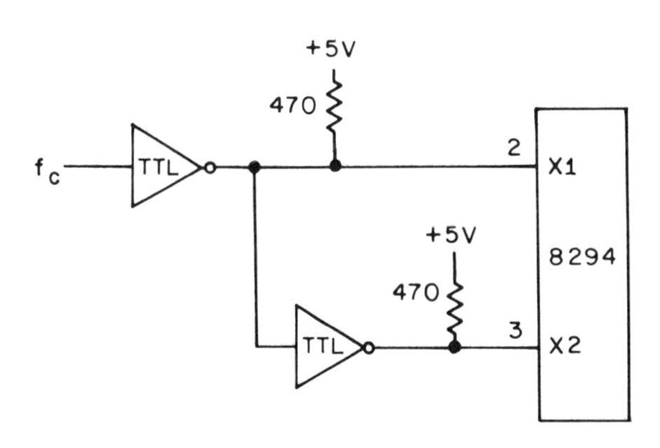

Figure 3b

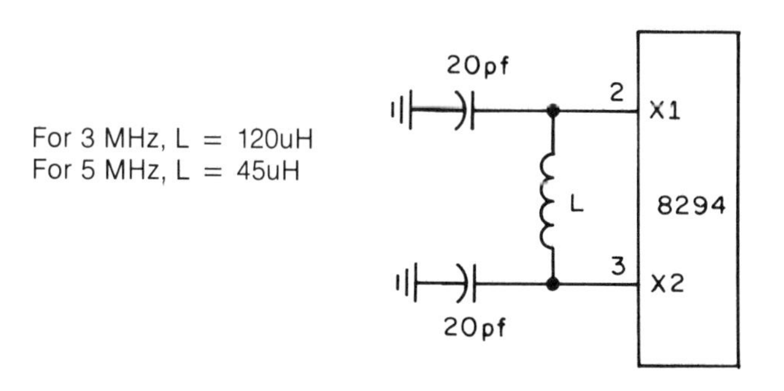

Figure 3c

MC6859

Data Security Device

The MC6859 Data Security Device, made by Motorola, is a programmable binary-coded scrambler designed to be incorporated in a wide variey of digital equipment requiring protection of data through cryptographic measures.

The cryptographic algorithm utilized by this chip is the Data Encryption Standard (DES) as adopted by the U.S. Department of Commerce, National Bureau of Standards (NBS), and described in publication FIPS PUB 46 (1-15-77).

Cryptographics

The MC6859 appears to the CPU as an interface adapter device which processes an intelligible message by scrambling it according to an algorithm. The CPU can then forward the encoded message to a modem or similar communications controller for transmission to a receiving station, assured that the message will not be understood if intercepted. Likewise, a received encoded message can be unscrambled by the MC6859 and presented to the CPU in its original text form.

The key to the algorithmic process is a 56-bit word (plus 8 bits of parity) called an Active Key code. The first of these keys is a Major Key, and it must be entered into the chip's *major key register* prior to data processing.

Data Processing

Data is transferred between the CPU and MC6859 over eight bidirectional DATA BUS lines (pins 17-22, 24, 2) in blocks of eight bytes at a time. The mode of the DATA BUS is determined by three control pins, defined as CONTROL ADDRESSes A0 (pin 3), A1 (pin 4), and A2 (pin 5). These control inputs are used in conjunction with the READ/WRITE (pin 8) input to direct the flow of data and its destination.

To encode an 8-byte block of data, the first seven bytes are written into the *write data/"C" key register* by properly configuring the CONTROL ADDRESS pins and pulling the READ/WRITE input LOW. The eighth byte, which is written into the *encipher data register*, automatically initiates the encryption process, as specified by the current Active Key. During enciphering, a Busy flag is set, and only a Read Status, hardware RESET (pin 7) or software Reset will be recognized; all other commands are ignored.

Upon completion of the algorithm, the Busy flag is removed and the READY INTERRUPT REQUEST output (pin 23) is driven LOW. The CPU then executes a read data operation by altering the control mode and raising READ/WRITE input HIGH. As each byte is read, zeroes are shifted into the *data register* to ensure data security.

The deciphering process is exactly the same as encoding, except that the eighth byte is loaded into the *decipher data register*, which effectively reverses the encoding procedure. During the algorithm process, either encode or decode, the bytes are constantly monitored for parity errors. If an error is detected, the PARITY ERROR INTERRUPT REQUEST (pin 1) goes LOW.

Chip Select

In order for the chip to be actively engaged, it must be properly addressed using the CHIP SELECT inputs. The CHIP SELECT

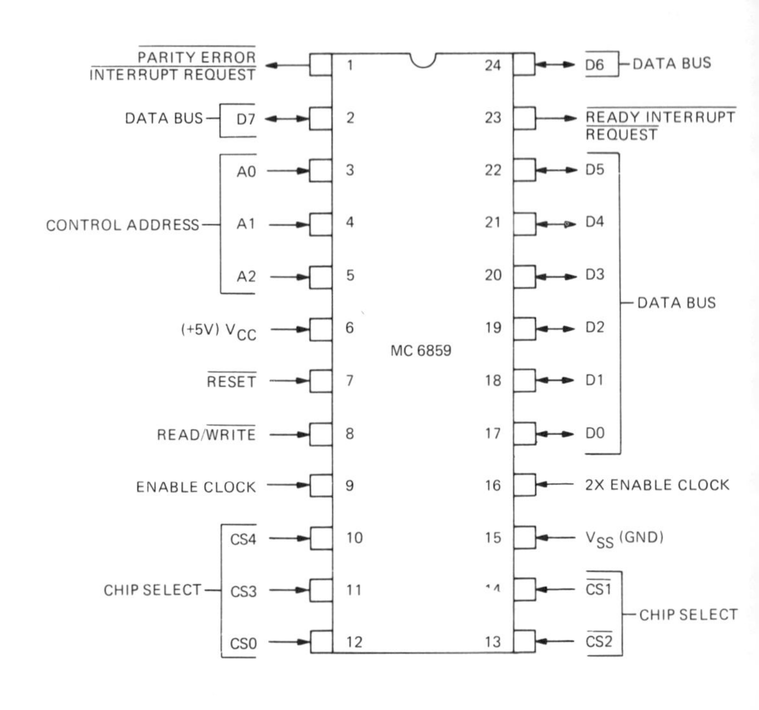

address is valid when CS0, CS3, & CS4 (pins 12, 11, & 10) are HIGH and CS1 & CS2 (pins 14 & 13) are LOW. When the device is not addressed, the DATA BUS ports assume their high-impedence state.

Clock

The rising edge of the ENABLE CLOCK (pin 9) initiates data transfer between the MC6859 and CPU, while the falling edge latches data into the chip during the write cycle. The 2X ENABLE CLOCK (pin 16) processes the encipher/decipher algorithm; a typical alogrithm operation requires 320 clock cycles.

Security Considerations

The security of a system employing NBS Data Encryption Standards depends only upon the cipher key used, and not the availability of the algorithm or equipment used to implement the algorithm. For greater security, a Secondary Encipher Key is available to the user, which is loaded into the MC6859 after the Major Key. The data is now double encoded. Since the DES algorithm utilizes a 56-bit Active Key, there are 2^{56} possible encrypted combinations, or about 7×10^{16} possibilities, available to the user.

If additional security is required, however, several techniques can be called upon to further increase data security. An easy way to increase security without reducing throughput is to perform the DES algorithm "in reverse." In other words, data is first deciphered at the sender by decoding the original message, and then unraveled at the receiver by enciphering the data to yield the original message. This technique works because the enciphering and deciphering algorithms are mirror images of each other.

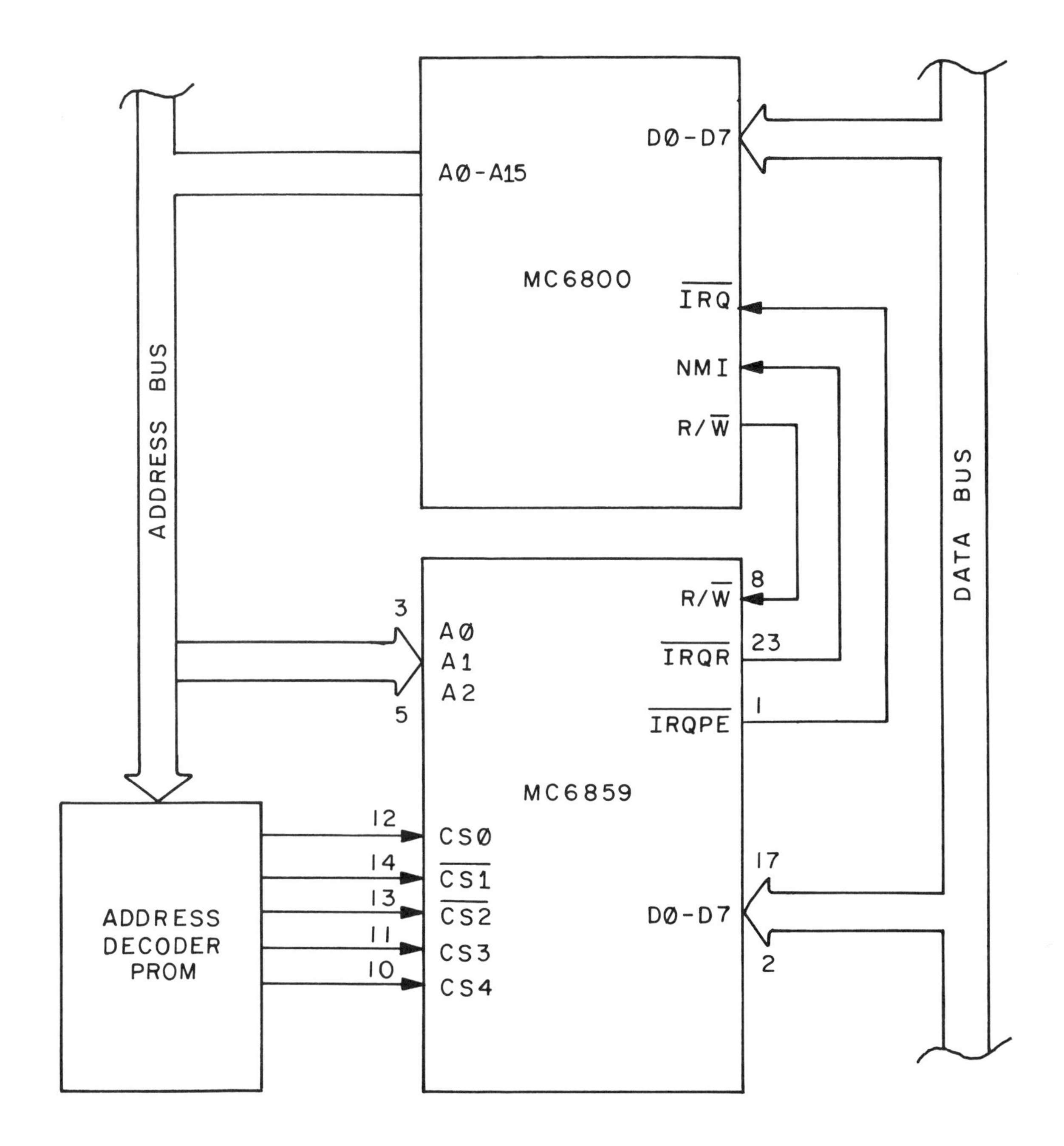

Figure 1

NOTICE

This product may not be exported without prior approval from the U.S. Department of State, Office of Munitions Control

CHAPTER THREE
INTERFACE

Z8036

Counter/Timer & Parallel I/O Unit

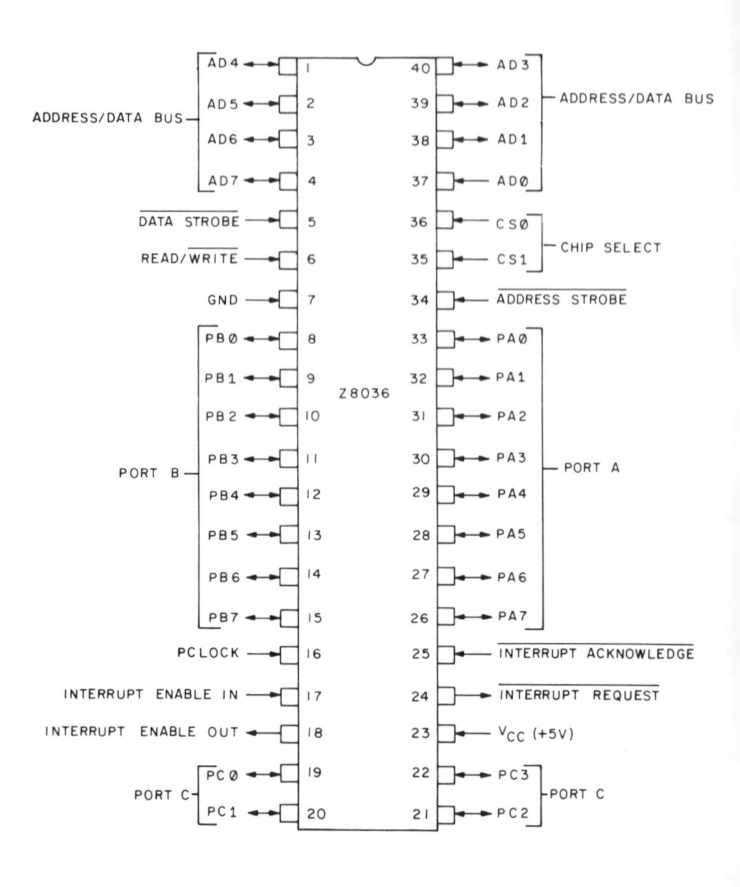

The Z8036 Counter/Timer and Parallel I/O element, made by Zilog, is a general purpose peripheral chip, satisfying most counter, timer, and parallel I/O requirements encountered in a microprocessor system. The versatile Z8036 chip contains three I/O ports and three 16-bit counter/timers. This particular version of the CIO device is specifically designed for use with Zilog's Z-Bus.

Operation

Basically, the Z8036 is a programmable, bidirection peripheral controller used to ease the monitoring and counting functions often done by the microprocessor. Programming the chip entails loading the control registers with bits to implement the desired operation.

Prior to programming, the device is reset by forcing the ADDRESS STROBE (pin 34) and DATA STROBE (pin 5) lines LOW simultaneously. Once reset, only the Rest bit can be read or written to; writing to all other bits is ignored and all reads return as zeros. Only after clearing the Reset bit can the other command bits be programmed.

Z-Bus Interface

The Z8036 interfaces to the CPU through a multiplexed ADDRESS/DATA BUS (pins 37-40, 1-4). The chip allows two schemes for register addressing. Both schemes use only six of the eight bits of the ADDRESS/DATA BUS.

To read the contents of a register, the CPU drives the READ/WRITE (pin 6) input HIGH and places an address on the ADDRESS/DATA BUS. The lower six bits of the least significant address byte are used to specify a register; the most significant address byte and status information are combined and decoded by external logic to provide two CHIP SELECT (pins 36, 35) signals that enable the chip. These six address bits, along with the CHIP SELECTS, are latched into the Z8036 by an ADDRESS STROBE.

The data contained in the selected register is then strobed onto the ADDRESS/DATA BUS when the CPU issues a DATA STROBE command. If the register indicated by the address does not exist, the bus simply remains in its high-impedence state.

Writing to a register is very much the same. The identical timing steps are required, with the READ/WRITE input pulled LOW to direct the flow of data into the chip instead of from it.

I/O Port Operations

Of the three I/O ports available, two are 8-bit general-purpose ports and the third is a special-purpose 4-bit port. The two general purpose elements, PORT A (pins 33-26) and PORT B (pins 8-15), can be configured as input, output, or bidirectional ports—all with handshaking capabilities. They can also be linked together to form a single 16-bit port that is input or output programmable on a bit-by-bit basis, provided the hand-shaking routine is eliminated. PORT B also provides access for Counter/Timers #1 and #2.

The function of the special-purpose 4-bit port, PORT C (pins 19-22), depends upon the roles of PORTs A and B. In many cases, PORT C is called upon to supply the handshake lines for the other two. Any bits of PORT C *not* used for handshaking can be used to provide access to the third internal counter, or as simple I/O lines on a bit-programmable basis.

Request/Wait

PORT C can also be programmed to provide a status signal in addition to its handshaking duties, which is either a REQUEST or a WAIT signal. A REQUEST output indicates when a port is ready to perform a data transfer via the Z-Bus. It is intended for use with a DMA-type device. The WAIT signal, on the other hand, provides synchronization for transfers with a CPU. Because the extra PORT C line is used, only one general purpose port can function with a handshake line and a REQUEST/WAIT line; the other port must be bit programmed.

Pattern Recognition

In all configurations, PORTS A and B can be programmed to recognize specific data patterns and generate interrupts when the pattern is encountered. Pattern recognition may be performed on all bits, including those used in I/O for the counter/timers.

The pattern can be independently specified for each and every bit as a "1", "0", rising edge, falling edge, or any transition. These may be intermixed in any order, and individual bits can be masked off the pattern altogether. A pattern match is defined as the congruent occurance of all unmasked bits. Unfortunately, PORT C does not contain pattern recognition.

With the exception of the CPU bus interface and programming methods, the Z8036 is identical in operation to the Z8536, and the reader will find additional details outlining their operation by reviewing that section.

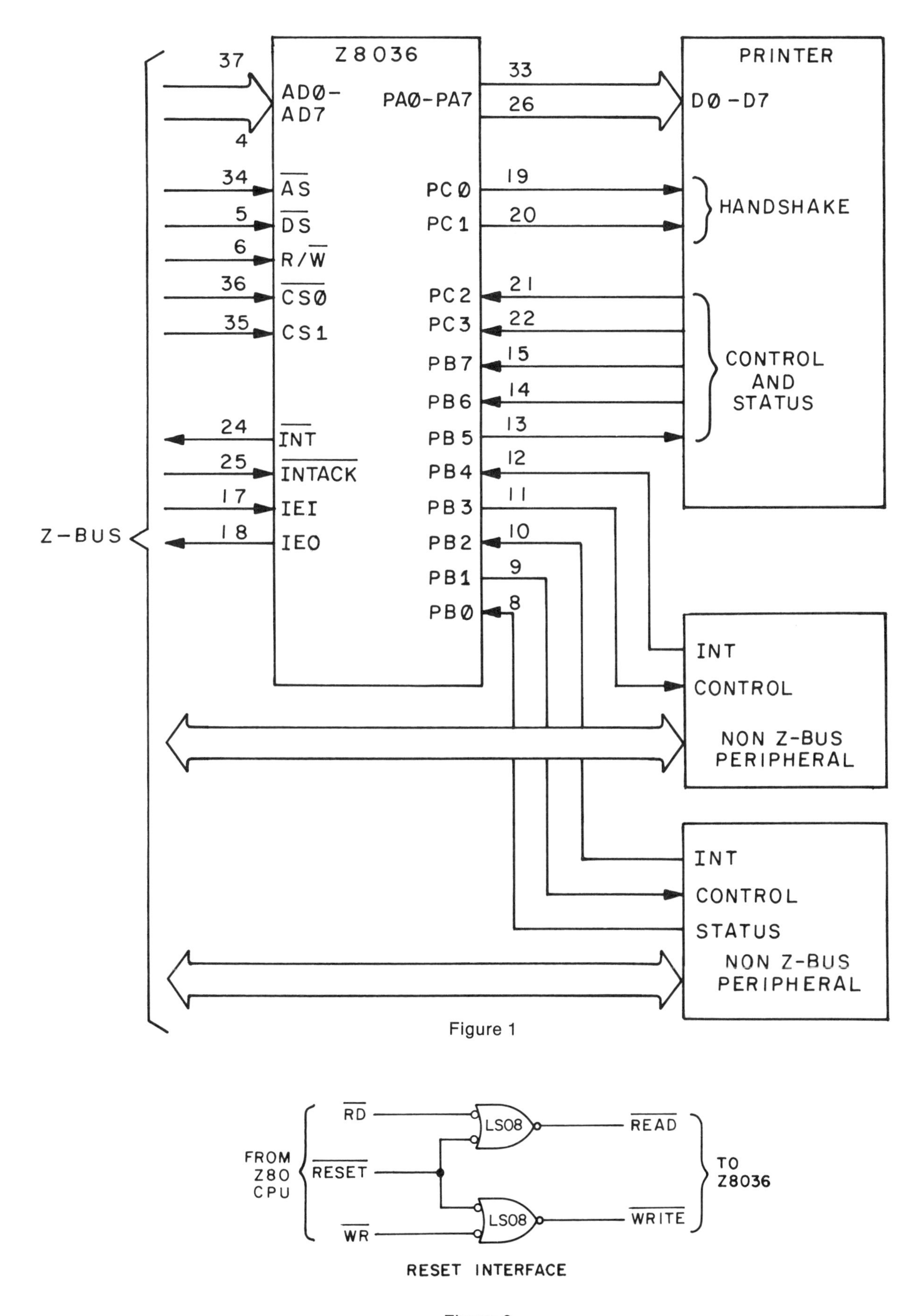

Figure 1

Figure 2

Z8536

Counter/Timer & Parallel I/O Unit

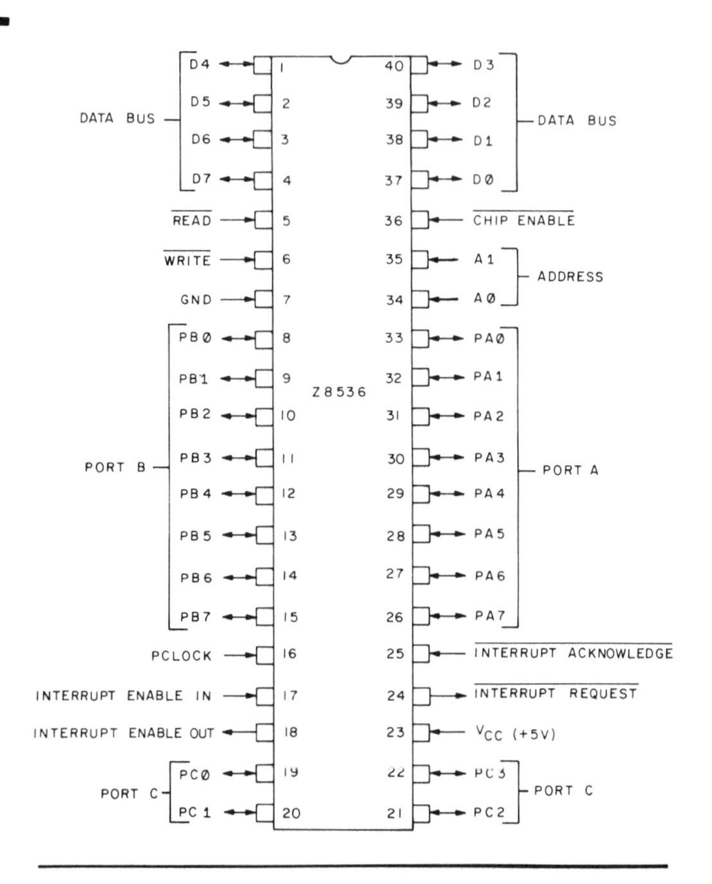

The Z8536 Counter/Timer and Parallel I/O element, by Zilog, is a general purpose peripheral chip, satisfying most counter, timer, and parallel I/O requirements encountered in a microprocessor system. The device is easily interfaced to most popular microprocessors, and contains many programmable operations aimed toward specific applications.

Programming

Programming the chip entails loading the control registers with command bits to implement the desired operation. Only the data registers within the Z8536, however, are directly accessible. This is accomplished by configuring the ADDRESS lines, A0 (pin 34) and A1 (pin 35). All other internal registers are accessed by the following two-step sequence, with the ADDRESS line specifying the control operation (see fig. 2).

First, the CPU places the address of the target register on the DATA BUS (pins 37-40, 1-4). Both the A0 and A1 inputs must be HIGH, and the CHIP ENABLE (pin 36) engaged, before programming can take place. When the WRITE input (pin 6) is strobed LOW, the register address is transferred into a 6-bit *pointer register*. The next READ (pin 5) or WRITE operation will be performed on the specified register. As long as only READ cycles are performed, the register can be read continuously without updating the *pointer register*. A subsequent WRITE, however, will clear the pointer and reset it for the next address. The entire chip is reset by forcing the READ and WRITE inputs LOW simultaneously, normally an illegal condition.

Although the data registers are usually accessed using the ADDRESS inputs, they can be accessed by the above two-step procedure.

Counter/Timer Operation

The counter/timer function consists of three identical 16-bit counter/timers. Each counter/timer contains a presettable down counter, a *time constant register* (storing the value to be loaded into the down counter), a *current count register*, and two registers for control and status.

The flexibility of the counter/timers is enhanced by providing up to four lines of direct control and status monitoring per function. They include: counter input, gate input, trigger input, and output. Timers #1 and #2 are accessed through the bit ports of PORT B; counter/timer #1 is accessible through pins 11 to 15, and counter/timer #2 is accessible through pins 8 to 11. Timer #3 is accessed through the four pins of PORT C (pins 19-22). The relationship between the pin numbers and the four control and status lines is shown in fig. 3.

When external counter/timer lines are to be used, the associated PORT pins must be vacant and programmed in the proper data direction. Lines for the counter/timer have the same characteristics as simple input lines: They can be specified as inverting or noninverting, they can be read, and will respond to pattern recognition logic.

Once a down counter is loaded with the value contained in the *time constant register*, the counting sequence begins and continues toward terminal count as long as all the hardware and software gate inputs are HIGH. If either of these gates goes LOW, the countdown halts until the HIGH level is re-established.

The counting rate is determined by the mode of the counter/timer. In the Timer Mode, the down counter is clocked internally by a signal that is half the frequency of the PCLOCK (pin 16) input. Used as a counter, the down counter is incremented on the rising edge of the counter input. Each time the counter reaches terminal count, an interrupt is generated (provided the interrupts are enabled).

Interrupt

The Z8536 has five possible sources of interrupt: the three counter/timers and PORTs A and B. An INTERRUPT REQUEST (pin 24) is asserted when an interrupt occurs, to which the CPU responds with an INTERRUPT ACKNOWLEDGE (pin 25). As part of the INTERRUPT ACKNOWLEDGE cycle, the Z8536 is capable of responding with an 8-bit interrupt vector that specifies the source of the interrupt. The internal interrupt priority is fixed within the chip, with counter/timer #3 having the highest priority and PORT A the next.

Pattern Recognition

In all configurations, PORTs A and B can be programmed to recognize specific data patterns and generate interrupts when the pattern is encountered. Pattern recognition may be performed on all bits, including those used as I/O for the counter/timers.

The pattern can be independently specified for each and every bit as a "1", "0", rising edge, falling edge, or any transition. These may be intermixed in any order, and individual bits can be masked off the pattern altogether. A pattern match is defined as the congruent occurance of all unmasked bits. Unfortunately, PORT C does not contain pattern recognition.

With the exception of the CPU bus interface and programming methods, the Z8036 is identical in operation to the Z8536, and the reader will find additional details outlining their operation by reviewing that section.

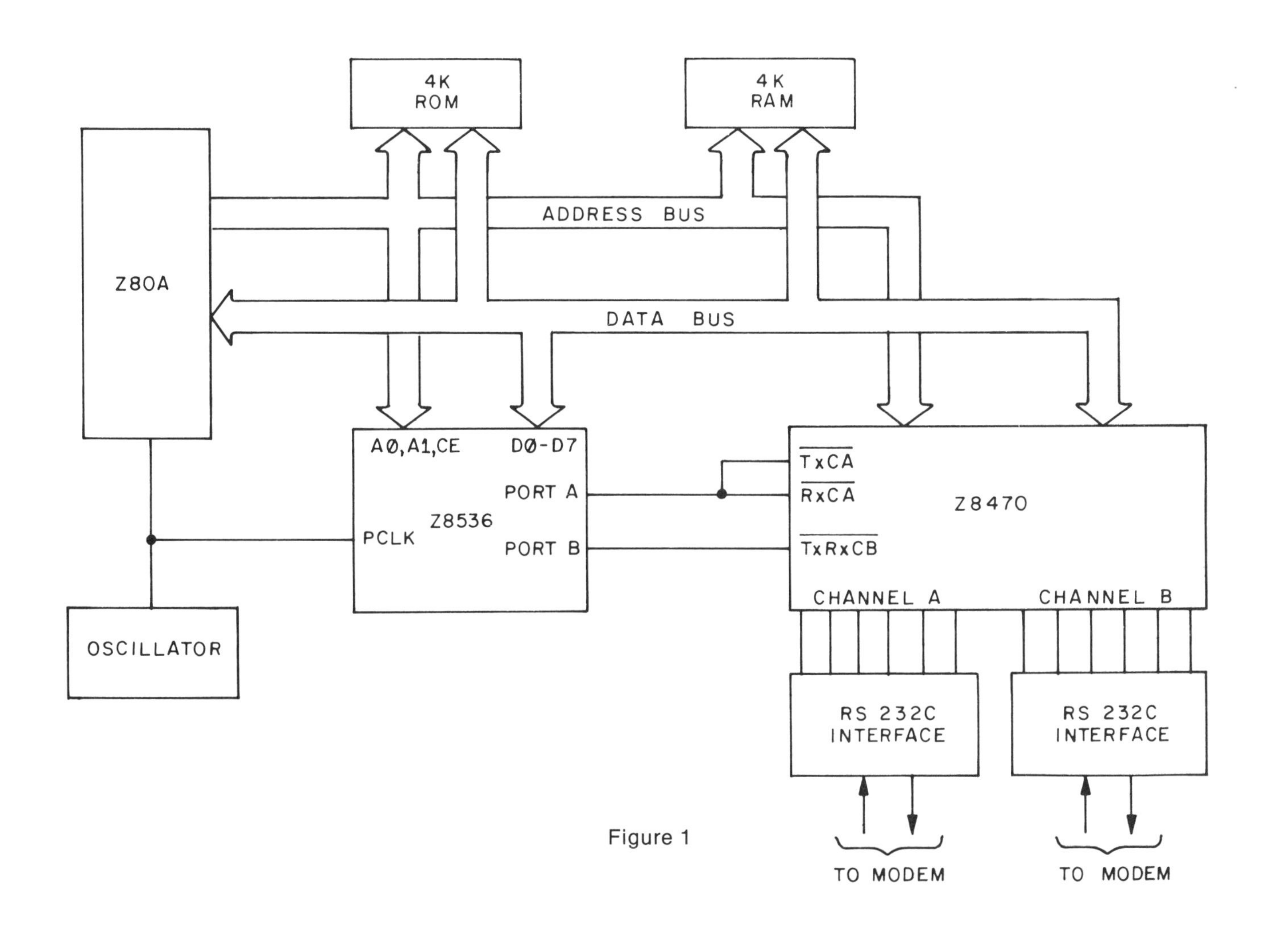

Figure 1

A1	AØ	REGISTER
0	0	PORT C DATA REGISTER
0	0	PORT B DATA REGISTER
1	1	PORT A DATA REGISTER
1	1	CONTROL REGISTERS

Figure 2

FUNCTION	TIMER #1	TIMER #2	TIMER #3
OUTPUT	PB4	PBØ	PCØ
COUNTER INPUT	PB5	PB1	PC1
TRIGGER INPUT	PB6	PB2	PC2
GATE INPUT	PB7	PB3	PC3

Figure 3

CDP1851

Programmable I/O Interface

The CDP1851, by RCA, is a Programmable Interface device designed for use with CDP1800 series microprocessors. Each port of the interface can be programmed with either byte or bit I/O pins for interfacing to peripheral devices such as printers and keyboards.

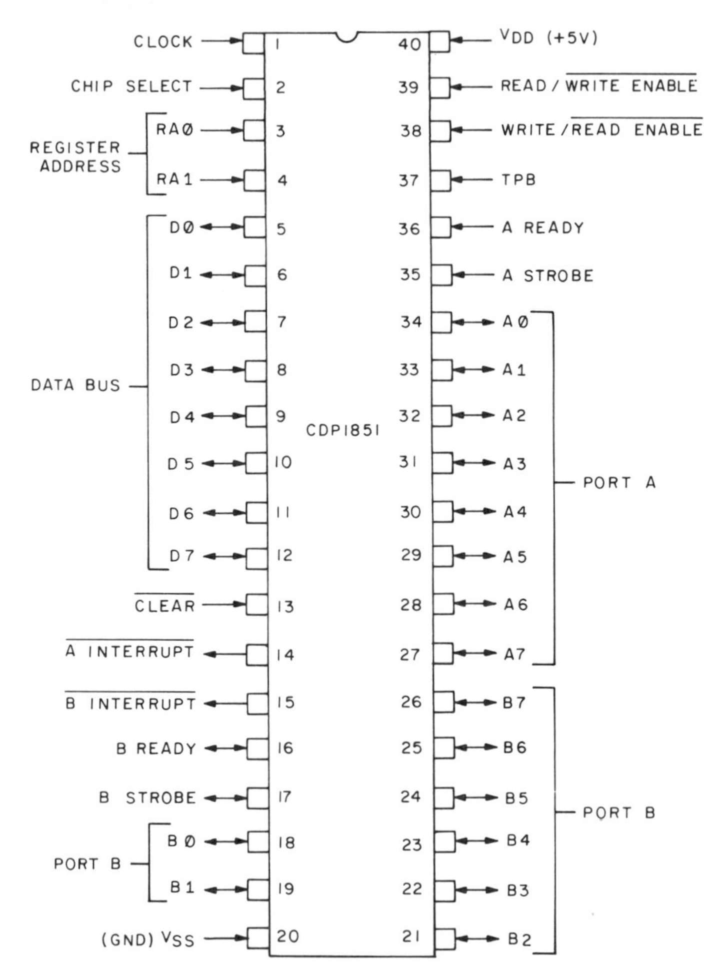

Functional Description

The 1851 is a two-port I/O interface with four programmable modes: Input, Output, Bidirectional, and Bit-Programmable. When programmed, these ports provide an interface between an external peripheral and the CPU. Prior to programming, however, the chip is initialized by a CLEAR (pin 13) command during the power-on phase. The program registers are then accessed through the DATA BUS (pin 5-12) by configuring the REGISTER ADDRESS (pins 3, 4) inputs, and the I/O port programmed.

PORT A (pins 34-27) is programmable in all four modes, while PORT B (pins 18-26) is programmable for the Input, Output, and Bit-Programmable modes. When PORT A is in the Bidirectional Mode, PORT B must be in the Bit-Programmable Mode. Except in this single restriction, all modes may be individually executed by either port.

Input Mode

Either port can be set for the Input Mode. Selection of the intended port is determined by the REGISTER ADDRESS inputs, which joins the proper PORT to the DATA BUS lines. Each port has two handshaking lines that sequence the operation of this mode.

A HIGH-to-LOW transition of the STROBE (pins 35 and 17, for PORT A and B, respectively) will input data to the chip and generate an interrupt while doing so to indicate that data has been received by the port. When the CPU reads the data, the interrupt is reset and the READY line (pin 36, 16; A, B) asserted to report that the channel is clear.

Output Mode

The STROBE and READY lines are also the handshaking lines for the Output Mode. Upon receipt of a STROBE pulse, data is output from the port to the peripheral and an interrupt set. After the CPU loads new data into the port, the interrupt is reset and the READY line goes HIGH to indicate that the port is full.

Bidirectional Mode

Only Port A can be programmed for the Bidirectional Mode. Since the data flow is bidirectional, four handshaking lines are required to synchronize operations—and PORT A borrows two from PORT B. For data inputs, the A STROBE input loads peripheral data into the 1851, generating an interrupt as it does. As soon as the CPU reads the data from the port, the A READY output is forced HIGH

The PORT B handshaking controls are used for output operations. Data is transmitted on a B STROBE signal, setting an interrupt to indicate that the port is now empty. When the CPU loads new data into the A port, the interrupt is cleared and the B READY line asserted.

The direction of the data flow through the DATA BUS is governed by two control lines. A positive input to the WRITE/READ ENABLE (pin 38) writes data from the CPU bus into the 1851. Alternately, data is read from the 1851 when the READ/WRITE ENABLE input (pin 39) is actively HIGH.

In the Bidirectional Mode, PORT B has no access to its handshaking lines, and must be programmed in the Bit-Programmable Mode. Furthermore, the *status register* must be read to determine whether the source of the interrupt was an A STROBE or a B STROBE.

Bit-Programmable Mode

In the Bit-Programmable Mode, each and every pin associated with the port can be individually programmed as an input or output data bit—including the READY and STROBE lines. The exception, of course, is when the Bidirectional Mode of PORT A must use the B READY and B STROBE lines.

The eight I/O pins of each port can generate interrupts. These interrupts can be programmed to occur on certain logic conditions, such as pattern recognition. An interrupt mask determines which bits are monitored and which are not. Any combination of masked or monitored I/O data bit lines is permissible, and any combination of inputs or outputs can be monitored for interrupts. The interrupt will be reset when the logic condition no longer exists. Handshaking lines cannot be monitored for pattern recognitions.

In all modes, interrupts can be made available to the external INTERRUPT pins through software enabling. The A INTERRUPT (pin 14) corresponds to PORT A; B INTERRUPT (pin 15) to PORT B. The INTERRUPT outputs are open-drain drivers that permit wired ORing, and must be used with pull-up resistors, typically 10k, tied up to V_{DD} (pin 40).

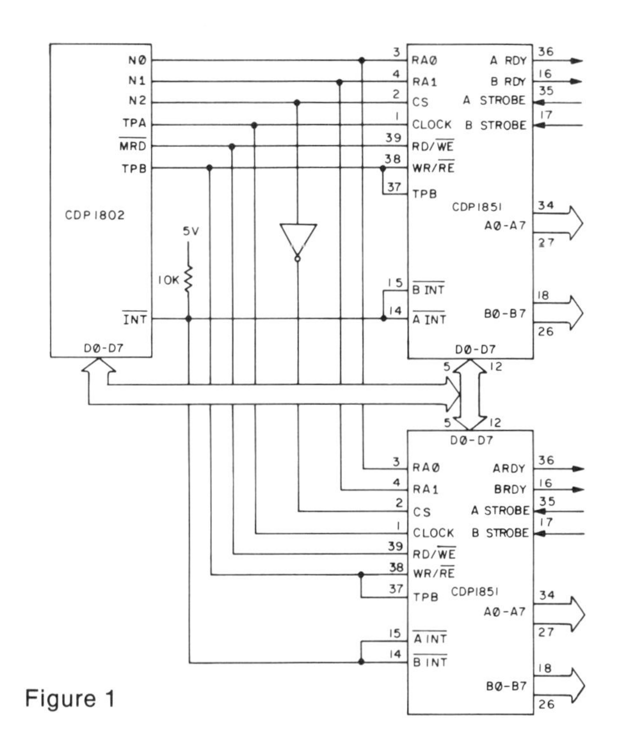

Figure 1

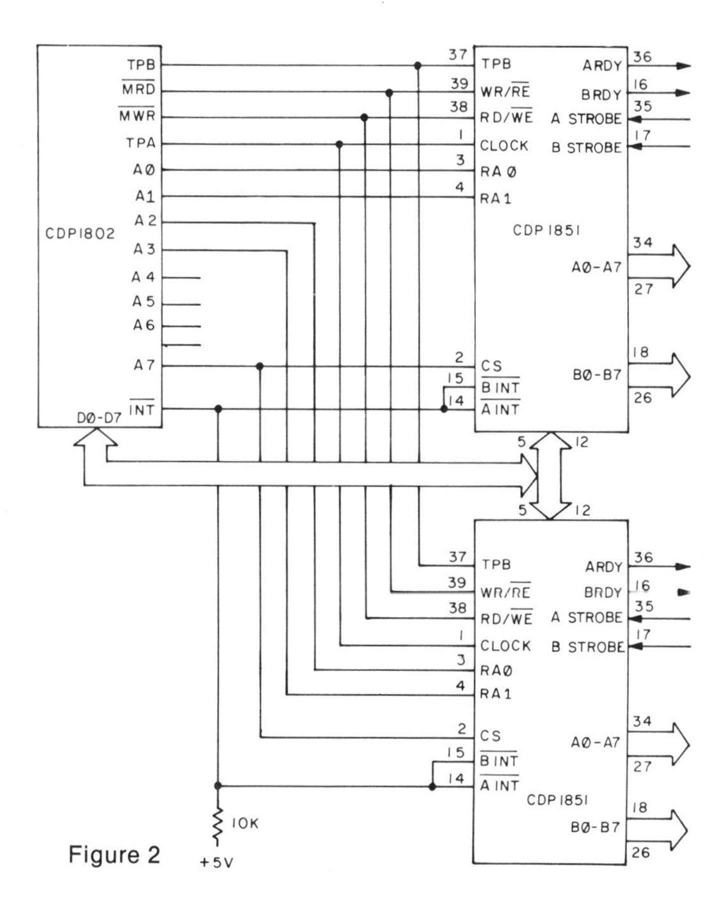

Figure 2

The memory addressed configuration allows up to four CDP1851s to occupy memory space with no additional hardware. Address lines A4, A5 and A6, A7 are used as RA0 and RA1 inputs to the third and fourth chips, respectively.

CY232

Parallel/Serial Interface Device

The CY232 Parallel/Serial Interface Device, by Cybernetics Micro Systems, is designed to be a communications interface between parallel TTL lines and a serial asynchronous communications system. The chip can address up to 256 parallel devices, and operates in full duplex mode using standard RS-232C format, making it compatible with existing systems.

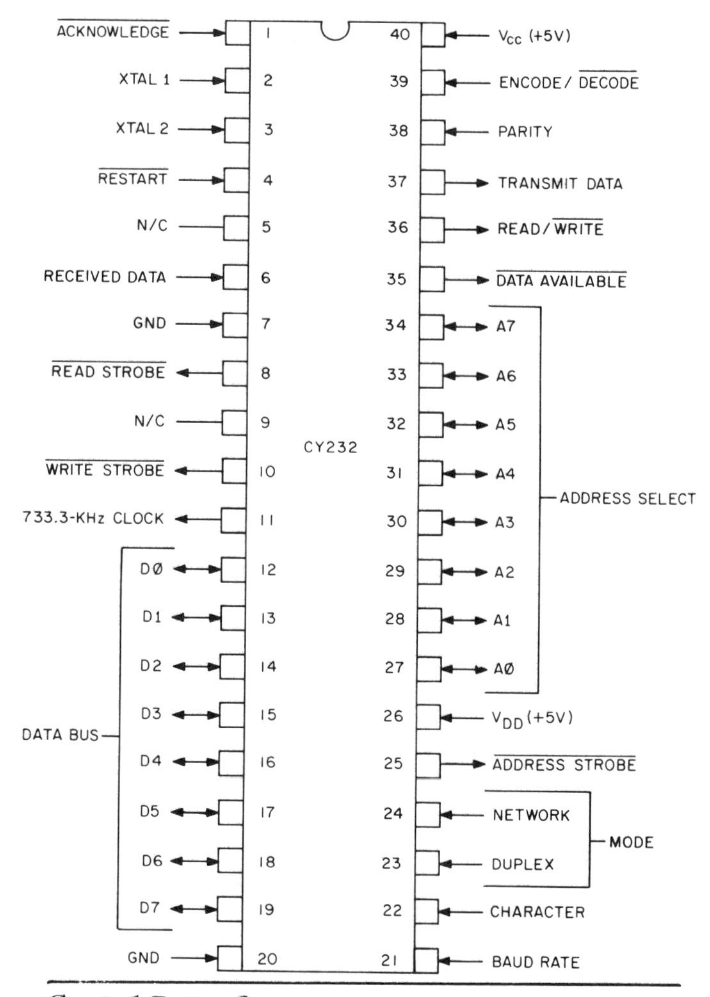

Parallel Peripheral Addressing

The CY232 will receive parallel data from as many as 255 8-bit devices, or distribute parallel data to 255 devices by simply addressing the intended peripheral. Locating the device in service is done with the bidirectional ADDRESS SELECT lines (pins 27-34), which are software defined (input or output) and can operate in one of two addressing formats—encoded or decoded.

When encoded, external devices 0 through 255 may be addressed with positive true logic. For example, 0001 0000 would address chip number 16. When decoded, however, only eight devices may be addressed. In this mode, one of the eight ADDRESS SELECT lines will be taken LOW as the chip select, and the other seven will remain HIGH. An example of this mode shows that address 1110 1111 will select device #4.

The address line format is determined by the ENCODER/DECODE input (pin 39), where a HIGH input represents true binary code (encoded), and a LOW selects the decoded mode.

Before each read or write operation, though, the CY232 checks to see if the address presented is a valid one. This allows the user to program accessible locations only, and prevents wasteful time on addresses which don't physically exist (unconnected devices). Although this is basically an internal software operation, it can be performed externally by recognizing the ADDRESS STROBE output (pin 25) when the new address is written into the chip. External circuitry can then decide if the address presented is valid or not by forcing the ACKNOWLEDGE input (pin 1) HIGH (valid) or LOW (disallowed).

Data Bus

Valid data is transferred either to or from the CY232 (as set by the software instructions) through eight DATA BUS lines (pins 12-19); the READ/WRITE output (pin 36) indicates the direction of the data flow. In addition, two separate outputs report data written *from* the CY232 (WRITE STROBE, pin 10) or data read *into* the chip (READ STROBE, pin 8) when either goes HIGH. Notice the data flow is expressed in CPU directives. These three outputs are intended to provide external circuitry with handshaking signals.

A second type of handshaking is available using the DATA AVAILABLE output (pin 35) and the ACKNOWLEDGE input. The handshake is accomplished by taking the ACKNOWLEDGE input HIGH before the DATA AVAILABLE output is strobed LOW. If a successful data transfer is made after the DATA AVAILABLE goes LOW, the external device notifies the CY232 by taking the ACKNOWLEDGE input LOW. Pulling the ACKNOWLEDGE pin LOW before the DATA AVAILABLE is asserted prevents the initiation of any read or write operations. If ACKNOWLEDGE and DATA AVAILABLE are not used in an application, they should be tied together.

Serial Interface

Serial communications is performed via two ports: RECEIVED DATA (pin 6) and TRANSMIT DATA (pin 37). The baud rate for the two is established by the BAUD RATE input (pin 21). A HIGH input will set the rate at 1200, a LOW sets it at 300 baud; a floating input assumes a 2400 baud rate.

Before the data can be converted from parallel form into serial (or vice versa for the receiver), the chip must be defined as to the serial data format. Three control pins define the serial network, and are sampled for all time (as is the BAUD RATE) with a RESTART (pin 4) pulse, and are as follows.

The protocol format is set by the CHARACTER input (pin 22). This input can provide Binary (input HIGH), ASCII Character (input LOW), or ASCII HEX (input open) symbols for communications over the serial interface.

The network configuration and operating mode is determined by a 2-bit address to the NETWORK (pin 24) and DUPLEX (pin 23) inputs. Altogether, there are three concepts involved: Master, Slave, and Echo. In the Master Mode, the CY232 will initiate transmission of data on its own accord by scanning the messages from the various parallel devices. If not in the Master Mode, then the chip is considered to be a Slave, and will not initiate messages. The Echo Modes (four in all) will retransmit the message it receives from the serial network.

The CY232 will operate in either a full duplex or half duplex mode, with automatic turnaround capabilities. Whatever mode is established, the chip will perform constant parity checks on the data flow as described by the setting of the PARITY (pin 38) input. A LOW input signifies an odd parity, a HIGH sets an even parity.

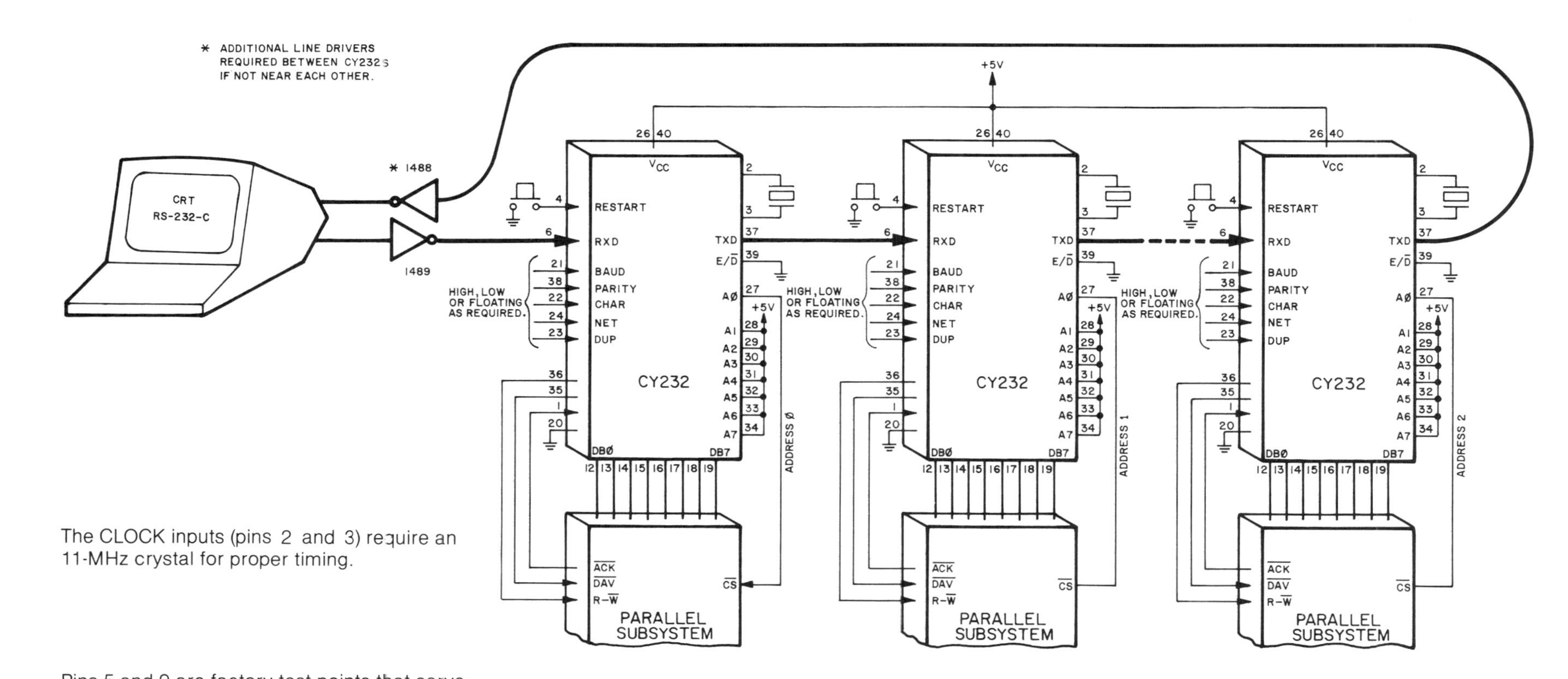

The CLOCK inputs (pins 2 and 3) require an 11-MHz crystal for proper timing.

Pins 5 and 9 are factory test points that serve no useful purpose, and should be left open during operation.

MC68122

Cluster Terminal Controller

The MC68122 Cluster Terminal Controller, by Motorola, relieves the CPU of the time-consuming chores related to communicating with peripheral devices, such as terminals and line printers, by establishing the data communications link. The cluster controller acts primarily as a "front end processor" and works with either synchronous or asynchronous bus lines.

To perform its duties, the MC68122 contains an enhanced version of the popular MC6800 CPU. All device functions—as well as actual program execution within the chip—are totally transparent to the user.

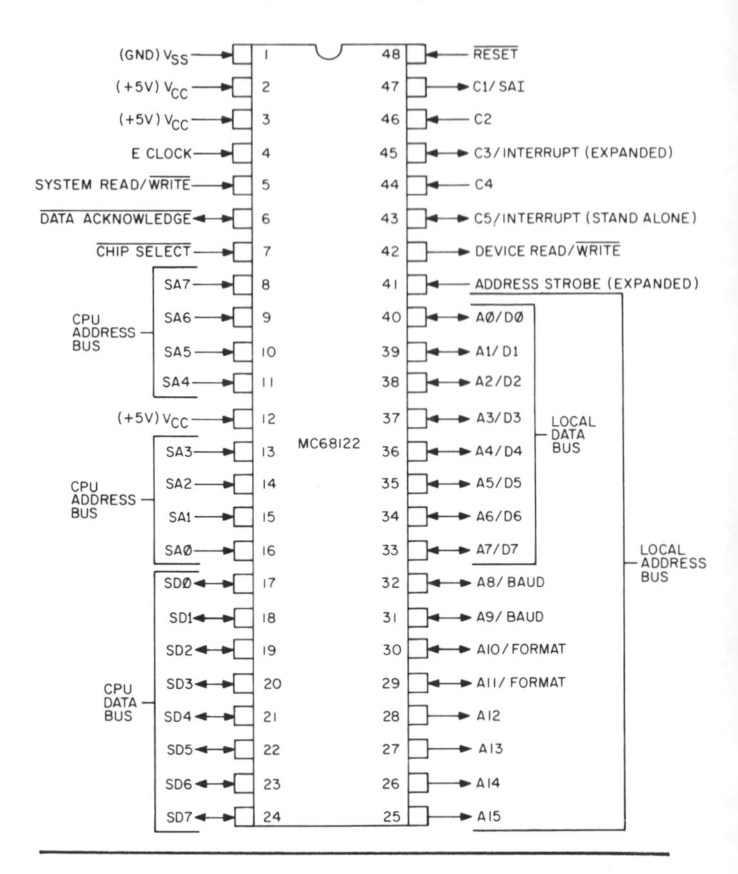

CPU Interface

Data transfer between the host CPU and the MC68122 is accomplished by a dual port RAM in the controller chip known as the *transfer area RAM.* Because it is dual ported, the *transfer area RAM* may be accessed by both the MC68122 execution unit (internal CPU) and an external CPU.

There are four *request registers* which control ownership of the *transfer area RAM.* These *request registers* are tools provided to the programmer for use in arbitrating simultaneous access of the same source.

The *transfer area RAM* is accessed externally by using eight CPU ADDRESS BUS lines (pins 16-13, 11-8) and eight bidirectional CPU DATA BUS lines (pins 17-24). Three control lines provide either synchronous or asynchronous operation. The DATA ACKNOWLEDGE pin (pin 6) is the control line used to configure the synchronous or asynchronous interface; grounding this input specifies a synchronous system and data will transfer during the positive level of the E CLOCK (pin 4), provided the CHIP SELECT (pin 7) is enabled.

In the asynchronous mode, the DATA ACKNOWLEDGE pin performs an output function. A LOW output indicates that valid data is on the bus for transfer during the read cycle, or that data has been written during a write cycle — otherwise, it remains HIGH. The asynchronous operation uses this signal to synchronize the data flow with the internal registers.

The direction of the data flow is controlled by the SYSTEM READ/WRITE (pin 5)input. A LOW on this line enables the input buffers, and the data is transferred from the host processor to the MC68122. When pin 5 is HIGH, the *data output buffers* are turned on and data flows from the bus ports to the CPU.

Peripheral Interface

Data transfer between the MC68122 and the peripheral can be done in one of two configurations. The two different operating modes are selected by hardware programming and are referred to as the Stand Alone Mode and the Expanded Mode. An example of the hardware programming can be seen in fig. 2. The operating mode is programmed by the level asserted on C5 during the positive edge of RESET (pin 48).

In the Stand Alone Mode, duplex serial communications is established between the controller and one peripheral. The serial interface is available through a single serial port made up of pins 45, 46, and 47—ports C3, C2, and C1, respectively. Pin C1 is the transmitted data output, pin C2 is the received data input. Once the mode has been programmed, C5 can serve as an interrupt output.

The serial communications has two data formats and a variety of data baud rates. The baud rate is defined by configuring pins 31 and 32; four rates are available for any given E CLOCK frequency. When available, pin 45 (C3) can be used as an external clock input, in which case the baud rate is equal to one-eighth the input frequency, otherwise it is programmed as the baud clock output.

The data format is selected by configuring three pins: pins 29, 30, and 45 (see above). Serial formats include standard mark/space (NRZ) and bi-phase; both provide one start bit, eight data bits, and one stop bit. Although the transmitter and receiver are functionally independent, both must use the same data format and baud rate.

Expanded Mode

For systems requiring high data throughout for more than one device, the MC68122 provides extra processing power in the Expanded Mode.

The expanded configuration provides servicing for up to 128 devices when coupled with asynchronous serial communications interface adapters, such as the MC6850. These adapters can be connected to any serial-type device, such as a CRT controller, provided you use one interface chip per channel.

The peripheral chips are addressed by sixteen LOCAL ADDRESS BUS lines (pins 40-25), which can select any one of the 128 devices for communication through the bidirectional LOCAL DATA BUS lines (pins 40-33). Notice that eight of these pins double as both ADDRESS and DATA ports. The direction of the data transfer is controlled by the READ/WRITE (pin 42) output. The multiplexed address outputs are enabled by the ADDRESS STROBE input (pin 41) and decoded by an address latch chip, as seen in fig. 3. In the Stand Alone Mode, all Expanded Mode related pins are left disconnected.

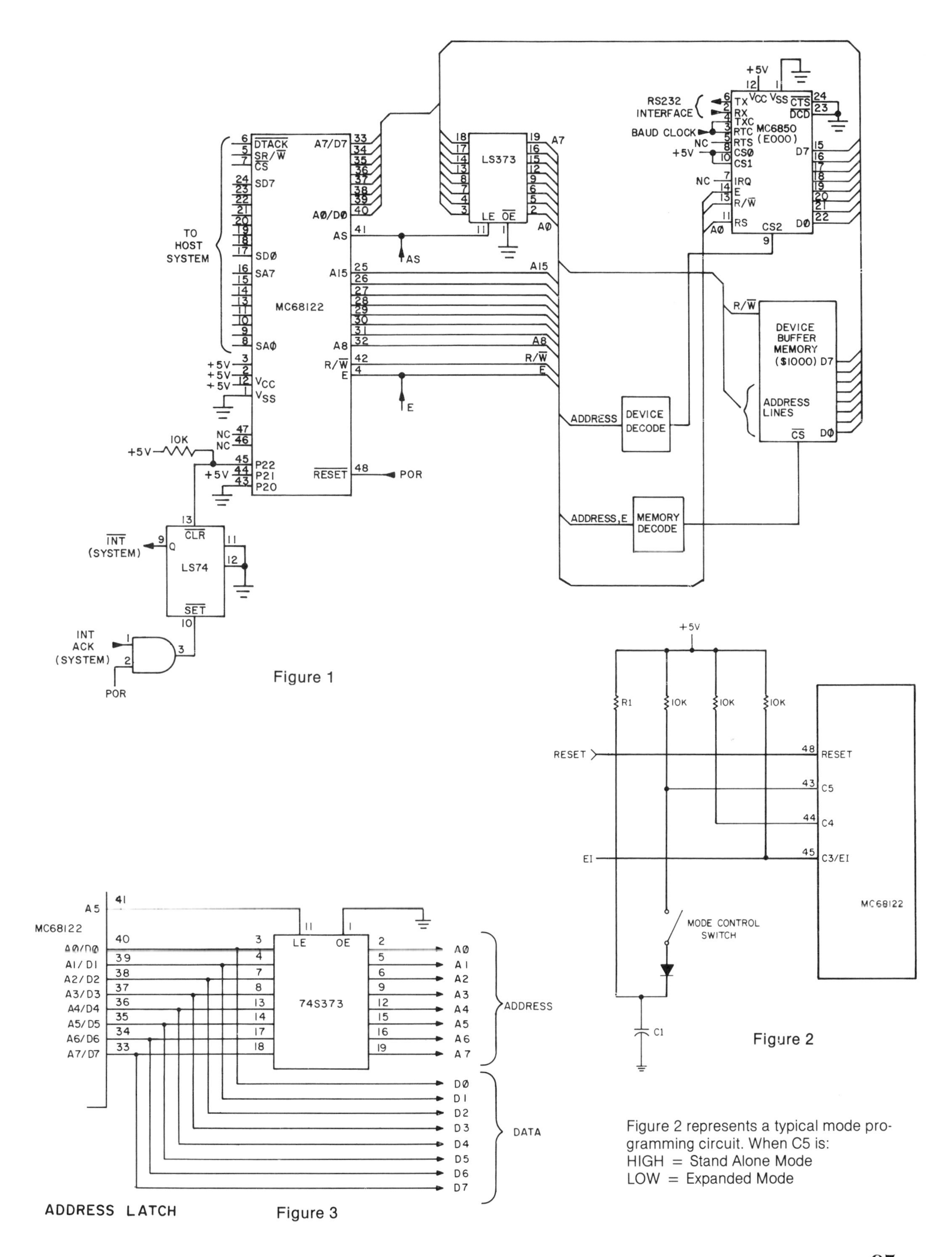

Figure 1

Figure 2

Figure 3

Figure 2 represents a typical mode programming circuit. When C5 is:
HIGH = Stand Alone Mode
LOW = Expanded Mode

8255A

Programmable Peripheral Interface

The 8255A, made by Intel, is a Programmable Peripheral Interface device designed for use in most microcomputer systems. Its function is that of a general-purpose I/O component to interface peripheral equipment to the microprocessor system bus.

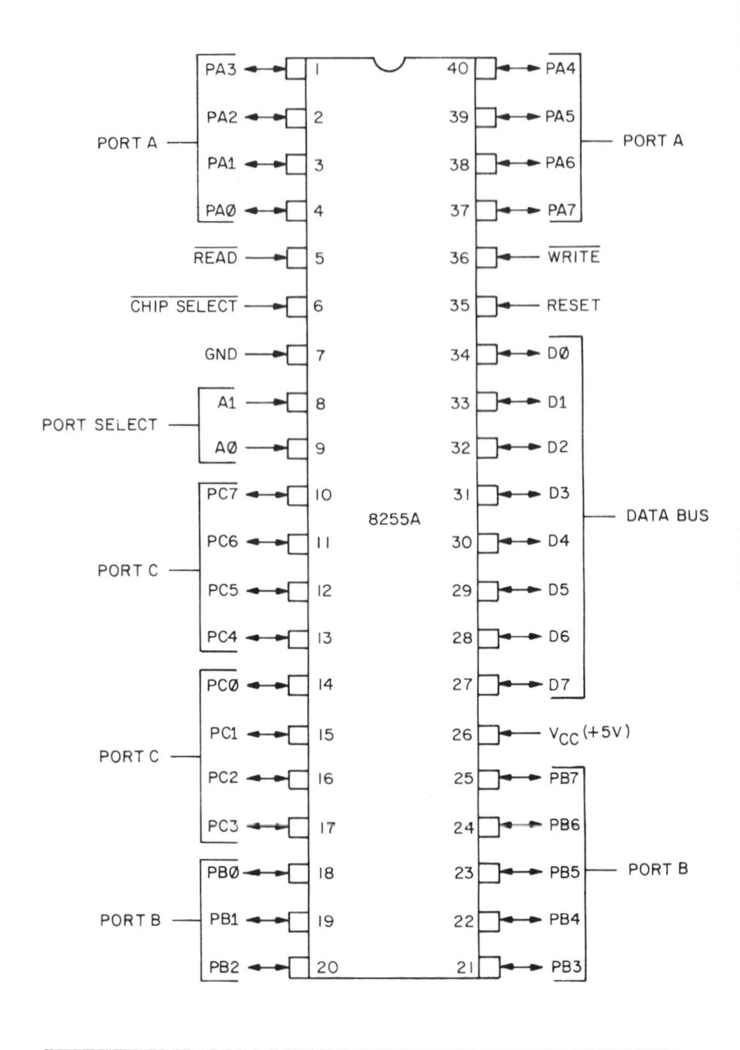

Chip Architecture

The 8255A is a powerful tool with enough flexibility to interface almost any I/O device to the CPU bus without the need for additional external logic. Each peripheral device normally has a unique service routine associated with it which manages the software interface between it and the CPU. The functional configuration of the 8255A is programmed according to this service routine, and actually becomes an extension of the system software.

Interfacing and control between the CPU and 8255A is performed using a DATA BUS (pins 34-27) and the standard READ (pin 5), WRITE (pin 36), and CHIP SELECT (pin 6) controls.

There are three modes of operation associated with the 8255A. The mode definitions and possible pin combinations may seem confusing at first, but after a cursory review of the complete device operation, a logical I/O pattern will surface.

Mode 0

This mode provides simple input and output operations when no handshaking (feedback control) is required. In this mode, data is simply written or read from a specified report.

The chip has 24 I/O pins—divided into three groups of eight—which may be programmed for peripheral interfacing. Any port (group) can be made input or output—provided all related pins perform the same function. In other words, all eight pins of PORT A (pins 4-1, 40-37) must be configured the same, either input or output, but not intermixed. The same is true of PORT B (pins 18-25). PORT C, however, is split into two 4-bit ports, upper (pins 13-10) and lower (pins 14-17), which can be managed separately. Altogether, 16 different input/output configurations are possible in Mode 0 when all port combinations are considered.

The decision to make any port an input or output is determined by a Control Word written into the *control word register*. Once the ports are configured, data is passed through the device by addressing the desired port, using the PORT SELECT (pin 9, 8) inputs, and enabling the DATA BUS. Only when used as outputs are the interface pins latched; inputs can not be latched.

Mode 1

Mode 1 is basically the same as Mode 0, except that the data is transferred under the control of handshaking (strobe) pulses. To accomplish this, the I/O pins are divided into two groups of 12 pins apiece. PORT A and the upper half of PORT C make up one group, while PORT B and the lower portion of PORT C comprise the other. The handshaking signals are derived from the PORT C affiliate. Therefore, each group consists of one 8-bit data port, either PORT A or PORT B, and one 4-bit control port.

As before, either port may be configured input or output, provided you don't intermix the two within the port. PORT A, however, can be the opposite of PORT B. Because data is transferred using a strobe, both inputs and outputs are latched.

Data is loaded and latched into the inputs (when configured) on the LOW level of the STROBE INPUT. This pin corresponds to PC4 (pin 13) for PORT A and PC2 (pin 16) for PORT B. Once the transfer takes place, it is acknowledged by taking the INPUT BUFFER FULL output HIGH (pins 12 and 15, A & B, respectively).

The output control handshaking signals consist of an OUTPUT BUFFER FULL (pins 10, 15; A, B) and an ACKNOWLEDGE INPUT (pins 11, 16; A, B), which acknowledge the transfer of data from the 8255A to the peripheral. Interrupts are available for both input and output configurations.

Mode 2

This functional configuration provides a means for communicating with a peripheral through a bidirectional I/O bus. However, only PORT A is applicable to Mode 2; PORT B cannot be configured bidirectionally.

Mode 2 is structured on a single 8-bit bus for both transmission and reception of data. Control signals are provided by a 5-bit control port (PORT C) to maintain proper flow discipline in a manner similar to Mode 1.

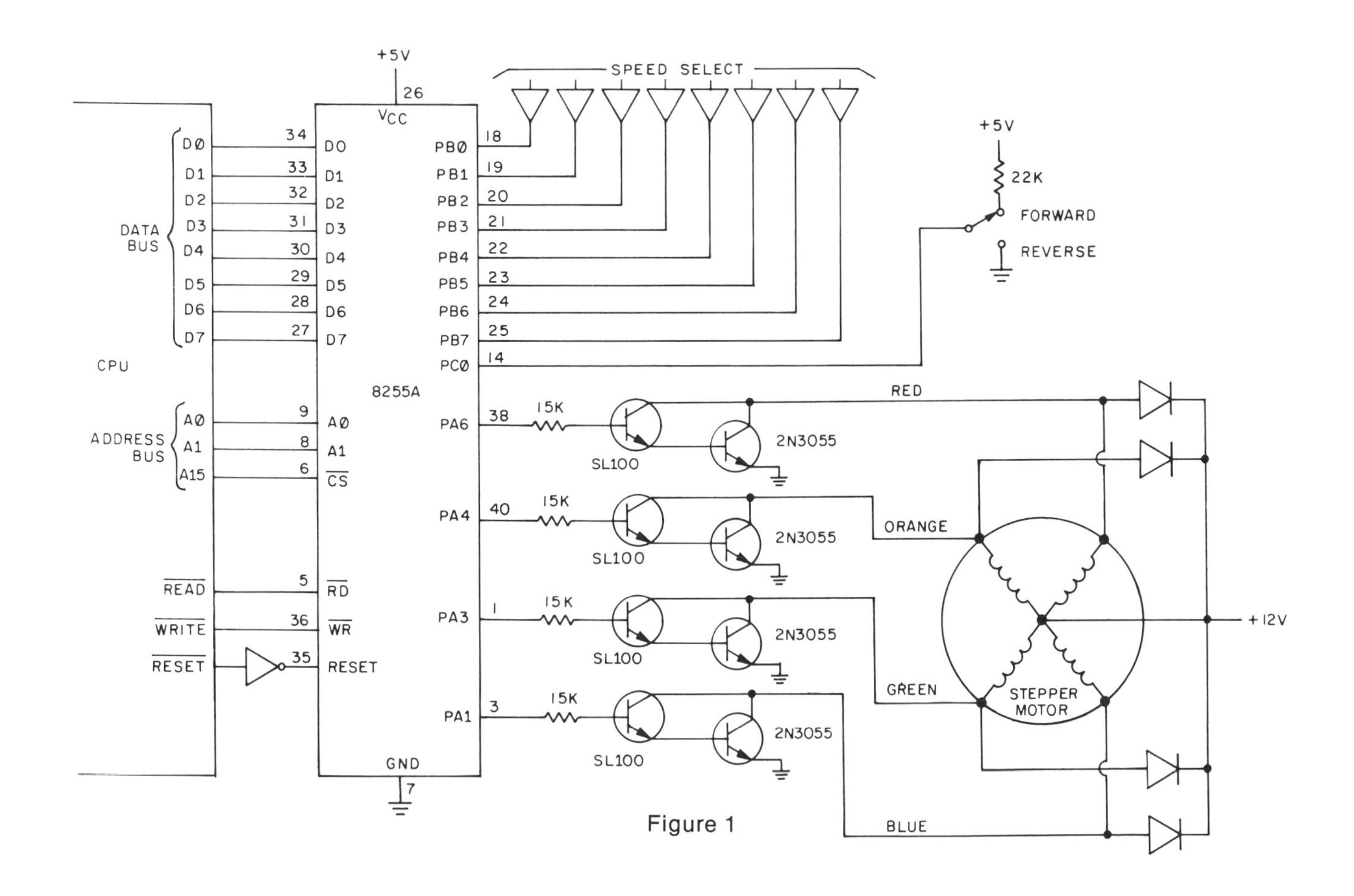

Figure 1

PORT A	MODE 0 INPUT	MODE 0 OUTPUT	MODE 1 INPUT	MODE 1 OUTPUT
PA 0	IN	OUT	IN	OUT
PA 1	IN	OUT	IN	OUT
PA 2	IN	OUT	IN	OUT
PA 3	IN	OUT	IN	OUT
PA 4	IN	OUT	IN	OUT
PA 5	IN	OUT	IN	OUT
PA 6	IN	OUT	IN	OUT
PA 7	IN	OUT	IN	OUT
PORT B				
PB 0	IN	OUT	IN	OUT
PB 1	IN	OUT	IN	OUT
PB 2	IN	OUT	IN	OUT
PB 3	IN	OUT	IN	OUT
PB 4	IN	OUT	IN	OUT
PB 5	IN	OUT	IN	OUT
PB 6	IN	OUT	IN	OUT
PB 7	IN	OUT	IN	OUT
PORT C				
PB 0	IN	OUT	$INTR_B$	$INTR_B$
PB 1	IN	OUT	IBF_B	$\overline{OBF_B}$
PB 2	IN	OUT	$\overline{STB_B}$	$\overline{ACK_B}$
PB 3	IN	OUT	$INTRA_A$	$INTRA_A$
PB 4	IN	OUT	$\overline{STB_B}$	I/O
PB 5	IN	OUT	IBF_A	I/O
PB 6	IN	OUT	I/O	$\overline{ACK_A}$
PB 7	IN	OUT	I/O	$\overline{OBF_A}$

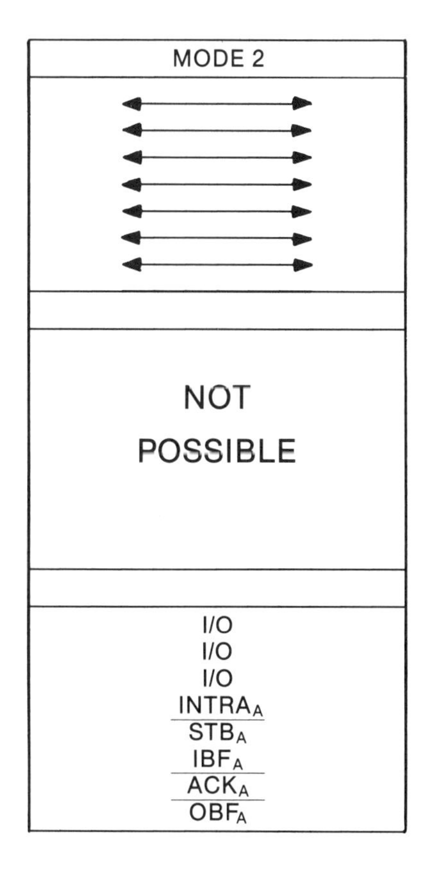

MODE 2
NOT POSSIBLE
I/O
I/O
I/O
$INTRA_A$
$\overline{STB_A}$
IBF_A
$\overline{ACK_A}$
$\overline{OBF_A}$

Figure 2

MC68230

Parallel Interface/Timer

The MC68230 Parallel Interface/Timer, by Motorola, offers versatile 8-bit and 16-bit parallel interfacing capabilities—with a computer oriented timer system—in one package. The chip has an asynchronous bus interface, primarily designed for use with the MC68000 microprocessor. With care, however, it can connect to most synchronous CPU bases.

Parallel Interface Ports

The MC68230 contains two 8-bit parallel interface ports, PORT A (pins 4-11) and PORT B (pins 17-24), that may be concantenated to form a single 16-bit interface port. Management of these ports is under the control of 25 internal registers, which are addressable through five REGISTER SELECT inputs (pins 29-25).

The *general control register* contains a 2-bit field that specifies a set of four operational modes. These modes govern the overall performance of the two ports and determine their interrelationships. The four modes are: unidirectional 8-bit, unidirectional 16-bit, bidirectional 8-bit, and bidirectional 16-bit interfaces.

Some modes, however, require additional information to further define operation. These instructions are contained in the *submode field registers.* Each port mode/submode combination specifies a set of programmable characteristics that fully defines the behavior of that port and of the two associated HANDSHAKE pins (pins 13 and 14 for PORT A; pins 15 and 16 for PORT B).

Programming

Programming begins by initializing the registers with the RESET input (pin 39, LOW). Data is entered through eight bidirectional DATA BUS lines (pins 44-48, 1-3). The direction of the data flow is controlled by the READ/WRITE input (pin 43), and enters the chip when this pin is LOW. The CHIP SELECT input (pin 41) enables the registers for the current bus cycle when pulled LOW. Address Strobe and Data Strobe of the master bus, along with the appropriate address bits, must be included in the Chip Select equation. Data transfer is acknowledged by a LOW indicator on the DATA ACKNOWLEDGE output (pin 42) at the completion of the bus cycle. A double fetch operation is required for 16-bit transfer.

Port A and B functions have an independent pair of active LOW INTERRUPT REQUEST output and INTERRUPT ACKNOWLEDGE input pins, located at pins 35 and 36, respectively. In addition, a DIRECT MEMORY REQUEST (pin 34) output may be associated with either—but not both—PORT A or PORT B.

Port C

The chip contains a third port, PORT C (pins 30-37), which is rather unique. This port can be used as eight general purpose I/O pins, that may be individually programmed as inputs or outputs by the *port C data direction register,* or as six special function pins and two general purpose I/O pins. All pins not used for special purposes can be designated general purpose I/O pins with each pin bit programmable.

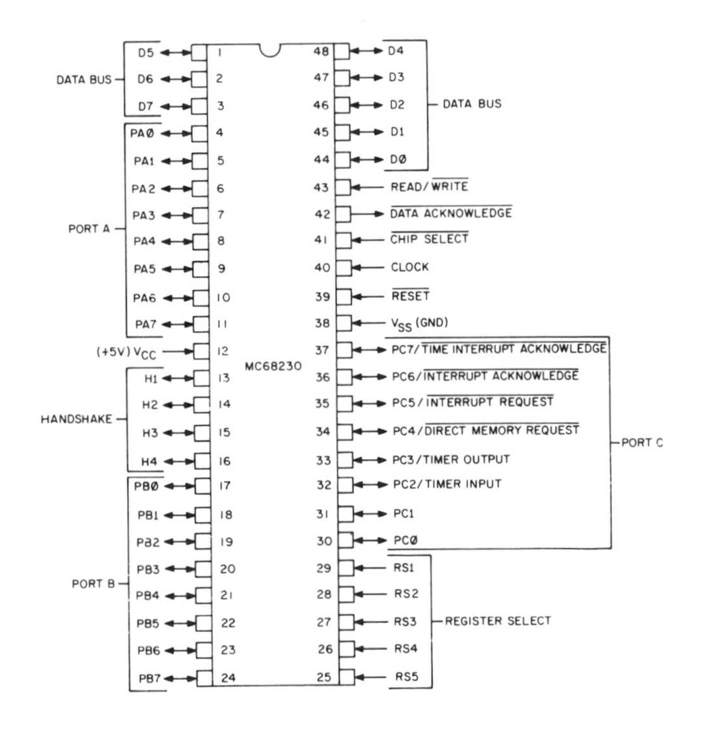

Timer

Three alternate functions of port C are timer input/output pins. The timer is a 24-bit synchronous down counter that can generate periodic interrupts, a square wave, or a single interrupt after a programmed time period. It can also be used for elapsed time measurements and as a device watchdog.

The timer value is loaded from three 8-bit *counter preload registers.* The *counter preload registers* are readable and writable at any time and this occurs independently of any timer operation. In other words, no protection mechanisms are provided against ill-timed writes, a task that must become the responsibility of the input controller.

The timer can be decremented by either the master CLOCK (pin 40) or by an external TIMER INPUT (pin 32). In addition, the timer may be clocked by the output of an internal 5-bit, divide-by-32 prescaler, driven by the internal or external clock. The input frequency to the 24-bit timer counter from the TIMER INPUT or prescaler output *must* fall between 0 and 1/32 of the CLOCK input frequency — regardless of the configuration chosen.

For configurations in which the prescaler is used (with either the CLOCK input or the TIMER INPUT), the contents of the *counter preload register* is transferred to the counter the first time the prescaler rolls over. For configurations in which the prescaler is not used, the contents of the *counter preload register* are entered on the first edge of the TIMER INPUT pin, which is used as a trigger.

The TIMER OUTPUT pin (pin 33) may provide an active LOW timer interrupt request output, in conjunction with the TIMER INTERRUPT ACKNOWLEDGE input (pin 37), or a general purpose square wave output.

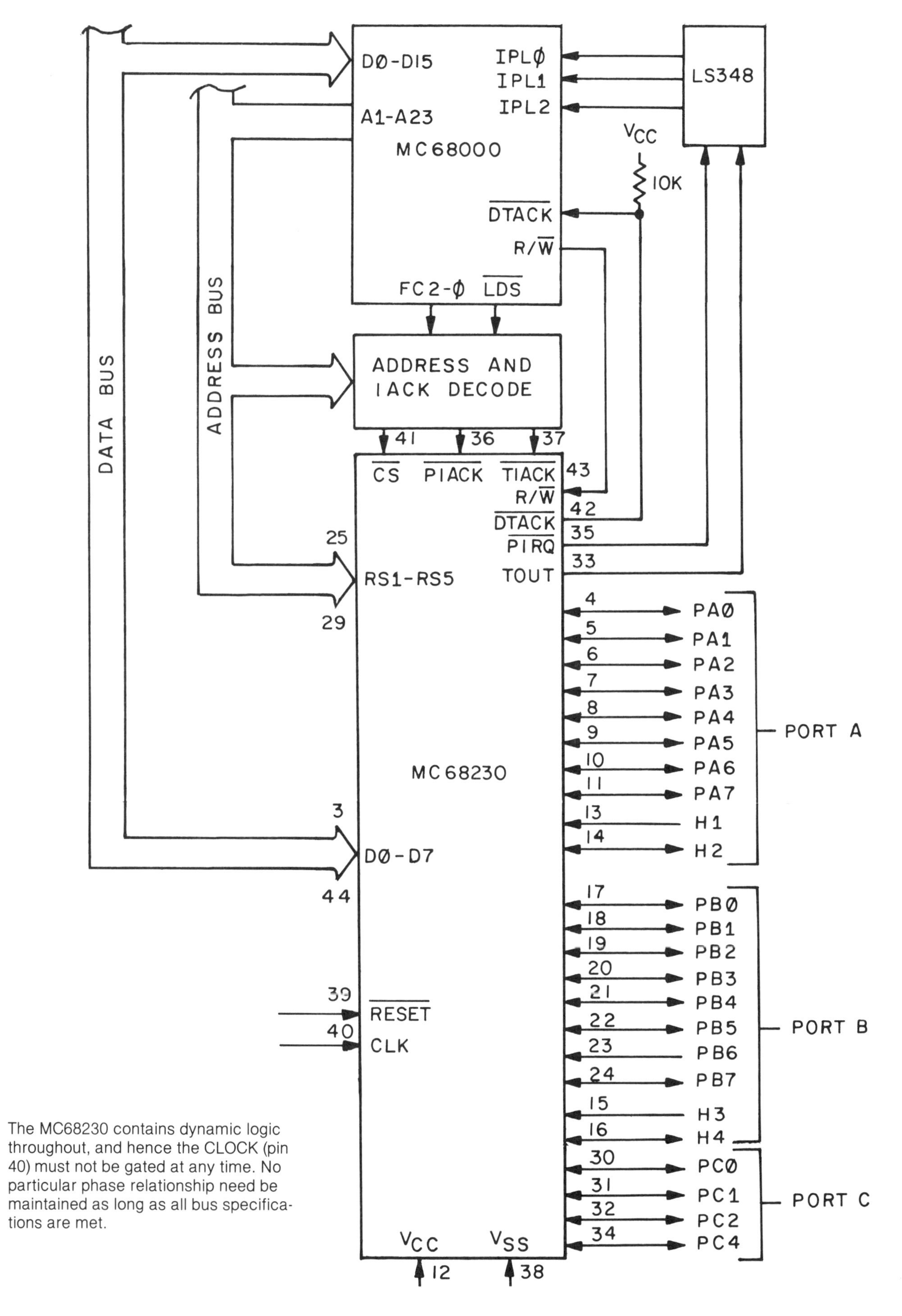

The MC68230 contains dynamic logic throughout, and hence the CLOCK (pin 40) must not be gated at any time. No particular phase relationship need be maintained as long as all bus specifications are met.

CHAPTER FOUR
CONTROL

CY500

Stored Program Stepper Motor Controller

The CY500, made by Cybernetic Micro Systems, is an ASCII programmable Stepper Motor Controller configured to control any 4-phase stepper motor. The chip not only contains the timing logic necessary for proper motor control, but also includes an internal 18-byte buffer capable of storing instructions and commands. In many applications, the user will find that the CY500 can function on its own—independent of the microprocessor—except for initial program loading.

Operation

Programs are entered into the chip through eight parallel DATA input lines (pins 12-19). The data is considered valid when strobing the WRITE (pin 1) input and transferred into memory on the falling edge of the READ (pin 8) output. Data can be loaded in either ASCII language or machine code; language selection is determined by the ASCII/BINARY (pin 33) control line. Once loaded, the chip self executes the program independently.

Timing Clock

The chip is regulated by an internal clock, which is frequency controlled using a resonant circuit across the XTAL (pins 2 and 3) inputs. Ranging in frequency from 1-MHz to 6-MHz, several timing controls are possible, as demonstrated by fig. 2; including crystal control, LC tank, and external clock input. The CLOCK/15 output (pin 11) provides a signal which is 1/15th the input frequency.

The stepping rate for the motor is basically controlled by the XTAL frequency. Timing values for the chip are scaled down to $6/f$, where f is the crystal frequency in MHz. Each motor step contains a precise number of clock intervals, as determined in software, and ranges from one step every five seconds to 3350 steps/second when using a 6-MHz crystal. The stepping rate is non-linear, however, with resolution increasing at the lower speeds.

Mechanical Synchronization

Most stepper motors are employed as parts of functional systems. These systems often must synchronize the behavior of the various subsystems to each other—or to real world occurrence, such as an operator input. The CY500 has been designed to provide both input and output signals for synchronization of interacting components.

There are two sources for controller output. The PULSE output (pin 35) gives an output pulse of 5-25-us in duration for every step command, while the TOGGLE output (pin 36) supplies an alternating signal which flips states with each consecutive step. In addition, related timing signals are available from PROGRAM out (pin 31), MOTION COMPLETE (pin 37), and PROGRAM COMPLETE (pin 32).

The input controls include a PROGRAM WAIT (pin 38) which effectively holds the program in waiting until certain external conditions are satisfied, and a TRIGGER input (pin 30) that is used to gate the steps one at a time. An ABORT command (pin 6) brings all stepping operations to an immediate halt.

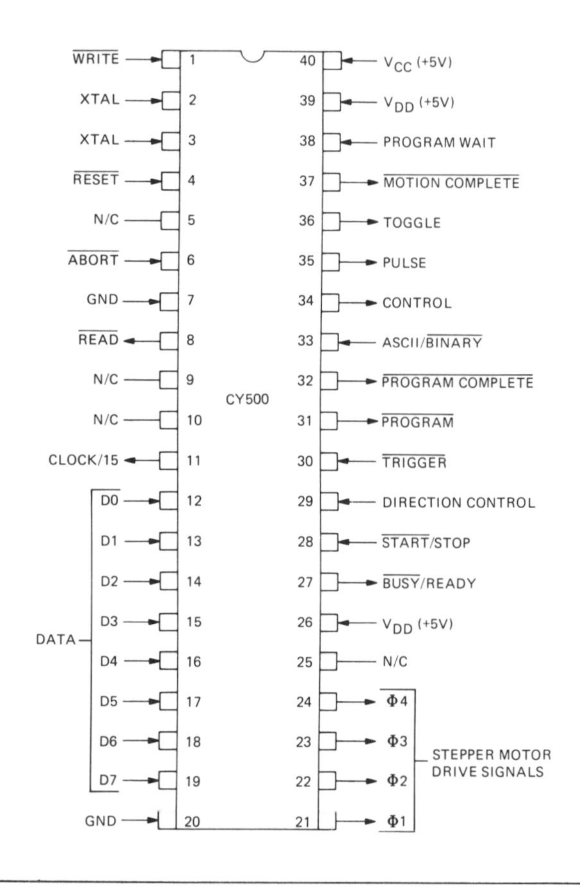

Manual Operation

The controller chip can also be manually operated by placing it in a software Command Mode and using an ASCII encoded keyboard to input the data. A WRITE signal on pin 1 will transfer the keyboard input into the chip, where the instruction is acted upon immediately. This mode makes it very simple to set up an initial program because you see the results of every move as you type it in. Once you are satisfied with the operation, a Program Storage command will allow loading of the program into the registers.

There are also two overriding input pins which control the motor regardless of mode; the DIRECTION CONTROL (pin 29) and the START/STOP (pin 28) controls will advance and stop the motor in either direction manually.

Outputs

Although the CY500 provides the timing and logic signals necessary to control a stepper motor, a driver circuit is needed to make a complete system. The user has two options available.

By using the four phase outputs from the chip (pins 21-24) you can drive a 4-phase motor directly. However, a power buffer that is capable of handling the power requirements of the motor must be used to interface the two. Your other choice is to employ a commercially available translator driver; only the PULSE output and CONTROL signals (pin 34) will be needed for operation. The rest of the motor's operations are carried out by the translator, which incidentally, permits the use of 3- and 5-phase stepper motors as well as the 4-phase.

The CY500 is very similar to the CY512, although not pin-for-pin interchangeable, and the reader will find data pertaining to the CY500 by reviewing that section.

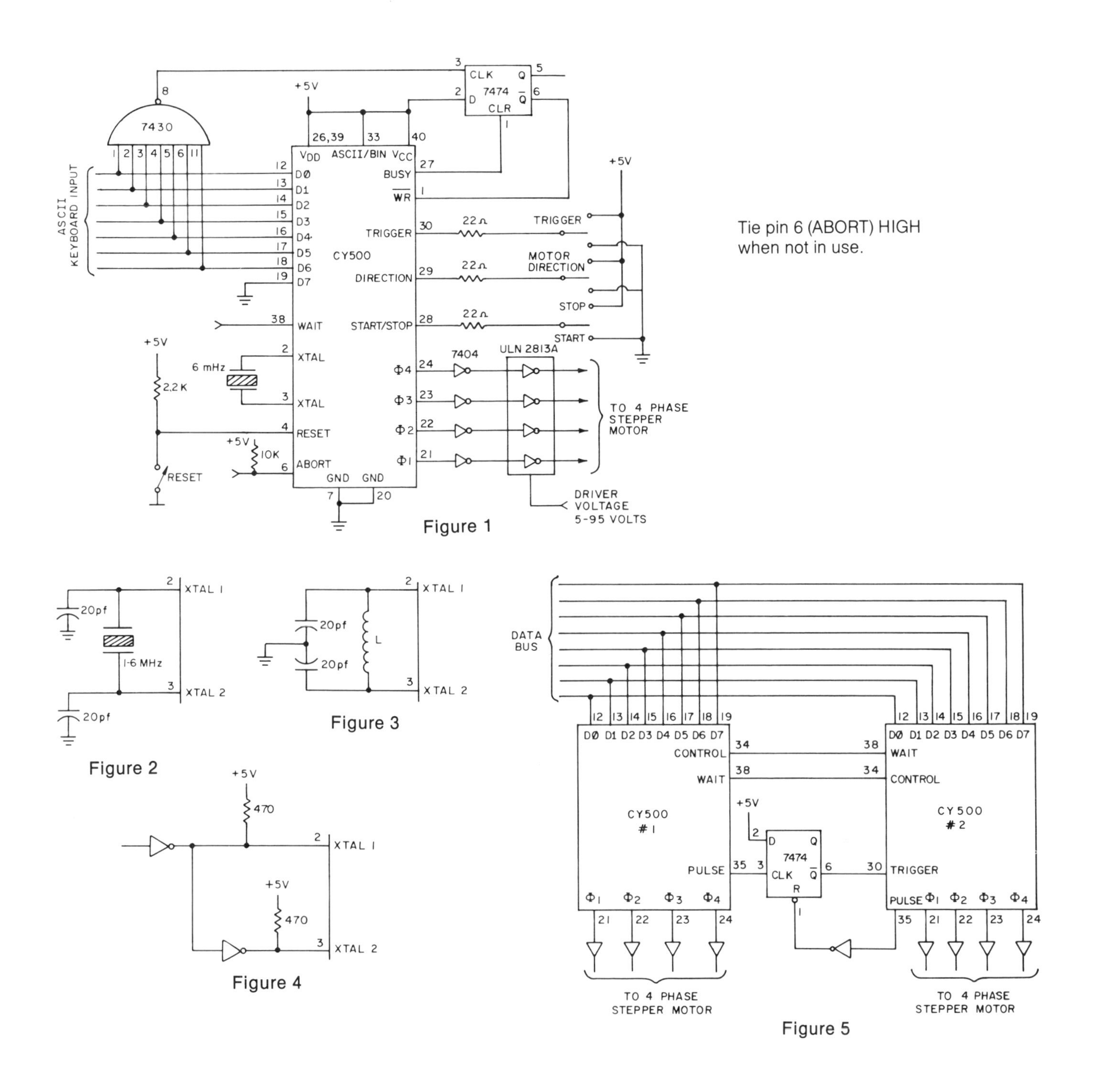

Figure 1

Figure 2

Figure 3

Figure 4

Figure 5

Color TV burst crystals, 3.58-MHz, may be used.

CY512

Intelligent Positioning Stepper Motor Controller

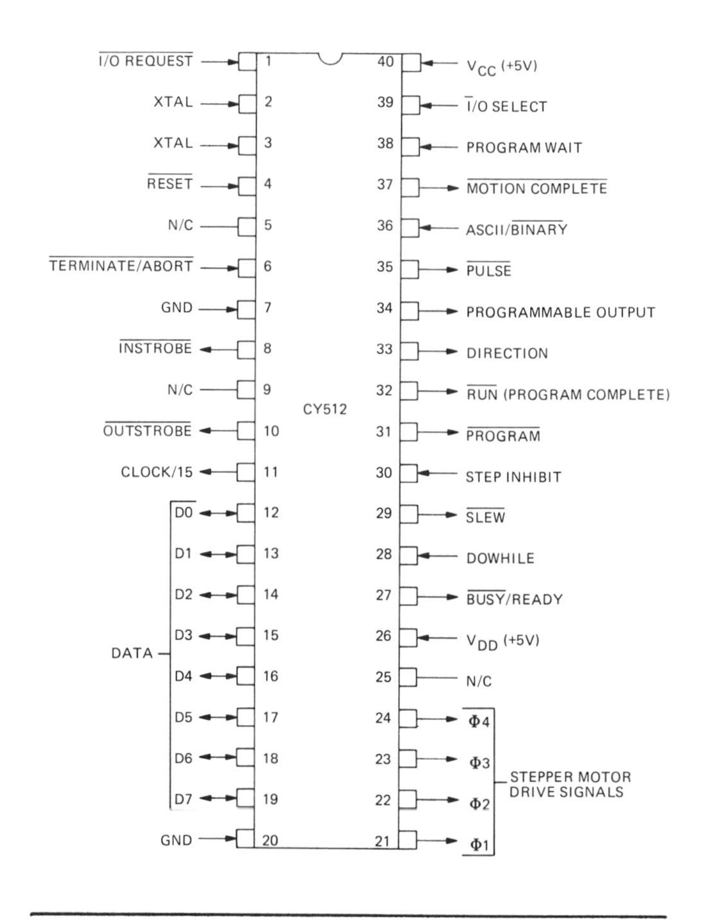

The CY512, by Cybernetic Micro Systems, is an ASCII programmable Stepper Motor Controller capable of directly controlling any 4-phase stepper motor. It is essentially an upgraded version of the popular CY500 chip. Among its improvements, the frequency range has been extended to 11-MHz, for more precise rate control, and the program memory expanded to 48 bytes—nearly three times the capacity of the CY500. Furthermore, improved software allows the controller to automatically determine CW and CCW motions without outside instruction, as indicated by the DIRECTION (pin 33) output.

Programming

Programs are entered into the chip's *program registers* through eight bidirectional DATA lines (pins 12-19). The I/O SELECT input (pin 39) governs the direction of data flow through the parallel DATA ports. When I/O SELECT is LOW, data is written into the registers by strobing the I/O REQUEST (pin 1) input LOW, with the transfer taking place on the trailing edge of the INSTROBE (pin 8) handshaking pulse. Data can be entered in either ASCII or machine code; selection of the language is determined by the ASCII/BINARY (pin 36) control line.

Once the data has been loaded, the chip will self administer the program completely independent of the microprocessor. During execution of the program, further instructions are prevented from interfering with the chip, as indicated by a LOW condition on the RUN (pin 32) output. New input is not recognized until the RUN output goes HIGH (program complete) or a RESET (pin 4) is initiated.

The contents of the internal registers can be examined when I/O SELECT is set HIGH. The OUTSTROBE (pin 10) output is the handshaking pulse associated with data transfers in the Read Mode.

In many applications, the user will find that the CY512 can function on its own—independent of the CPU, except for initial program loading. However, even the CPU is replaceable with a program storage ROM, which feeds the program to the controller during the power-up phase of operation.

Slew Rate Control

It is often desirable, particularly during high-speed operations, to provide a controlled acceleration and deceleration ramp for the motor, since it is virtually impossible to go from a dead stop to maximum velocity in one step, and vice versa. This sequence is provided for in the Slew Mode program command.

Although the stepping rate is in software, three control functions aid the system in maintaining synchronization. During ramp sequencing, the SLEW output (pin 29) remains HIGH. Only after maximum speed has been achieved is it driven LOW; the output returns HIGH as soon as deceleration begins. The DOWHILE input (pin 28) forces the program into a conditional program loop, repeating a sequence over and over, until certain external requirements are satisfied. In addition, the CY512 provides an uncommitted PROGRAMMABLE OUTPUT (pin 34) which the user may apply as needed.

The ramp operation is generally satisfactory at low and medium speeds; but due to non-linear relationships, the chip may require an external timing circuit at higher slew rates, as seen in fig. 3. Particularly attractive in the CY512 is an ABORT function (pin 6) that can be logically wired to a subroutine which ramps the motor down safely in the shortest possible time.

Feedback Synchronization

Although slew rate control allows higher step rates than would be otherwise achieved with non-ramped signals, at higher step rates the next step command may occur before the motor has had a chance to complete the last step, throwing the entire system out of sync.

Using existing interrupter technology, though, it is possible to re-establish synchronization using feedback control signals to the CY512. The circuit in fig. 2 accomplishes this by triggering the STEP INHIBIT (pin 30) input using a light source and a slotted disc on the motor shaft. Until an "in position" signal is received from the light monitor, the program is held in wait, slowing the CY512 from a rate that is too fast for the motor to follow.

Synchronous Operations

Provisions have been made for synchronous operations of more than one controller chip. The chips can contain either identical programs for multiple machine control, or individual programs for sequenced operation. An example of two controllers in synchronous arrangement as shown in fig. 5, page 103.

Many of the descriptions pertaining to the operation of the CY500 apply to the CY512, and the reader can gain additional insight into the controller's performance by referring to that section. However, the two chips are not interchangeable, and differences must be noted.

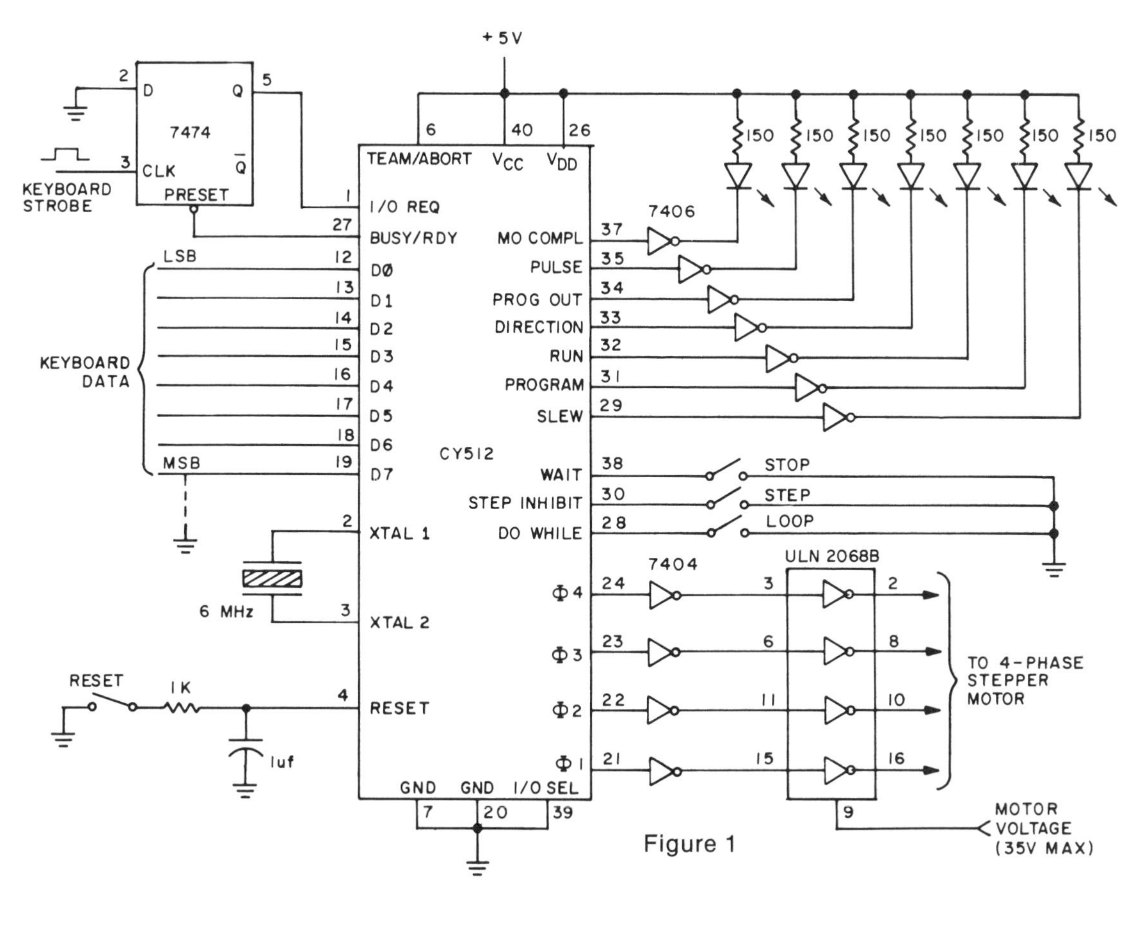

Figure 1

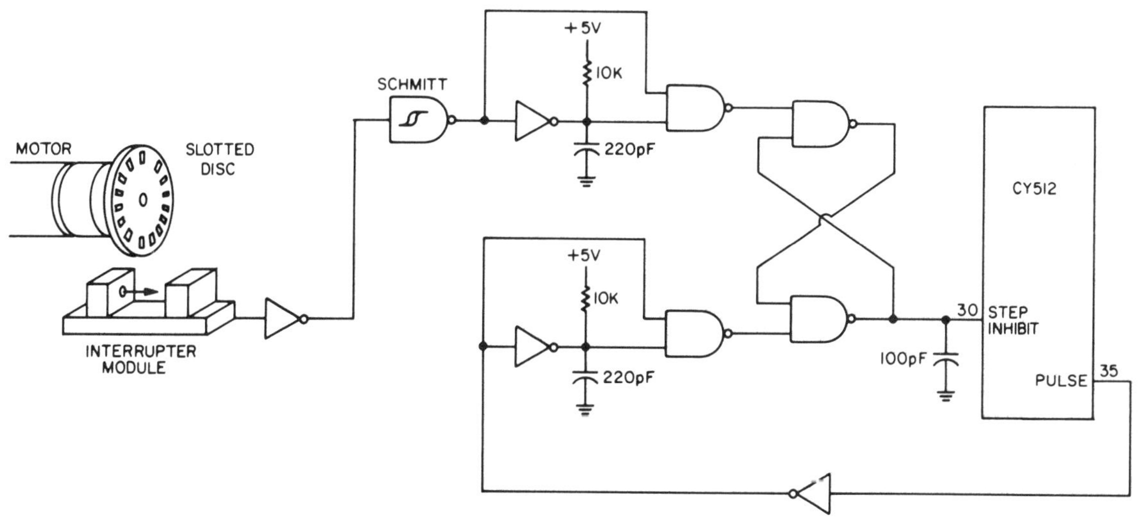

Figure 2

An interrupter module with slotted disk provides position feedback for precise control at maximum step rates.

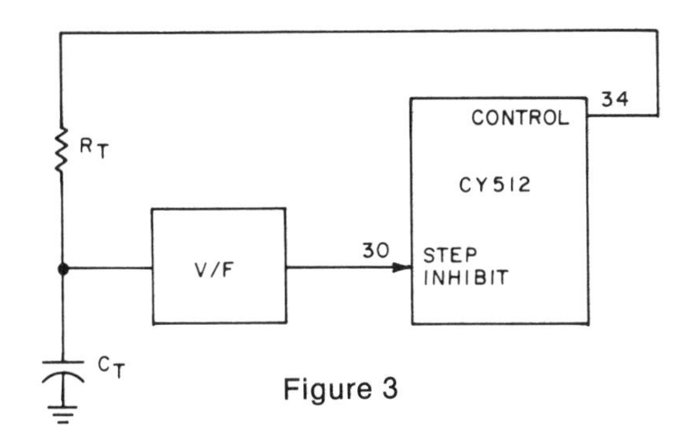

Figure 3

8295

Dot Matrix Printer Controller

The 8295 Dot Matrix Printer Controller, by Intel, is a device specifically designed to interface microprocessors to the LRC 7040 series of dot matrix impact printers, and other similar printers. If offers complete solenoid and motor drive timing, and contains an on-chip character generator accommodating 64 ASCII characters. Character density, width, and print intensity are all software programmable.

Printer Operations

The 8295 is designed to operate an impact printer whose head consists of seven solenoids driving seven stiff wires that impact the paper through an inked ribbon. Arranged so that the wires form a vertical column at the ribbon impact point, characters are generated by scanning the head across the page while the controller activates the appropriate solenoids for each 5x7 or 7x7 dot matrix character in the line.

For solenoid drive, the 8295 supplies seven SOLENOID DRIVE outputs (pins 27-33) plus a SOLENOID STROBE (pin 34). The SOLENOID STROBE modulates the SOLENOID DRIVE outputs externally to control the actual solenoid ON time, which determines the print intensity. This time is software programmable. These SOLENOID DRIVEs cannot drive the printer solenoids directly, and should be buffered through external solenoid driver circuits.

Printing automatically begins when a Carriage Return command is received. The print head is moved across the paper by the main drive motor, which is engaged when the MOTOR (pin 35) output is LOW. The controller locates the head position by monitoring the HOME (pin 39) input. This input is activated only when the head is at its initial starting position. Once the head sweep begins, the 8295 assumes complete control over the printing. Character density is programmable from 10 to 12 characters per inch, generating between 32 and 40 characters per line.

Paper feed is accomplished with a second synchronous motor and a PAPER FEED (pin 1) microswitch. The motor is actuated by the PAPER FEED MOTOR DRIVE (pin 21) output. The paper feed operation is independent of the print cycle, allowing the two to run simulaneously.

The 8295 also offers two GENERAL PURPOSE output pins, GP1 (pin 24) and GP2 (pin 23), which can be set or cleared by the CPU. They may be used with various printers to implement such functions as ribbon color selection, enabling form release solenoid, and reverse document feed. A TOP OF FORM input (pin 22) is provided for the type T printer.

Data Input

The controller will internally store up to 40 ASCII characters for printing. The 8295 is a command-oriented device, with the commands occupying the lower ASCII codes. Software control of the chip is very straightforward; the CPU simply issues ASCII commands to the controller along with the ASCII characters.

The user has the option of using serial or parallel interface to the main processor. However, the choice between them must be done early in the system design since it is a hardware—not a software—decision. Thus, the two modes cannot be intermixed in a single controller application.

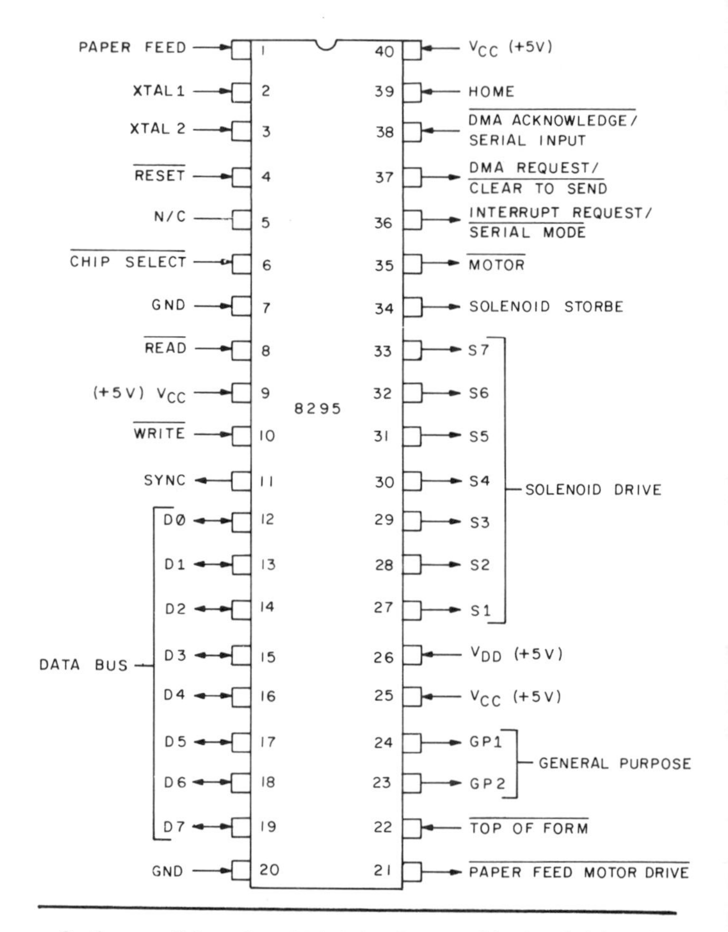

In the parallel mode, which is implemented by hardwiring pin 36 HIGH, the 8295 offers the traditional microprocessor bus interface. The bidirectional DATA BUS (pins 12-19) transfers data to and from the chip, while the READ (pin 8) and WRITE (pin 10) strobes control its direction when the CHIP SELECT (pin 6) input is enabled.

To further enhance bus efficiency and reduce CPU overhead, the 8295 provides for DMA operations. To initiate DMA transfers, the CPU merely loads the DMA controller, typically an 8257, with the print buffer starting address and writes the Enable DMA command into the 8295. The controller does the rest. A DMA transfer begins with a DMA REQUEST (pin 37) and is performed when a DMA ACKNOWLEDGE (pin 38) response is received. The controller keeps track of the number of operations performed, and continues DMA until all the enlisted characters are transferred.

Serial Interface

In addition to the parallel interface options, the 8295 supports "stand alone" serial interface. In this mode, the only communications with the CPU is via a serial data link. This configuration is perfect for remote printer applications, where three wires may replace as many as twelve.

The serial mode is imposed by simply grounding the SERIAL MODE (pin 36) input. Establishing a baud rate is also accomplished in hardware; by arranging the D0, D1, and D2 (pins 12, 13, 14) pins, all standard baud rates from 110 to 4800 are accommodated. Note that grounding CHIP SELECT and WRITE, with READ tied HIGH, is required to establish the proper data direction flow into the chip. SERIAL INPUT (pin 38) data is requested by LOWering the CLEAR TO SEND (pin 37) output. The CLEAR TO SEND must be asserted for each character, which enter the chip one at a time.

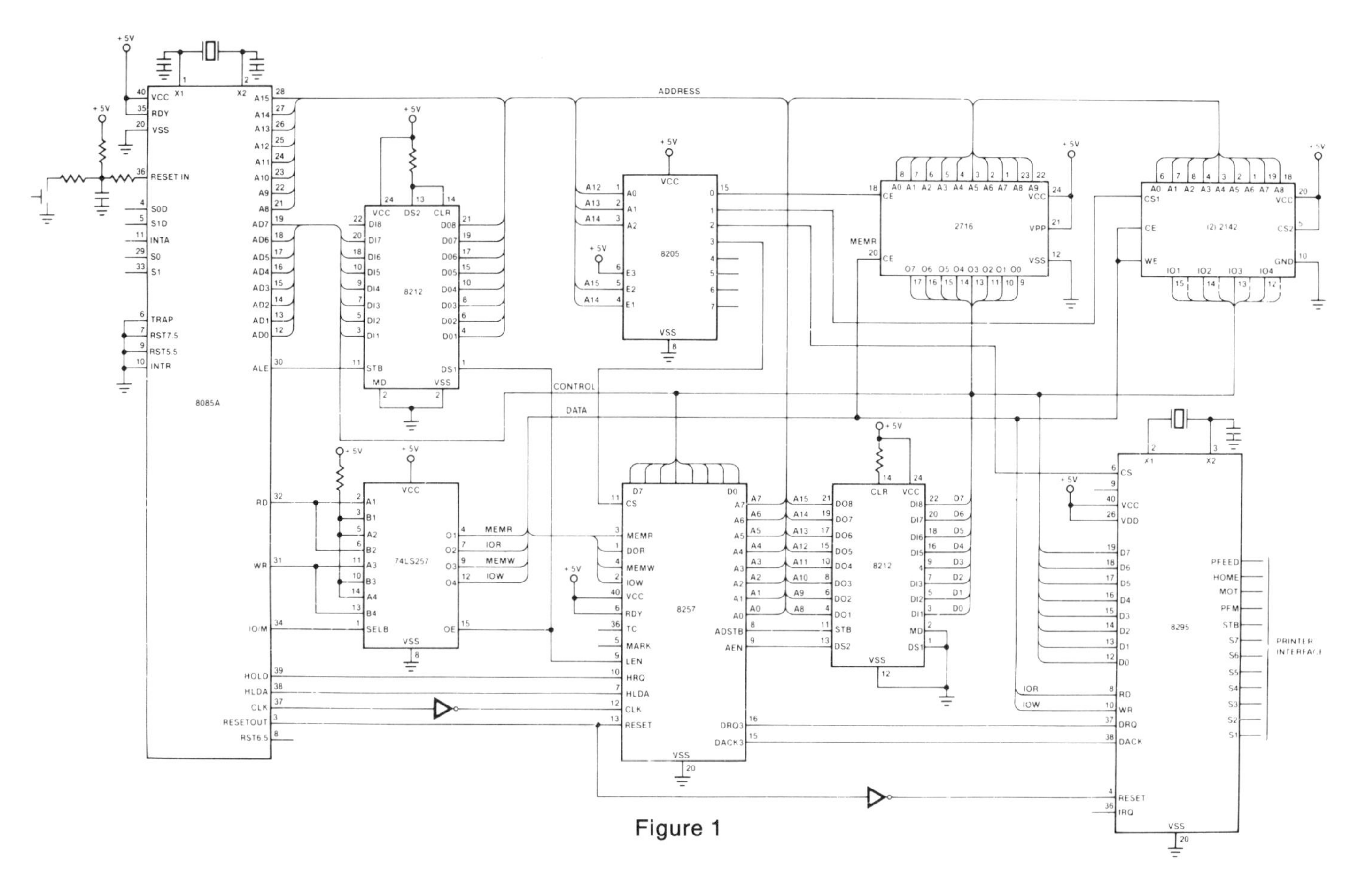

Figure 1

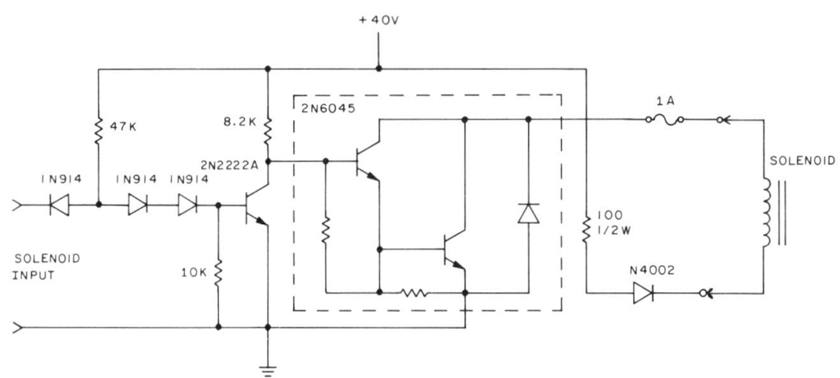

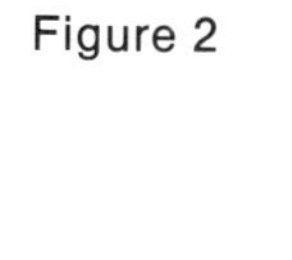

Figure 2

The 8295 is controlled by a 6-MHz crystal connected across the XTAL 1 (pin 2) and XTAL 2 (pin 3) inputs.

A 2.5-us SYNC (pin 11) pulse is provided as a strobe for external circuitry.

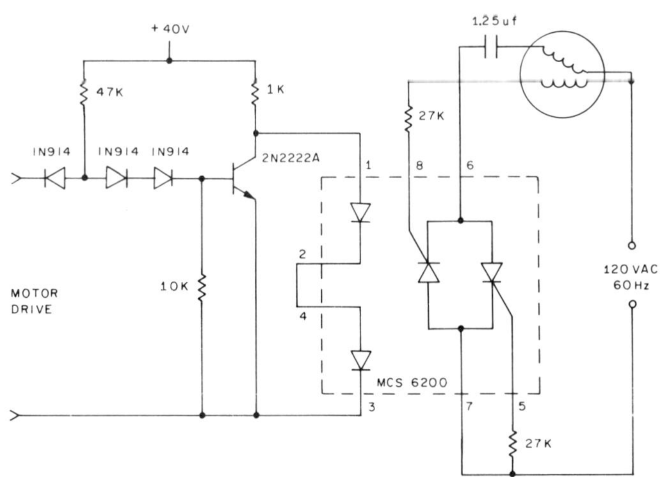

Figure 3

CY480

Universal Printer Controller

The CY480, introduced by Cybernetic Micro Systems, is a complete 5×7 dot matrix Printer Controller able to operate with a wide range of commercially available printers. Whether the choice be impact, thermal, or electrostatic, the chip easily interfaces with parallel 8-bit ports or serial TTL/RS-232 ASCII data. The device contains a character generator, timing circuits, printer controller, and handshaking routines.

Operation

A typical print cycle for the CY480 goes as follows: Data and instructions are received on the DATA lines (pins 3-6, 19-17) and stored in an internal *line buffer* until a carriage return or form feed command is received. Upon receipt of a carriage return, the chip flags a BUSY signal (pin 15, HIGH) and begins the task of printing.

This is done by first activating the printer motor with the MOTOR ON-FWD (pin 33) output, then engaging the clutch, if needed, using the CLUTCH ON (pin 32) output. As the head sweeps across the page, printer HAMMERs (pins 8-14) are fired, drawing a vertical column of up to seven dots. The head is incremented, and another column printed. This is repeated until the line is finished, at which point the carriage returns and awaits a new line.

Clock

The CY480 contains its own internal clock, and when left to its own devices, the chip will run at roughly 3-MHz by simply grounding the two CLOCK pins (pins 1 and 2).

However, there are more stable methods in which the clock can be operated. For instance, it can be externally synchronized using a TTL buffered input, or crystal controlled. The schematic in fig. 3 shows several different clocking arrangements, each with its advantage. For consistent high-quality printouts, though, a crystal controlled clock is recommended. You'll notice that C_{ext} is labeled optional since stray capacitance is often sufficient.

Line Width

Although a single chip is capable of storing and printing up to 48 characters per line, it is possible to expand this capacity to any number of characters by daisy chaining two or more chips. When interconnecting multiple chips, however, a communications link must be established between them. This is accomplished by paralleling pins 22, the INTERSYSTEM COMMUNICATION pin, and establishing the chip's position in the chain by configuring the POSITION 0 and POSITION 1 (pins 37, 36) inputs. Smooth control from one chip to another is maintained by linking the READY OUT (pin 34) pin of the first chip to the READY IN (pin 35) input of the next chip, and so forth on down the line. In the single chip configuration, these pins are grounded.

Data Input

Data can be input in either parallel or serial form. To communicate in the serial mode, the SERIAL DATA input (pin 16) is tied directly to the STROBE pin (pin 38); otherwise, SERIAL DATA is left open. The baud rate is determined by grounding DATA lines D1, D2, or D3 (pins 4, 5, 6). The rate is selected by an assigned code, which can be established by the terminal or hardwiring.

Parallel interfacing is established by simply connecting DATA lines 0 through 6 directly to the standard 8-line bus. If the printer chip is powered by a voltage source independent from the host interface, a buffered input is recommended.

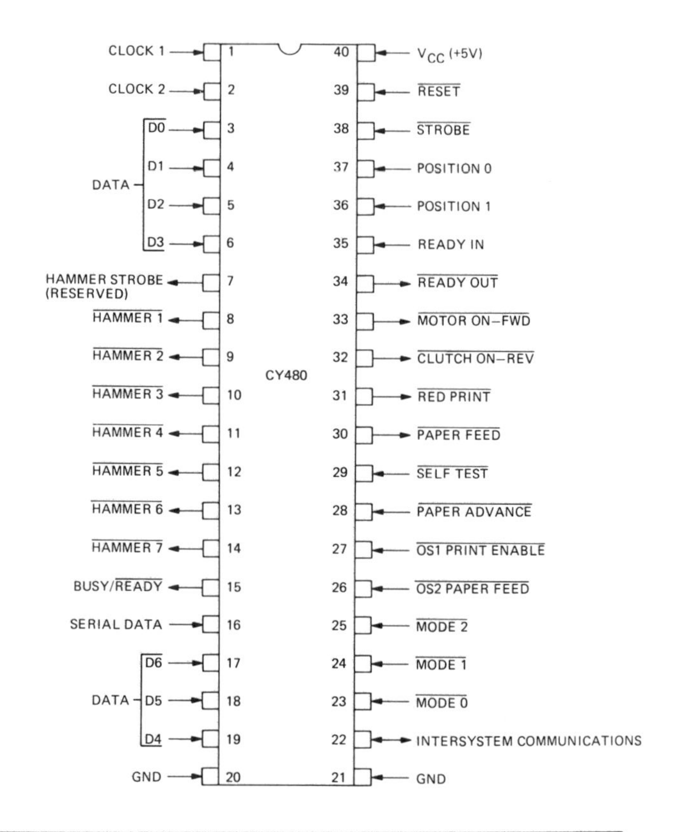

Printer Control

In order to faithfully print the data, it is imperative that the chip be synchronized to the printer. There are two feedback functions which prompt the chip to respond with the proper command. An input to OS1 (pin 27) tells the CY480 when to start printing and whether or not the print head is properly oriented; OS2 (pin 26) monitors the paperfeed. The printer is controlled by output pins 28 through 31.

The MODE inputs (pins 23-25) establish the proper control modes for the printer involved. In addition to those printers already programmed into the chip, printers with similar characteristics can also be controlled through manipulation of these MODE inputs and software instructions.

Software Instructions

The greatest feature of this chip is its ability to cope with almost any given situation. This is done using software programming which can alter the internal standards of the chip. Proper use of these instructions allows you to reverse or invert the print, stretch the print either horizontally or vertically, and even change the impact time of the hammer signals.

Although too extensive to list in this summary, these commands occupy the first 33 ASCII input codes. All commands are latched into the instruction register and remain effective until replaced or RESET (pin 39).

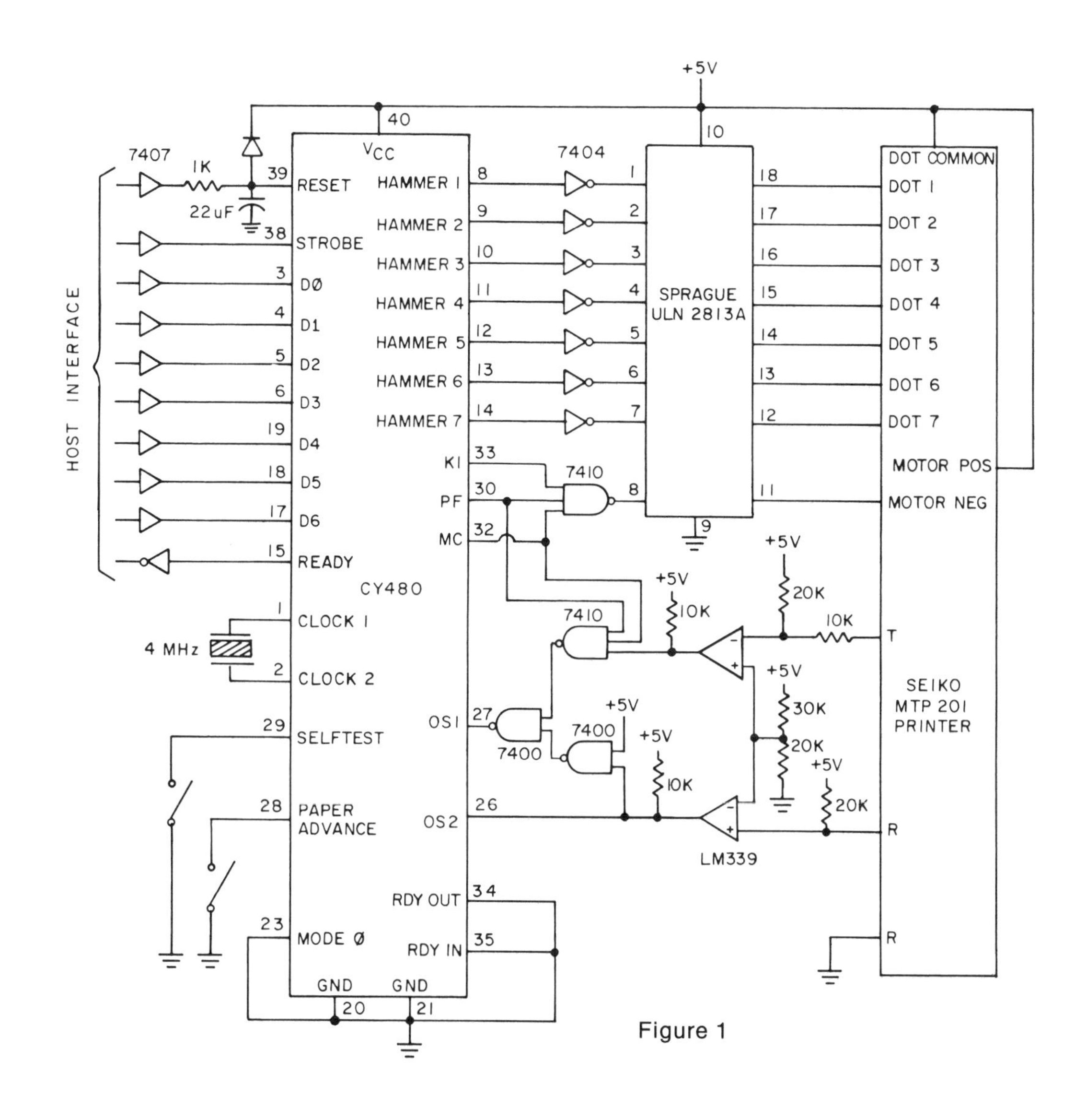

Figure 1

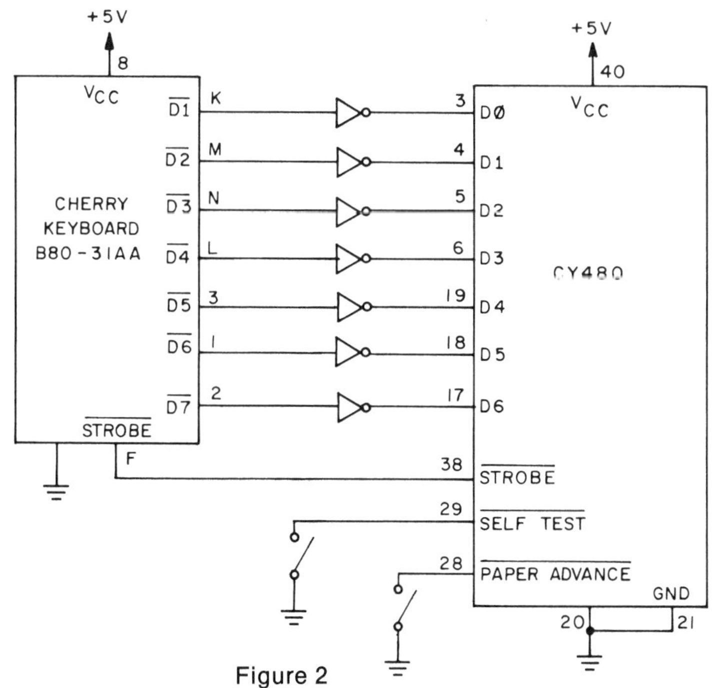

Figure 2

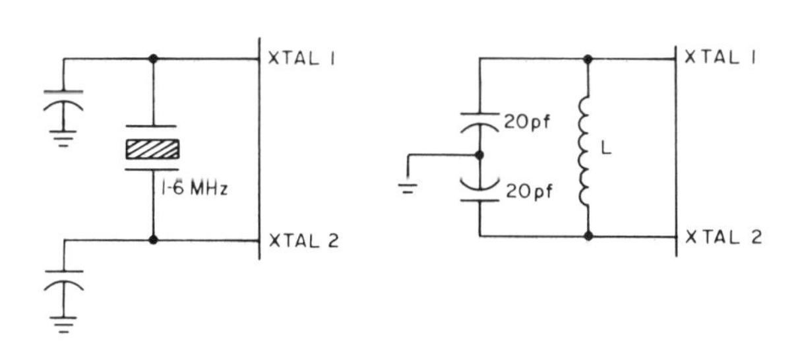

Figure 3

FD1792

Floppy Disk Formatter/Controller

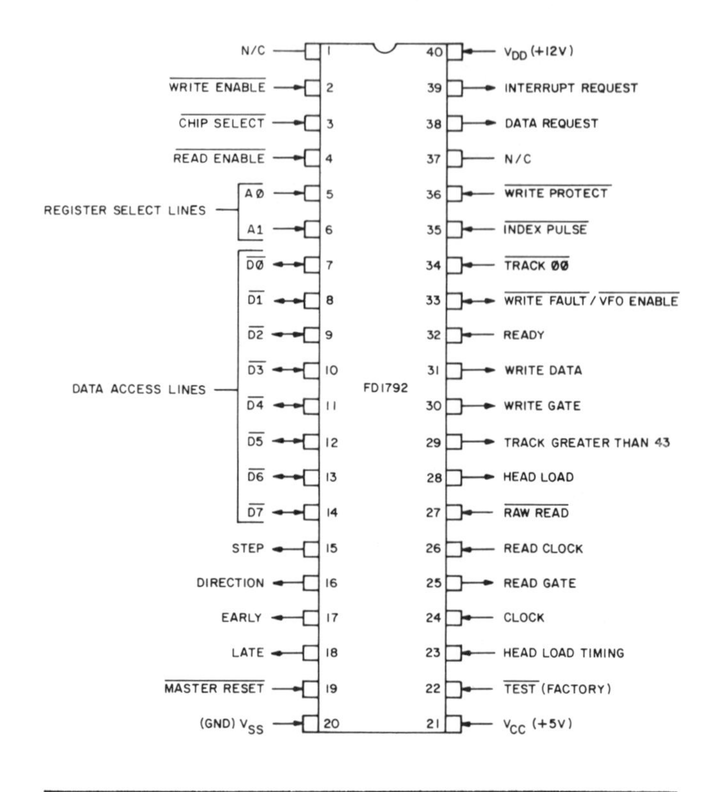

The FD1792, introduced by Western Digital, is an LSI device which performs the functions of a Floppy Disk Formatter and Controller in a single chip. The device is designed to interface with the disk drive electronics of a floppy disk memory system, and is compatible with a large variety of data systems, including the IBM 3740 and System 34 data formats.

The 179X family of floppy disk controllers consists of six devices, of which the FD1792 is a member. Since the functions of the controller chip are too numerous to list in one page, the reader will gain a total picture of its capability by reviewing all members of the 179X family in the following pages.

General Information

The floppy disk has become one of the most prolific forms for mass memory storage in small- and medium-sized computer systems. The floppy disk itself is a circular piece of thin plastic coated with a magnetic film, which is enclosed in a square, protective jacket. The plastic disk spins within its jacket at a fixed speed (usually 360 RPM) in much the same manner as a stereo record revolves on a turntable.

The disks come in two physical sizes. The 8-inch disk, originally introduced by IBM and measuring 8-inches *square* (protective jacket and all), is the larger of the two; the smaller version of the same disk, the 5¼-inch disk, is often referred to as a mini-floppy. With complete compatability as a goal, the 1792 has been designed to accommodate both sizes.

The data is magnetically stored on the disk in a circular path called a track. Unlike the spiral path of an audio disk, though, the tracks of the floppy disk are completely circular and concentric. Altogether, there are 77 tracks on a standard disk, whether it be the 8-inch or 5¼-inch version. The difference between the two disk sizes lies in the length of the track for each. The 8-inch tracks are obviously much longer, and capable of storing more data.

Motor Controls

A magnetic head assembly comes in contact with the disk for writing or reading information on the disk. The head assembly is usually positioned using a stepper motor, which in turn, must be controlled by the 1972's controller electronics. The head is incremented across the disk using the STEP (pin 15) and DIRECTION (pin 16) control signals. The STEP control moves the head one track per pulse, while the DIRECTION output determines which direction it will move: step in (toward the spindle) or step out (to the outer edge)

The head assembly control logic also includes a *track register* which contains the number of the data track presently being serviced. An optoelectronic sensor initializes this register by sending a signal to the TRACK 00 (pin 34) input when the head is positioned over the outermost track. The *track register* allows the controller to respond to a Seek Command by comparing the contents of the register to the desired track number and stepping in the required number of pulses in the appropriate direction.

Due to the fact that the inner races are somewhat shorter than the outer ones, the pulses become more crowded in these lanes, and it is often necessary to reduce the recording current to keep them from bleeding together. The controller indicates when the head is positioned between tracks 44 and 76 (the ones most frequently affected) by setting the TRACK GREATER THAN 43 output (pin 29) HIGH.

To save wear and tear on the floppy disk, the magnetic head doesn't come in contact with the disk until a read or write operation is to be performed. The HEAD LOAD control (pin 28) engages the head just prior to data transfer, lowering it into position. Since there is a time delay between the time the command is issued and the actual contact takes place, the controller holds off the data operation until it receives a "ready" signal from the HEAD LOAD TIMING (pin 23) input. The HEAD LOAD TIMING pulse is a timeout signal, usually supplied by a one-shot trigger, which allows the head time to move into position. After the external circuit has timed-out, the head is assumed to be engaged and data operations can proceed.

The transfer of data to or from the disk is dependent upon the status of three control lines. The READY line (pin 32) can be used for a number of functions—such as "door open", drive motor on, etc.—and must be actively HIGH before operation can begin. To synchronize track timing of the disk to the 1792 controller, a small hole is punched through the plastic disk near its center. Optoelectronic sensors supply a pulse to the INDEX PULSE input (pin 35) each time the hole is detected. This pulse identifies the beginning of the data track.

To prevent the accidental erasure of permanently stored material, a notch is cut in the disk envelope which, when detected, sends a signal to the WRITE PROTECT (pin 36) input, disabling the write function altogether.

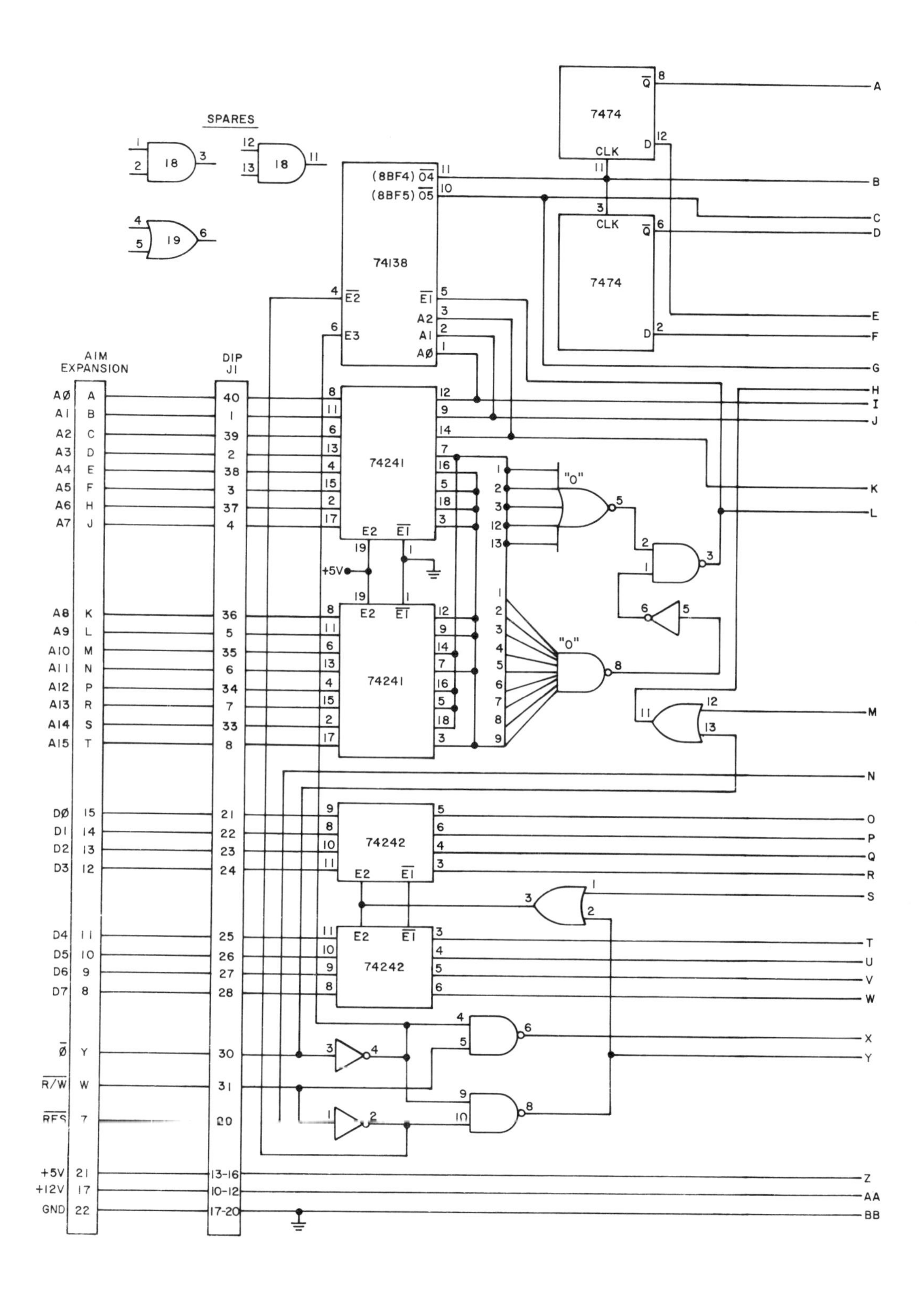

Interfacing the 1791 to the Rockwell AIM 64 computer. A complete schematic is represented by pages 111, 113, and 115.

Figure 1

FD1791

Floppy Disk Formatter/Controller

The FD1791, by Western Digital, is a Floppy Disk Formatter/Controller belonging to the 179X family. The following discussion pertains to that chip group.

Disk Format

The standard magnetic recording disk is divided into 77 data tracks. Each track is subdivided into a number of sectors. A sector can contain any number of data bytes, but they are generally limited to 128, 256, 512, or 1024 bytes long. Obviously, the number of bytes each sector contains determines the number of sectors one track will support.

Two standard data recording techniques are used to encode data information for storage on a floppy disk. The single-density method adheres to the IBM 3740 protocol, and uses a frequency modulated (FM) encoding technique to distinguish the clock pulses from the data. Each data bit, whether it be a "0" or "1," is bracketed by a clock pulse, and is referred to as a bit cell. In FM encoding, the data bit is written at the center of the bit cell, with the presence of a signal representing a logic "1" and its absence indicating a "0." Generally, it takes eight bits to make a byte.

The Standard 34 method of data recording increases the data storage capacity of a disk by two fold. Basically, it requires that the bit cell time slot be reduced from 4-us to 2-us. The encoding technique used for double-density recordings, however, is modified FM (MFM). In MFM encoding, the data bits are again written at the center of each bit cell. However, a clock pulse is written at the leading edge of the bit cell *only* if no data pulse was written in the previous cell and no data bit is to written in the present cell. In other words, a clock pulse is present only when two consecutive "0"s appear in the word, otherwise it is deleted. Using this scheme, twice as much data can be recorded without increasing the frequency rate.

The 1792 and 1794 devices of the 179X family record in the Single Density Mode only, while the other members of the 179X family can record in either mode. For double density recordings, the DOUBLE DENSITY input (pin 37) is pulled LOW.

Write

Once the mechanics of engaging the floppy disk have been resolved (see FD1792), the 1791 is ready to enter into its second phase of operation: data transfer.

To engage the disk drive in the write mode, the WRITE GATE output (pin 30) is activated, allowing current to flow into the Read/Write head. As a precaution against writing erroneous data, the *data register* must be loaded with the first word byte prior to activating the WRITE GATE. The memory write bytes are loaded into the *data register* when a WRITE ENABLE (pin 2, LOW) operation is performed by the CPU.

The data is encoded into a series of 500-ns pulses for FM or 200-ns pulses for MFM, and serially shifted out the chip through the WRITE DATA (pin 31) output. The appropriate clock pulses are also generated by the write logic and placed on the disk at the same time.

To verify that all operations are proceeding properly, the 1791 provides a WRITE FAULT input (pin 33) that queries the disk drive electronics as to the recordings process. Should a fault

occur, such as the failure to detect head write current, this input terminates the write operation and sets an interrupt.

Disk drive manufacturers often specify signal precompensation before writing data on the disk. This is particularly true in the Double Density Mode. Two signals are provided for write precompensation, EARLY (pin 17) and LATE (pin 18), which are used with external logic to achieve the necessary precompensation.

Read

Data is collected from the disk during a read operation, conditioned, and fed into the RAW READ (pin 27) input. This data stream contains clock pulses as well as data pulses. In order for the 1791 to separate the two, the original clock frequency must be derived from the data stream and presented to the READ CLOCK (pin 26) input. Phasing of the READ CLOCK signal relative to the RAW READ data is important, whereas polarity is not.

The READ CLOCK frequency is established using external circuitry, usually taking the form of a phase-locked loop. To initially synchronize the PLL with the clock signal, the controller chip supplies a timing READ GATE (pin 25) pulse whenever two bytes (single density) or four bytes (double density) of zeros are encountered within the RAW READ data stream. A VFO ENABLE (pin 33) output enables the PLL oscillator.

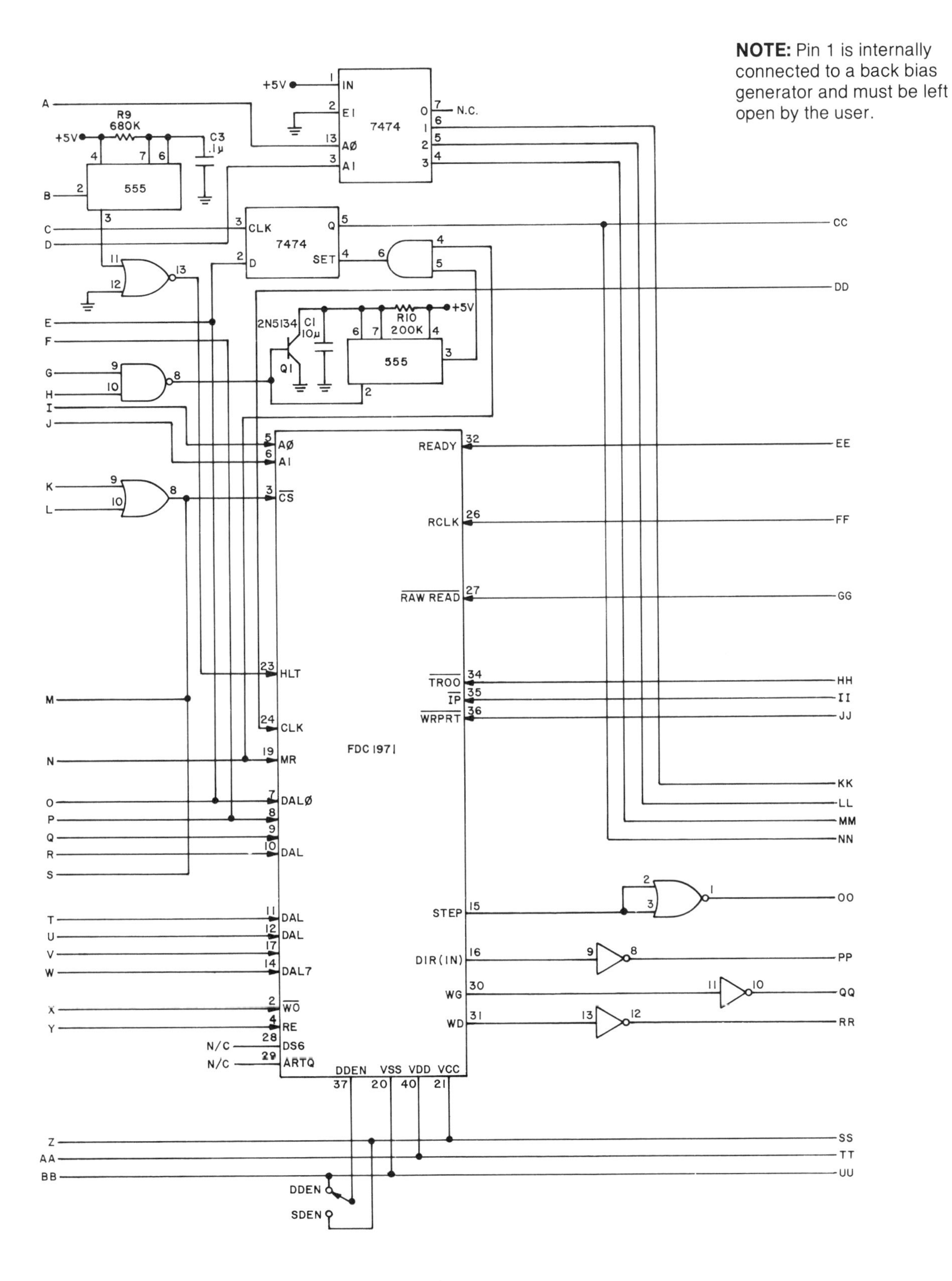

Interfacing the 1791 to the Rockwell AIM 65 computer. A complete schmatic is represented by pages 111, 113, and 115.

Figure 1

FD1795

Floppy Disk Formatter/Controller

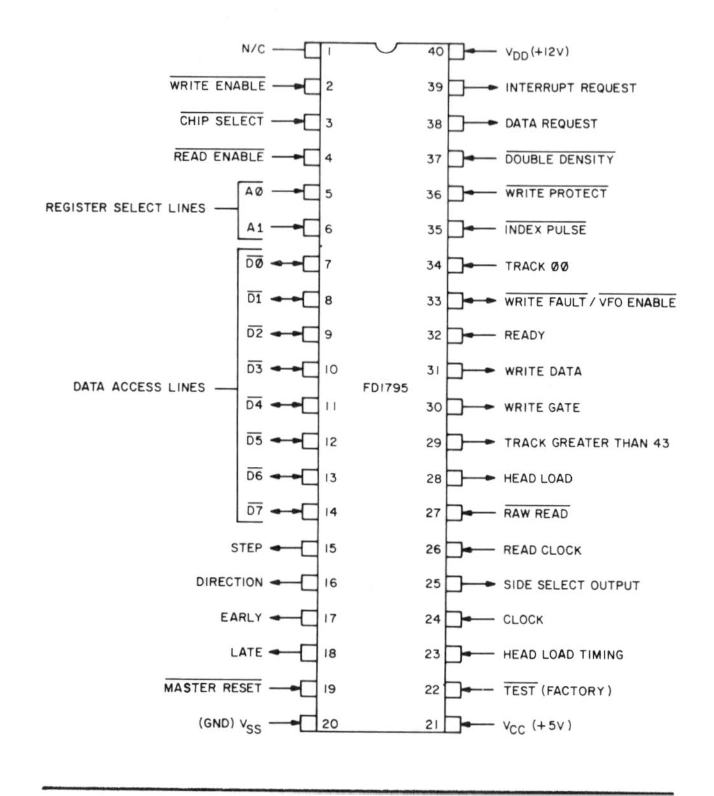

The FD1795, by Western Digital, is a Floppy Disk Formatter/Controller belonging to the 179X family. The 179X series can be considered the end result of both the FD1771 and FD1781 designs, and is IBM 3740 compatible in Single Density Mode (FM) and System 34 compatible in Double Density Mode (MFM). In order to maintain consistency between chips, all designs were made as close as possible in both software and hardware, with the pin assignments varying only slightly from one type to another.

Memory Expansion

The FD1791 and FD1792 floppy disk controller chips are limited to placing data on only one side of the disk. However, there is no reason that, with proper design, both sides of the disk can't contain information, much like a phonograph record does.

The 1975 has been designed to do just that. In systems so equipped, the 1795 controller will utilize the disk's dual-sided trait, with the SIDE SELECT OUTPUT (pin 25) determining which side of the disk is presently being serviced. Moreover, the 1795 still retains FM and MFM capabilities.

The controller may also operate as many disk drives as practical, provided proper interfacing techniques are observed. Normally, provisions are made in the hardware cabinet for up to four disk drives to share the same interface cable.

Computer Interface

To relieve the CPU of the software overhead normally associated with floppy disk drive operations, the 1795 contains five registers which respond to eleven commands. These eleven commands direct all controller operations.

The registers are accessed through eight bidirectional DATA ACCESS LINES (pins 7-14), and are selected by configuring the REGISTER SELECT LINES (pins 5, 6). The REGISTER SELECT LINES may actually tie to the CPU address lines, either directly or decoded, and become a memory-mapped location no different that conventional RAM.

When data transfer with the floppy disk memory is required by the host processor, the device address is decoded and the CHIP SELECT (pin 3) made LOW. The actual transfer of data is associated with only one register, the *data register.* This register is used for both reading and writing.

In the READ ENABLE (pin 4, LOW) mode, data bits are collected from the disk, accumulated, and transferred to the *data register.* When a complete word (byte) has been assembled, the 1795 activates an open-drain DATA REQUEST (pin 38) output. This indicates to the CPU that a data word is ready for transfer. The CPU responds by taking the CHIP SELECT LOW and assimilating the information. If the *data register* is not read by the time a new word is ready for transfer, the previous word is overrun and the Lost Data bit is set. The read operation continues until the end of the sector is reached.

Interrupts

At the completion of every command, an INTERRUPT REQUEST (pin 39) is generated. The INTERRUPT REQUEST is reset by either reading the *status register* or by loading the *command register* with a new instruction. In addition, an interrupt is generated if a Force Interrupt command condition is met.

Other CPU interface lines include CLOCK (pin 24), MASTER RESET (pin 19), and DOUBLE DENSITY (pin 37). The CLOCK frequency is 2-MHz for an 8-inch drive or 1-MHz for a 5¼-inch disk, with a 50% duty cycle. Accuracy must be held to ±1% since all internal timing, including stepping rates, are based on this clock.

The MASTER RESET line should be strobed for a minimum of 50-us after each power-on condition. This line clears and initializes all internal registers, and issues a Restore command on its rising edge.

Examining the 179X Family

The Western Digital 179X family of floppy disk controller chips consists of six chips altogether. Three types—the 1791, 1792, and 1795 — are examined in this book. They differ mainly in the methods by which the data is stored on the floppy disk itself. The 1792 is only a single-density chip; the 1791 expands that capacity by including both FM and MFM recording techniques, while the 1795 allows the user access to both sides of the disk.

Each of the above three chips have counterparts. The 1792 is related to the 1794, the 1791 to 1793, and the 1795 to 1797. Each chip is identical to its equivalent, with the exception of the DATA ACCESS LINES. While the first three chips interface to the CPU using inverted logic, its counterpart performs the same duties with true (positive) logic.

Another entire family of chips also exists which are related to the 179X series. They are the 176X chips, which are the 1-MHz mini-floppy equivalent of their relatives. All other parameters and functions remain the same.

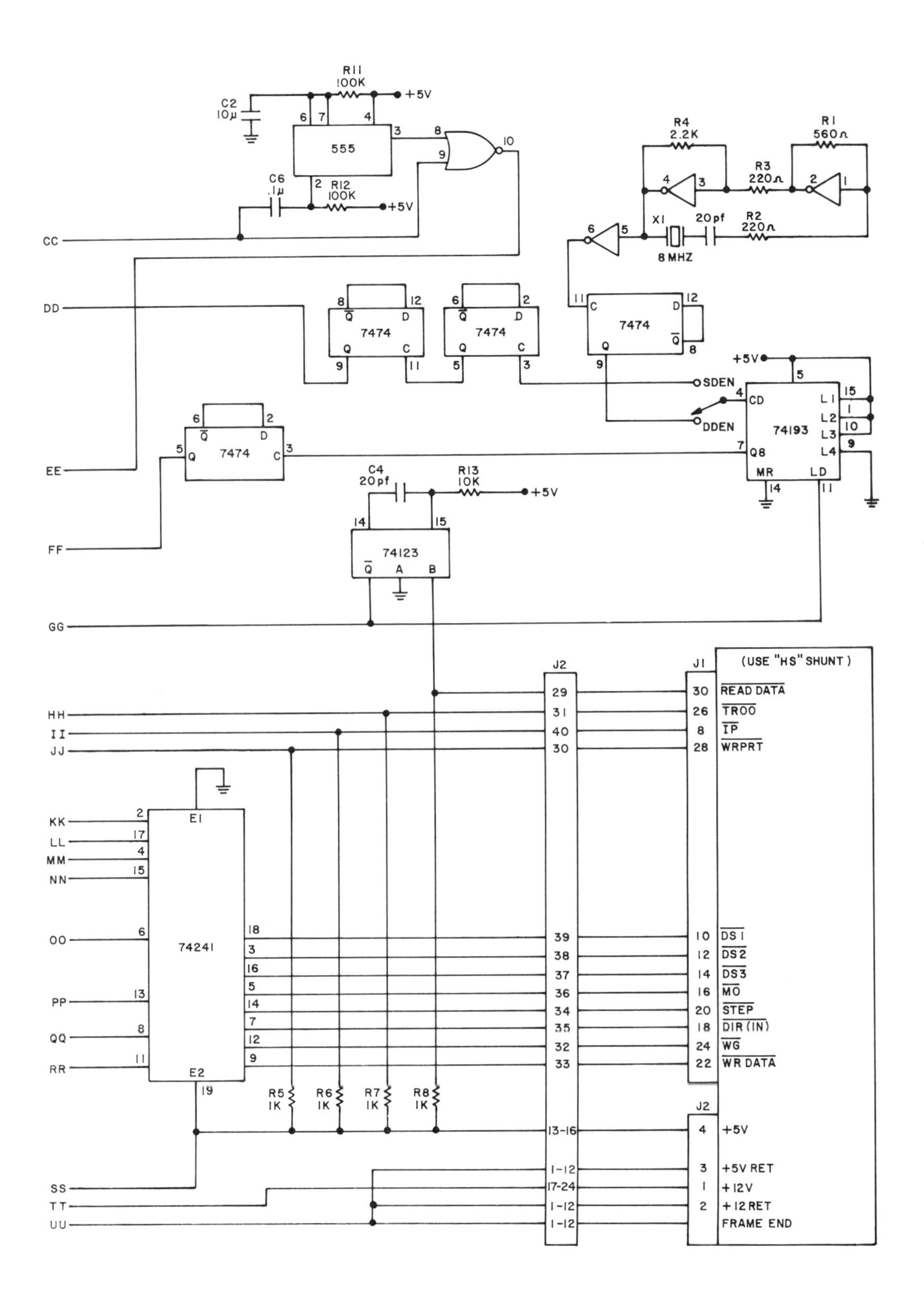

Interfacing the 1791 to the Rockwell AIM 65 computer. A complete schematic is represented by pages 111, 113, and 115.

Figure 1

WD2791

Floppy Disk Formatter/Controller

The WD2791 is a Floppy Disk Formatter/Controller made by Western Digital. The WD2791 is related to the 179X family of floppy disk controllers, also introduced by Western Digital, and is essentially an upgraded version of its predecessor. In fact, the two families virtually mimic each other when compared one to one.

However, the 279X series extends the performance of the previous chips by incorporating an on-chip phase-locked loop data separator and write precompensation logic, thereby sizably reducing the amount of required external logic. The entire 279X series is software compatible with the 179X, and software generated for the 1791 can be transferred to the 2791 system without modification.

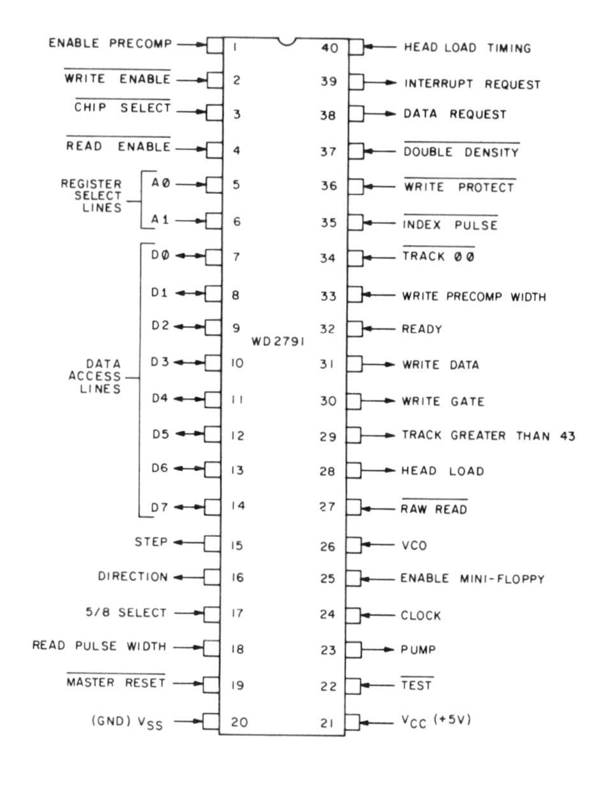

Data Separation

The 2791 is designed to operate with either its own internal data recovery circuit or with external data separator logic. When using an external PLL (phase locked loop) type separator, the 2791 supplies control signals for VCO (voltage controlled oscillator) synchronization.

To program the 2791 for external data separation, a MASTER RESET (pin 19) pulse of at least 50-us must be applied while the TEST (pin 22) line is held LOW. For operation, the TEST input is returned HIGH and a clock frequency equivalent to eight times the data rate (4-MHz for an 8-inch Double Density disk) is applied to the VCO (pin 26) input. Providing feedback for the external VCO, the PUMP output (pin 23) maintains a control voltage that tracks the phase difference between the raw data rate and the VCO to keep the two signals in step with each other.

For internal VCO operation, the TEST line is held HIGH during the MASTER RESET pulse, then set to logic "0" for the adjustment procedure.

Adjustment Procedure

A 50-k potentiometer tied to the READ PULSE WIDTH input (pin 18) is used to set the internal VCO clock pulse for proper phasing. With a scope on pin 29 (TG43), adjust the READ PULSE WIDTH control until it equals 1/8th the data rate. This adjustment would be 250-ns for an 8-inch Double Density recording.

Next, the VCO center frequency is adjusted using a 5-60-pf variable capacitor connected to the VCO (pin 26) input. With a frequency counter monitoring pin 16 (DIRC), adjust the trimmer capacitor to yield the appropriate data rate for the floppy disk involved. This frequency is 500-kHz for an 8-inch Double Density floppy disk. When making this adjustment, the user should be aware that the DOUBLE DENSITY line (pin 37) must be LOW while the 5/8 SELECT input (pin 17) is held HIGH, otherwise the adjustment time will be doubled. After the adjustments have been made, the TEST pin is returned to logic "1" and the device is ready for operation.

When operating in the internal mode, the PUMP signal is internally connected to the VCO for phase control. However, an external filtering circuit is usually necessary to avoid hunting by the VCO. A simple first-order lag-lead filter, similar to the one shown in figure 2, is all that is required for this purpose.

Write Precompensation

When operating in the Double Density Mode (pin 37, LOW), the 2791 is capable of providing precompensation for the WRITE DATA (pin 31) signal. Because the amount of precompensation varies from one disk drive to another, with typical values ranging from 100- to 300-ns, an external 10-k potentiometer connected to the WRITE RECOMP WIDTH pin (pin 33) allows the user to adjust the precompensation timing.

Adjusting the write precompensation timing is accomplished by pulling the TEST input LOW. A stream of pulses can then be seen on the WRITE DATA output when viewed with an oscilloscope. Now, tune the WRITE PRECOMP WIDTH pot until the precompensating pulse width equals the value specified by the disk drive manufacturer. This adjustment can be performed in-circuit since the WRITE GATE (pin 30) enable is inactive during the test.

Forcing the TEST input HIGH returns the device to normal operations. Precompensation will be performed on the WRITE DATA whenever the ENABLE PRECOMP input (pin 1) is set HIGH.

Disk Selection

The 2791 controller has been equipped to handle both 5¼-inch and 8-inch floppy disks. For data reading purposes, the 5/8 SELECT input (pin 17) will choose the internal VCO frequency necessary to decode the data for each of the two sizes. A HIGH input to this pin selects an 8-inch disk.

The controller will also write both formats without the need to change the CLOCK (pin 24) frequency. When the ENABLE MINI-FLOPPY input (pin 25) is used in conjunction with pin 17 (LOW), an internal divide-by-two circuit splits the master clock frequency in half. This allows both 5¼-inch and 8-inch drive operation with a single 2-MHz CLOCK.

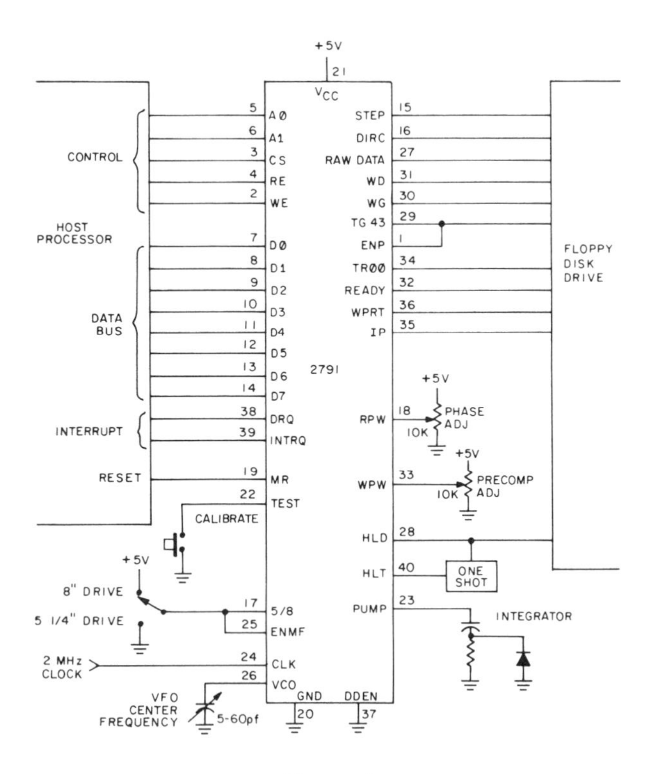

Figure 1

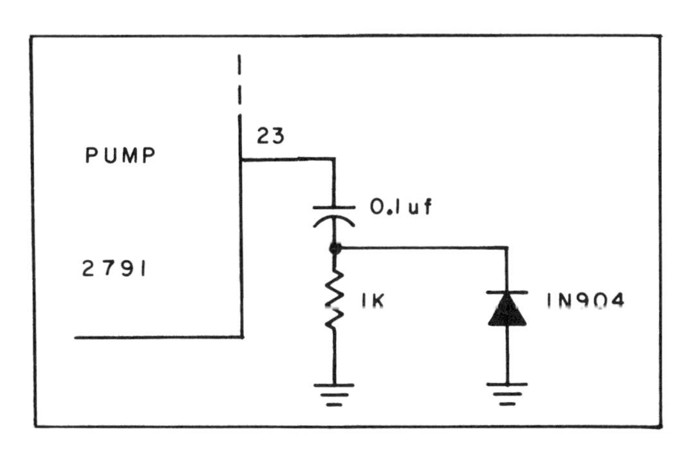

PUMP FILTER

Figure 2

When operating with 5¼-inch drives, increase the timing capacitor to 0.2 or 0.22uf.

FD1771

Floppy Disk Formatter/Controller

The FD1771, by Western Digital, is an LSI device that performs the functions of a Floppy Disk Controller and Formatter according to the IBM 3740 standard of single-density recording. Primarily intended for use within the disk drive electronics, the FD1771 generates control signals that can directly drive a three-phase motor. The end result is a simple CPU interface in lieu of a complicated disk interface.

Head Positioning

Four commands position the read/write head over the desired track on the floppy disk. By simply grounding the 3-PHASE MOTOR SELECT (pin 18) input, these commands are translated into control signals which can directly drive a 3-phase stepper motor. Three output signals—PHASE 1 (pin 15), PHASE 2 (pin 16), and PHASE 3 (pin 17)—actuate the stepping sequence. When stepping the head in, the sequence is PHASE 1-2-3-1; stepping out is PHASE 1-3-2-1. Note that PHASE 3 is positive logic, and requires an inverter for this application.

Additionally, the 1771 will also control a decoder-style head positioner. Forcing pin 18 HIGH allows the controller to issue 4-us HEAD STEP (pin 15) and 24-us DIRECTION (pin 16) commands to the disk drive.

The HEAD LOAD (pin 28) output places the magnetic head against the floppy disk medium. After 10-ms, the HEAD LOAD TIMING input (pin 23) is sampled, and if it is found to be HIGH the head is assumed to be in position. A LOW to this input recycles the head timing circuit.

Read Disk

The 1771 operates with an internal data separator. A 2-MHz CLOCK (pin 24) is internally divided by 4 to produce a 500-kHz clock for the separator logic. This clock synchronizes itself to the FLOPPY DISK DATA (pin 27) input from the disk drive raw data signal. From here, the clock is further divided by two and an internal detector extracts the data from the clock pulses. This serial data is accumulated and presented to the CPU on the DATA ACCESS LINES (pins 7-14) in parallel form. An open-drain DATA REQUEST output (pin 38) indicates when the *data register* has assembled a complete word. Should the CPU fail to read the contents of the *data register* before new data is received from the disk, the register is overrun and an error bit is set.

Since the CLOCK frequency is 2-MHz, it permits 500-ns resolution. On the inner tracks, however, the bit shift is often severe, especially without write precompensation, and the data or clock bits may occasionally fall outside the 500-ns window. In order to maintain an error rate better than one part in 10^8, therefore, an external data separator, such as a phase-locked loop, is recommended. External separation is accomplished by grounding the EXTERNAL DATA SEPARATION (pin 25) input.

Separated clock pulses from the PLL are now received at the FLOPPY DISK CLOCK (pin 26) input and data is input to the FLOPPY DISK DATA pin. Furthermore, these two lines, pins 26 and 27, may be reversed; the 1771 will determine which pin is clock and which is data when an Address Mark is detected.

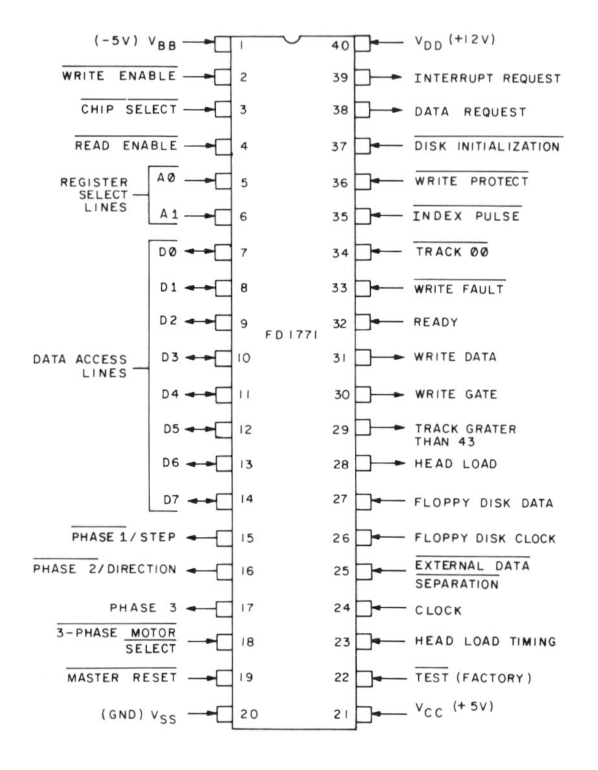

Write Disk

When writing onto the diskette, the WRITE GATE output (pin 30) is first engaged, allowing current to flow into the recording head. The data is then mixed with the clock pulses and serially shifted through the WRITE DATA (pin 31) output. The WRITE FAULT input (pin 33) is alerted when a writing fault has occurred in the drive unit, such as the failure to detect head current upon the receipt of a WRITE GATE signal, and terminates writing operations.

As a precaution against erasing permanent data, writing is prohibited when the WRITE PROTECT input (pin 36) is a logic LOW. Writing is also inhibited if the DISK INITIALIZATION input (pin 37) is LOW.

CPU Interface

All functions of the 1771 chip are controlled by five internal registers. These registers are accessed by configuring the REGISTER SELECT LINES (pins 5, 6).

Data traverses between the CPU and 1771 via the DATA ACCESS LINES using a WRITE ENABLE (pin 2) or READ ENABLE (pin 4) signal in conjunction with the CHIP SELECT (pin 3) input. Whenever a command or operation is concluded, the INTERRUPT REQUEST (pin 39) output is set. This interrupt is cleared when a new command is loaded into the *command register.*

Disk Drive Interface

Several control signals are necessary to coordinate the 1771 to the disk drive. The TRACK 00 input (pin 34) determines the location of the outermost track, while the INDEX PULSE (pin 35) locates the beginning of each track. The READY line (pin 32) is a general-purpose input that can be used for any number of indicators, such as a "door open" condition. This pin must be HIGH before READ or WRITE commands are performed.

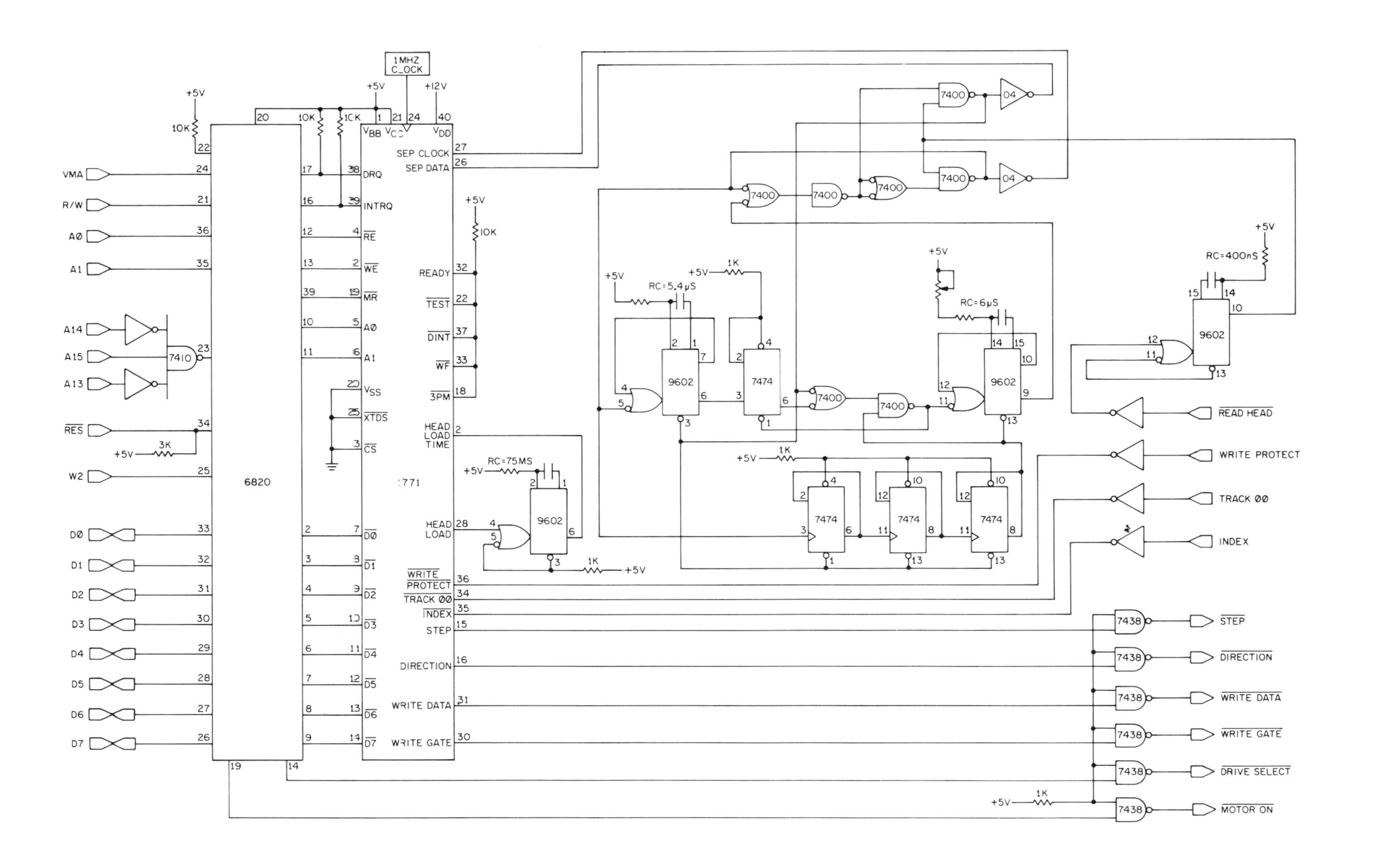

1MHZ CLOCK
6820
1771
SEP CLOCK
SEP DATA
DRQ
INTRQ
RE
WE
MR
AØ
A1
XTDS
CS
READY
TEST
DINT
WF
3PM
HEAD LOAD TIME
HEAD LOAD
WRITE PROTECT
TRACK ØØ
INDEX
STEP
DIRECTION
WRITE DATA
WRITE GATE
VMA
R/W
AØ
A1
A14
A15
A13
RES
W2
DØ
D1
D2
D3
D4
D5
D6
D7
RC=5.4µS
RC=6µS
RC=400nS
RC=75MS
READ HEAD
WRITE PROTECT
TRACK ØØ
INDEX
STEP
DIRECTION
WRITE DATA
WRITE GATE
DRIVE SELECT
MOTOR ON

8272

Floppy Disk Controller

The 8272, by Intel, is a Floppy Disk Controller which contains the circuitry and control functions necessary for interfacing a processor to four floppy disk drives. The device is capable of supporting either IBM 3740 single density format (FM) or IBM System 34 double density format (MFM), including double-sided recording.

In order to facilitate the use of the controller with the large number of commercially available disk drive systems, the 8272 permits software programming of the track stepping rate, the head load time, and the head unload time. In addition, handshaking signals are provided for DMA control and phase-locked loop data separation.

Disk Drive Selection

The 8272 generates all the signals necessary for the control and management of up to four floppy disk drives. Each drive is selected for operation by decoding the DRIVE SELECT (pins 29, 28) outputs.

The readiness of the disk drive mechanism is ascertained by monitoring the status of the drive's READY (pin 35) line prior to operation. This line must be HIGH before the 8272 will send or receive data from the disk drive. At power on, and after a power-on RESET (pin 1), the 8272 automatically enters a drive status polling mode.

In this mode, the controller sequentially scans each of the disk drive units looking for a change in the status of its READY line, usually due to the opening or closing of a door. In this manner, the 8272 notifies the processor when a floppy disk is inserted, removed, or changed by the operator. Only during actual data transfer or command inputs is this scanning process halted. Once the operation is complete, scanning resumes.

Disk Drive Signals

The 8272 controls the movements of the magnetic recording head(s) for proper track location. When more than one head is involved, as is the case with double-sided disks, the TWO SIDE input (pin 34) is set HIGH and the HEAD SELECT output (pin 27) determines which head is required for the operation.

The recording head is normally resting just above the disk when no data transfers are taking place, thus saving wear to the floppy disk surface. The head is lowered to the magnetic medium when HEAD LOAD (pin 36) command is issued. The amount of time it takes for the head to position itself on the disk surface is called the settling time. Since settling times vary from one disk drive to another, it is user programmable within the 8272 chip. After the settling time-out has expired, the head is assumed to be in position and data operations may begin. If, after a predetermined time, no further data transfers occur, the HEAD LOAD signal returns LOW, raising the head to its standby position.

Seek Mode

Locating the desired data track on a disk is accomplished by placing the controller in the Seek Mode. The Seek Mode is set by a software command and represented by a HIGH level on the SEEK (pin 39) output.

To keep tabs on the head's position, the 8272 stores the number of the current track (the one over which the head is residing) in a *track number register*. This register is initialized, and reset to zero, with a Recalibrate command. Basically, recalibration retracts the head until it is over the number 0 track, as indicated by a signal from the disk drive to the TRACK 0 (pin 33) input.

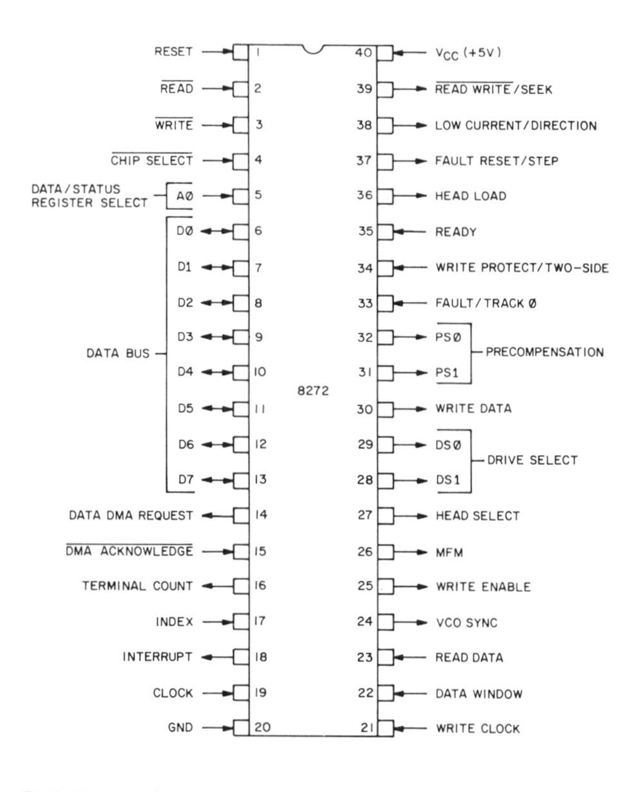

After the Seek command has been ordered, the controller begins searching for the specified track by comparing its number to the number contained in the *track number register*. Internal logic then determines in which direction the head must move using the DIRECTION (pin 38) output, and how far it moves by the STEP (pin 37) output. One STEP pulse is issued for each track; the rate at which they are sent is controlled in software using the Specify command.

To indicate the beginning of a disk data track, a hole is punched in the disk itself. An optoelectronic sensor feeds a pulse to the INDEX (pin 17) input every time this hole is detected, synchronizing the chip to the rotating disk.

Disk Formatting

The 8272 will also format tracks on a floppy disk. The particular disk format to be written is laid down according to the values programmed into the CPU during the Command Phase. The format must contain a Track Address (0 to 76), Head Address (side 0 or 1), Sector Address, and Sector Size. After each sector of a track is formatted, these parameters must be re-established, with the process continuing until an entire track is formatted.

The 8272 is second-sourced by the NEC uPD765A Floppy Disk Controller, and additional operating information is contained in that section.

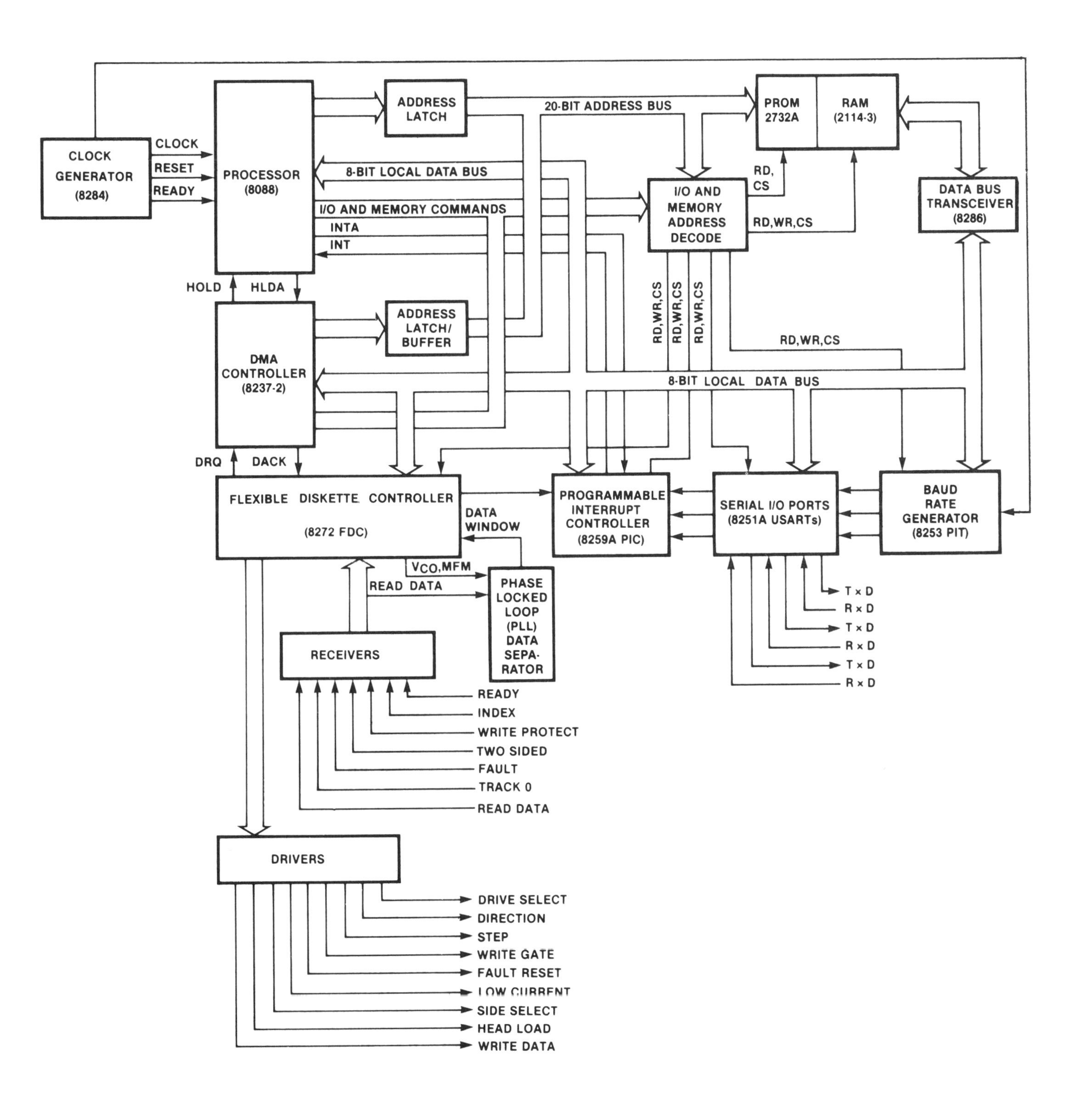

Using the uPD765A or 8272 Floppy Disk Controller, the system described in the following pages—122, 123, 126, and 127—can easily expand to 16 megabytes of data stored on four disk controllers.

Figure 1

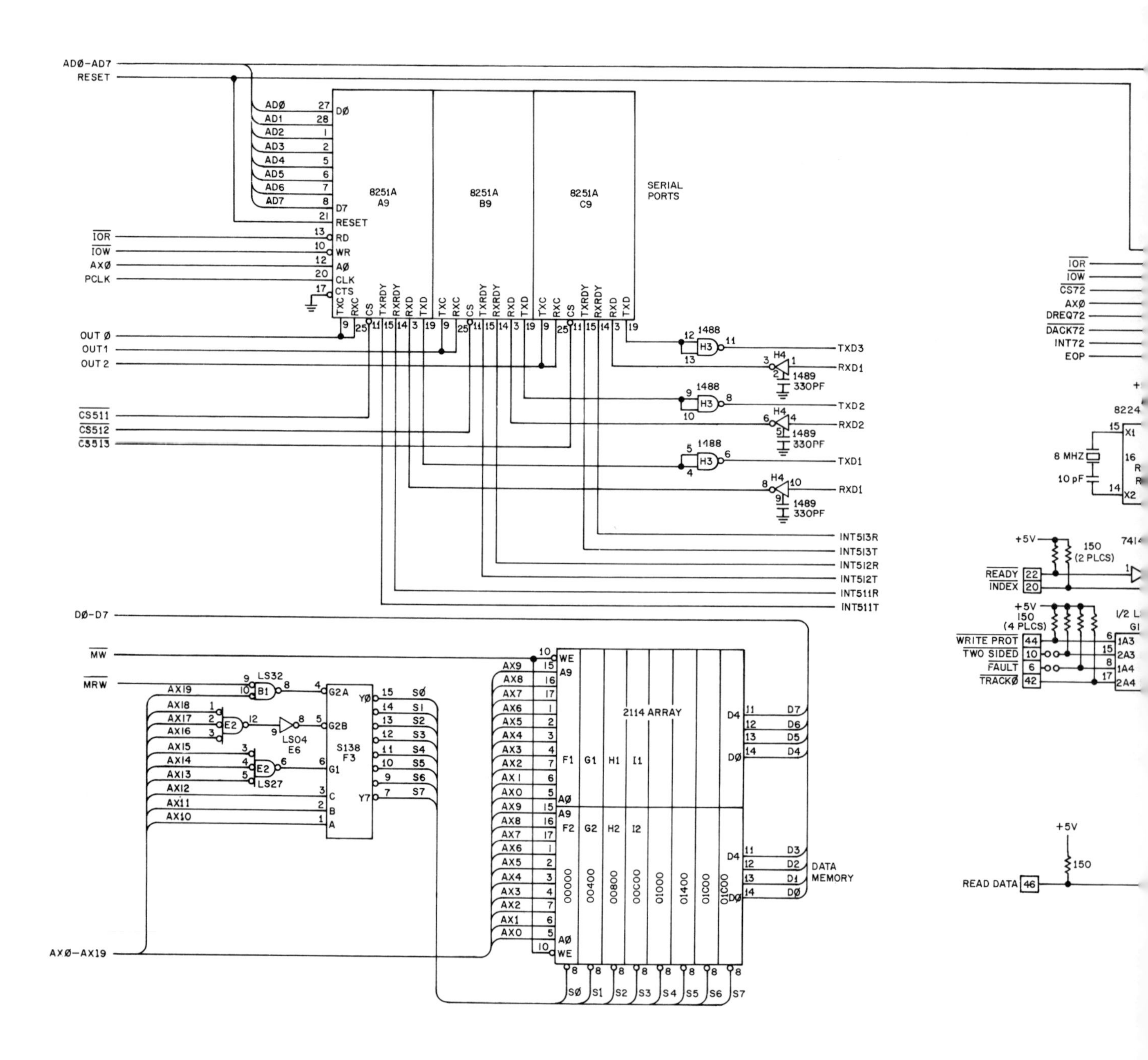
AD∅–AD7
RESET
8251A
A9
8251A
B9
8251A
C9
SERIAL
PORTS
IOR
IOW
AX∅
PCLK
OUT ∅
OUT1
OUT2
CS511
CS512
CS513
1488
H3
H4
1489
330PF
TXD3
RXD1
TXD2
RXD2
TXD1
RXD1
INT513R
INT513T
INT512R
INT512T
INT511R
INT511T
D∅–D7
MW
MRW
LS32
LS04
E6
LS27
S138
F3
2114 ARRAY
DATA
MEMORY
AX∅–AX19
CS72
DREQ72
DACK72
INT72
EOP
8224
8 MHZ
10 pF
+5V
150
(2 PLCS)
READY
INDEX
150
(4 PLCS)
WRITE PROT
TWO SIDED
FAULT
TRACK∅
READ DATA

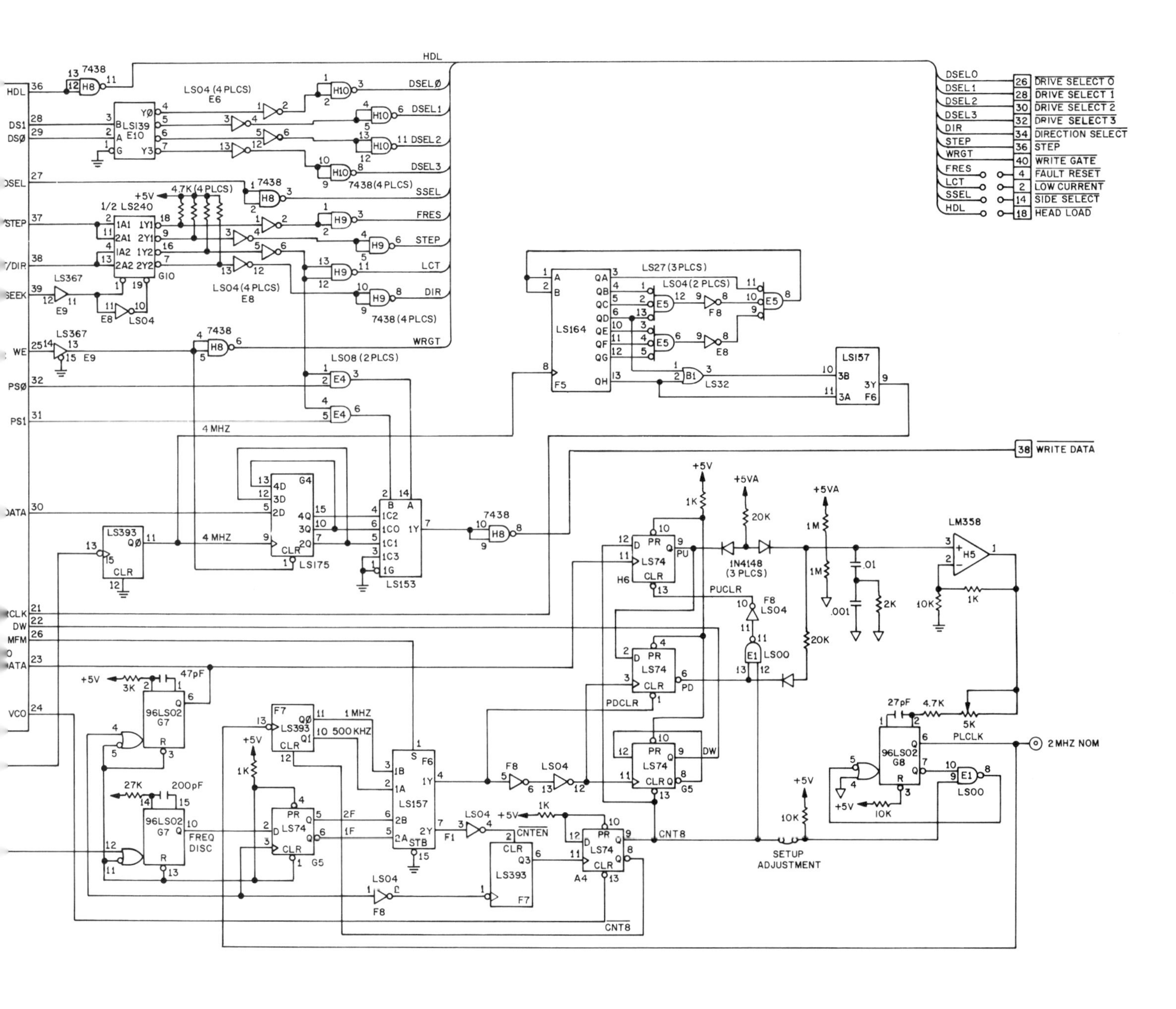

HDL
DSEL0
DSEL1
DSEL2
DSEL3
DIR
STEP
WRGT
FRES
LCT
SSEL
26 DRIVE SELECT 0
28 DRIVE SELECT 1
30 DRIVE SELECT 2
32 DRIVE SELECT 3
34 DIRECTION SELECT
36 STEP
40 WRITE GATE
4 FAULT RESET
2 LOW CURRENT
14 SIDE SELECT
18 HEAD LOAD
38 WRITE DATA
7438
LS04 (4 PLCS)
E6
LS139
E10
7438 (4 PLCS)
4.7K (4 PLCS)
+5V
1/2 LS240
G10
LS367
E9
E8
LS04 (4 PLCS)
LS08 (2 PLCS)
4 MHZ
LS393
LS175
LS153
LS164
F5
LS27 (3 PLCS)
LS04 (2 PLCS)
LS32
LS157
F6
+5VA
20K
1N4148
(3 PLCS)
LM358
H5
PUCLR
PDCLR
LS74
H6
G5
LS00
SETUP
ADJUSTMENT
96LS02
G7
G8
FREQ
DISC
47pF
3K
27K
200pF
27pF
4.7K
5K
PLCLK
2 MHZ NOM
1 MHZ
500 KHZ
CNTEN
CNT8
A4
F7
F8
HDL
DS1
DSØ
DSEL
STEP
DIR
SEEK
WE
PSØ
PS1
DATA
DW
MFM
VCO

uPD765A

Floppy Disk Controller

The uPD765A, by NEC Electronics, is a Floppy Disk Controller capable of interfacing up to four floppy disk drives to a processor. The controller will support both IBM 3740 single-density and System 34 double-density MFM recordings.

Processor Interface

The 765A contains two registers which may be accessed by the main processor; a *status register* and a *data register.* All controller functions are performed through these two registers.

The *data register*—which is actually several registers in a stack, with only one register presented to the CPU at a time—stores all data, commands, and parameters necessary to successfully operate the device. Access to this registers is through eight bidirectional DATA BUS (pins 6-13) lines when the DATA/STATUS REGISTER SELECT input (pin 5) is HIGH. The direction of the data flow is determined by the READ (pin 2) and WRITE (pin 3) signals, which are enabled by the CHIP SELECT (pin 4). The *status register* is a read-only register used to display internal chip conditions, and can be accessed by pulling the DATA/STATUS REGISTER SELECT pin LOW.

The 765A is capable of executing 15 different high-level commands. Each command is initiated by a multi-byte word from the CPU. After the command is executed, the results of its operation are transmitted back to the CPU as a multi-byte message. To avoid errors, the contents of the *status register* must be read after each byte transfer, and before the next byte can be sent. A nine-byte word requires nine readings of this register.

Read Data

Nine bytes are required to place the 765A into the Read Data Mode. Basically, the controller has two functions to perform: Controlling the floppy disk drive electronics and transferring data. These two functions are separate and distinct.

During the first phase of a disk reading operation, the drive electronics must be engaged. Internal routines locate the desired track, load the head, and find the proper sector. The chip is then switched over to the read function. Since several of the controller's pins are multiplexed, and do double duty, it is important to distinguish between the control and data function. The READ WRITE/SEEK (pin 39) output is HIGH when the 765A is guiding the mechanics through its paces, and LOW when data transfers occur.

Raw data from the floppy disk is input to the READ DATA (pin 23) pin, where the data pulses are sorted from the clock pulses. The 765A requires a PLL oscillator to accomplish this. The PLL supplies a timing pulse, which is synchronized to the original recording on the disk, to the DATA WINDOW.

The VCO SYNC output (pin 24) essentially turns the PLL off when no valid data is being received from the disk drive, allowing it to idle at its center frequency. For double-density recordings, the MFM (pin 26) output doubles the PLL frequency.

After a byte has been assembled, the 765A sets the INTERRUPT output (pin 18) HIGH. During disk data transfers, the controller must be serviced by the CPU every 27-us, or every 13-us in the MFM mode, to prevent data overrun.

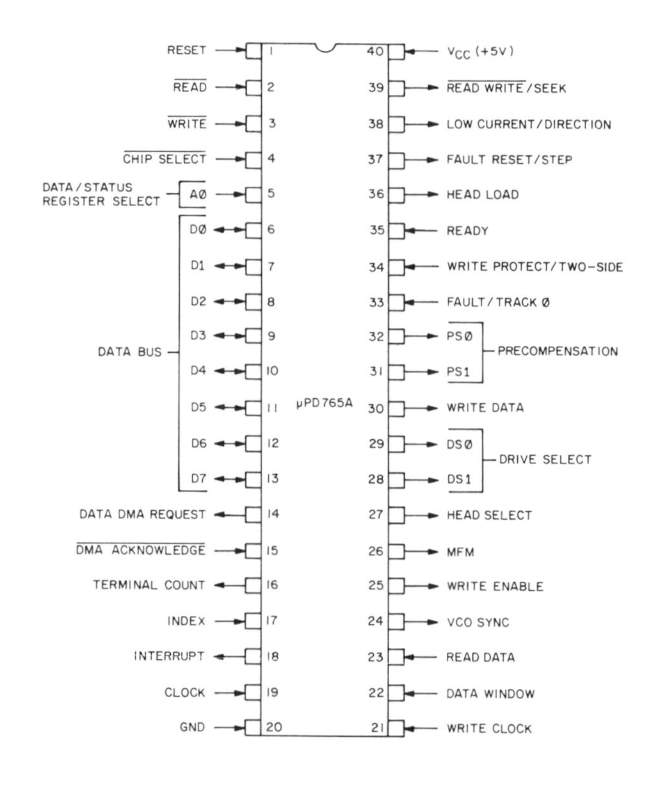

Write Data

A set of nine bytes is also required to set the chip into the Write Data Mode. Once the disk is set up (sector selected, etc.), the WRITE ENABLE output (pin 25) goes HIGH and data is serially output through the WRITE DATA (pin 30) port at the rate set by the WRITE CLOCK (pin 21). This clock must be enabled for all operations, both read and write. Precompensation is performed on the data stream using external circuitry steered by the PRECOMPENSATION (pins 32, 31) outputs.

Should a fault condition exist, such as the absence of head current after the WRITE ENABLE is activated, the disk drive sets the FAULT input (pin 33) HIGH, aborting the write operation. The fault indicator is cleared by a FAULT RESET (pin 37) output pulse. Writing is prohibited if the WRITE PROTECT input (pin 34) is HIGH.

DMA Operations

If the 765A is in the DMA mode, no interrupts are generated during the transfer of data. Instead, the controller asserts a DATA DMA REQUEST (pin 14) when each byte is available for transfer. The DMA controller responds to this request with both a DMA ACKNOWLEDGE (pin 15) and a READ (pin 2) signal. If the 765A is in the Write Data Mode, however, the DMA ACKNOWLEDGE will be accompanied by a WRITE (pin 3) signal instead. After the execution phase of data transfer is complete, the TERMINAL COUNT input (pin 16) is raised HIGH, and DMA operations are terminated.

The uPD765A is identical to the 8272 Floppy Disk Controller made by Intel. Refer to that section for additional information.

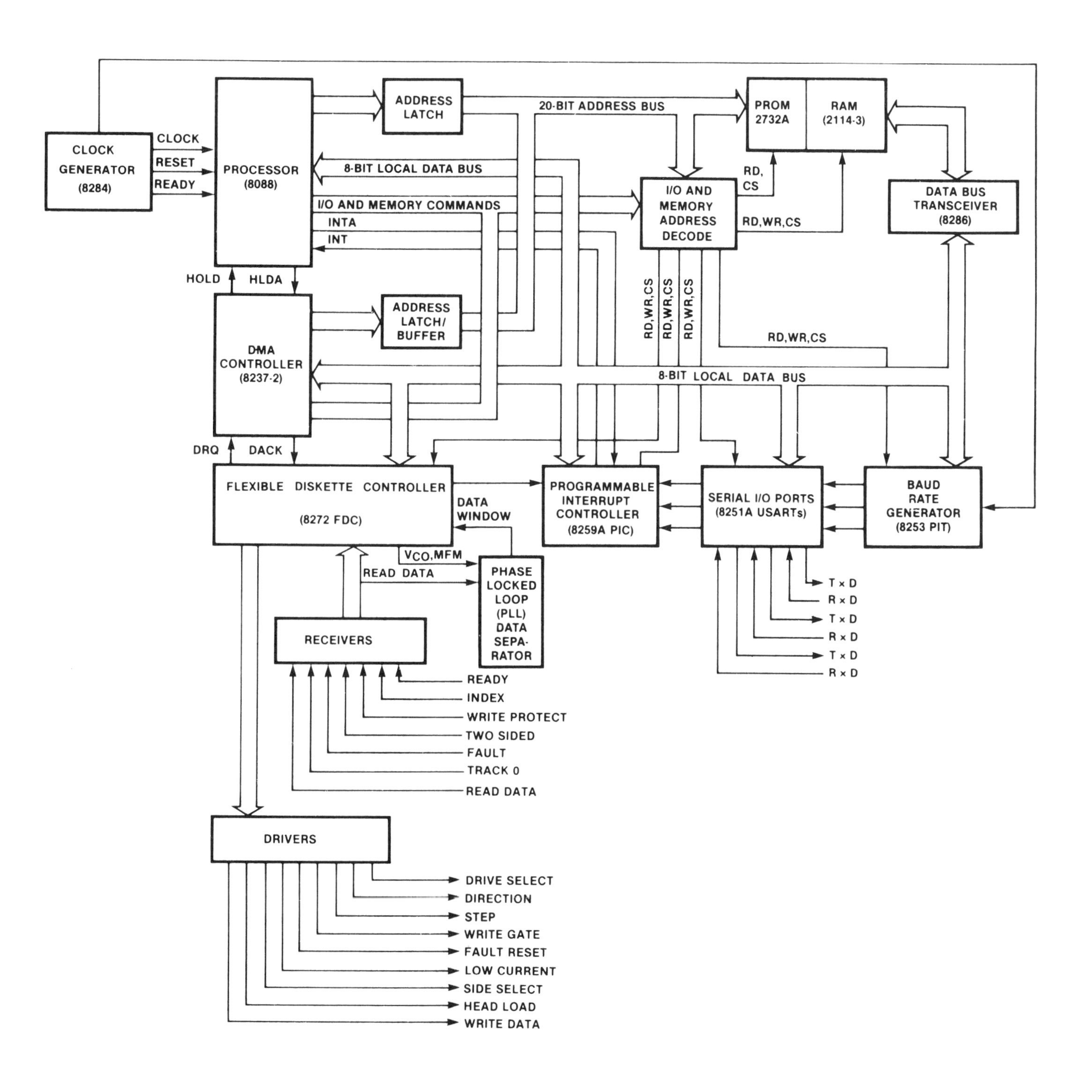

Using the uPD765A or 8272 Floppy Disk Controller, the system described in the pages—122, 123, 126, and 127—can easily expand to 16 megabytes of data stored on four disk controllers.

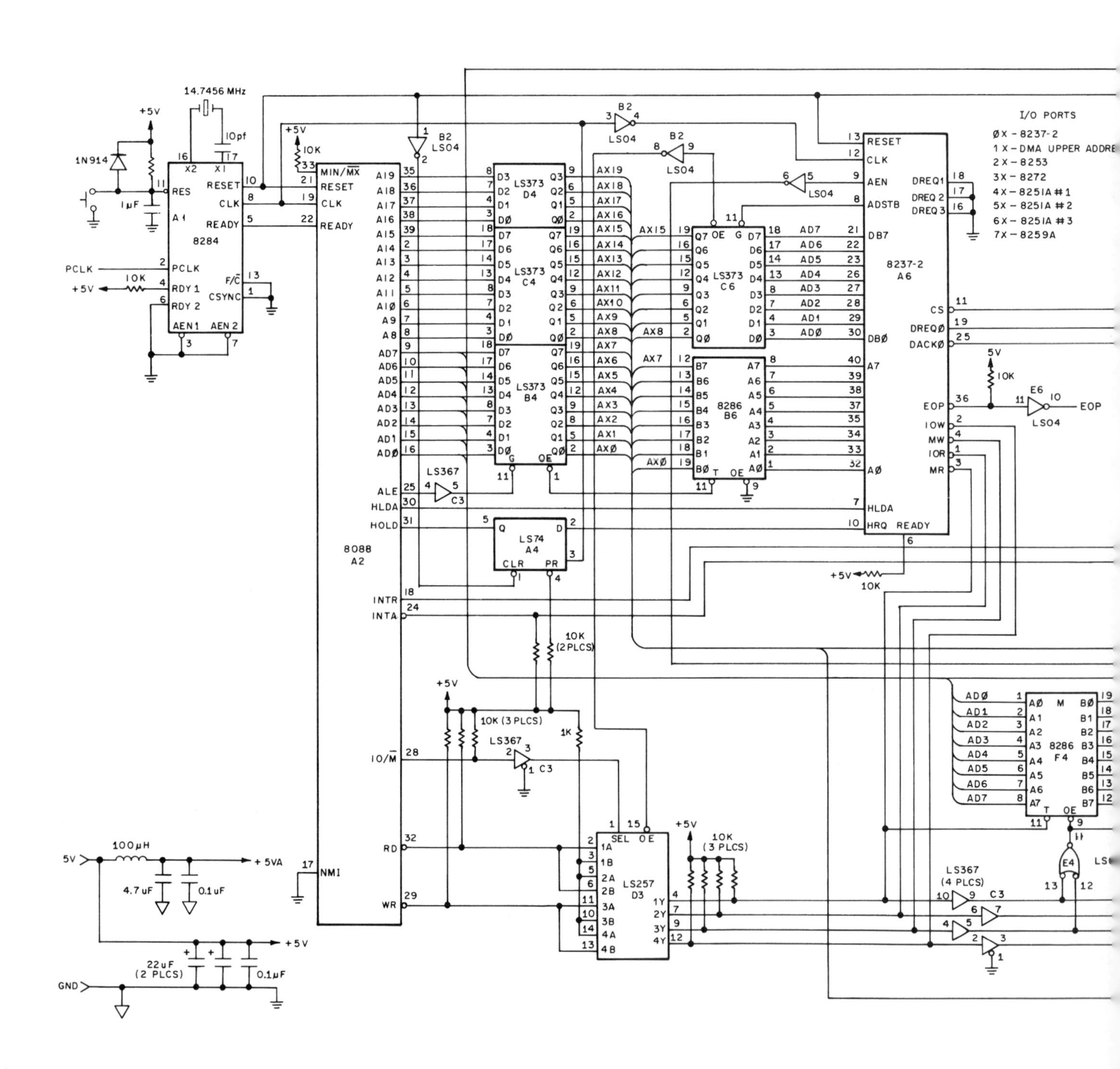
I/O PORTS
ØX - 8237-2
1X - DMA UPPER ADDRE
2X - 8253
3X - 8272
4X - 8251A #1
5X - 8251A #2
6X - 8251A #3
7X - 8259A
14.7456 MHz
10pf
+5V
1N914
1μF
8284
RES
RESET
CLK
READY
PCLK
RDY 1
RDY 2
CSYNC
AEN 1
AEN 2
10K
LS04
MIN/MX
8088
A2
ALE
HLDA
HOLD
INTR
INTA
IO/M
RD
WR
NMI
LS373
D4
C4
B4
C6
LS367
C3
LS74
A4
CLR
PR
8286
B6
F4
8237-2
A6
ADSTB
AEN
DREQ 1
DREQ 2
DREQ 3
CS
DREQØ
DACKØ
EOP
IOW
MW
IOR
MR
HRQ
E6
10K (2 PLCS)
10K (3 PLCS)
1K
LS257
D3
SEL
OE
10K (3 PLCS)
LS367 (4 PLCS)
E4
100μH
+5VA
4.7 uF
0.1 uF
22uF (2 PLCS)
GND

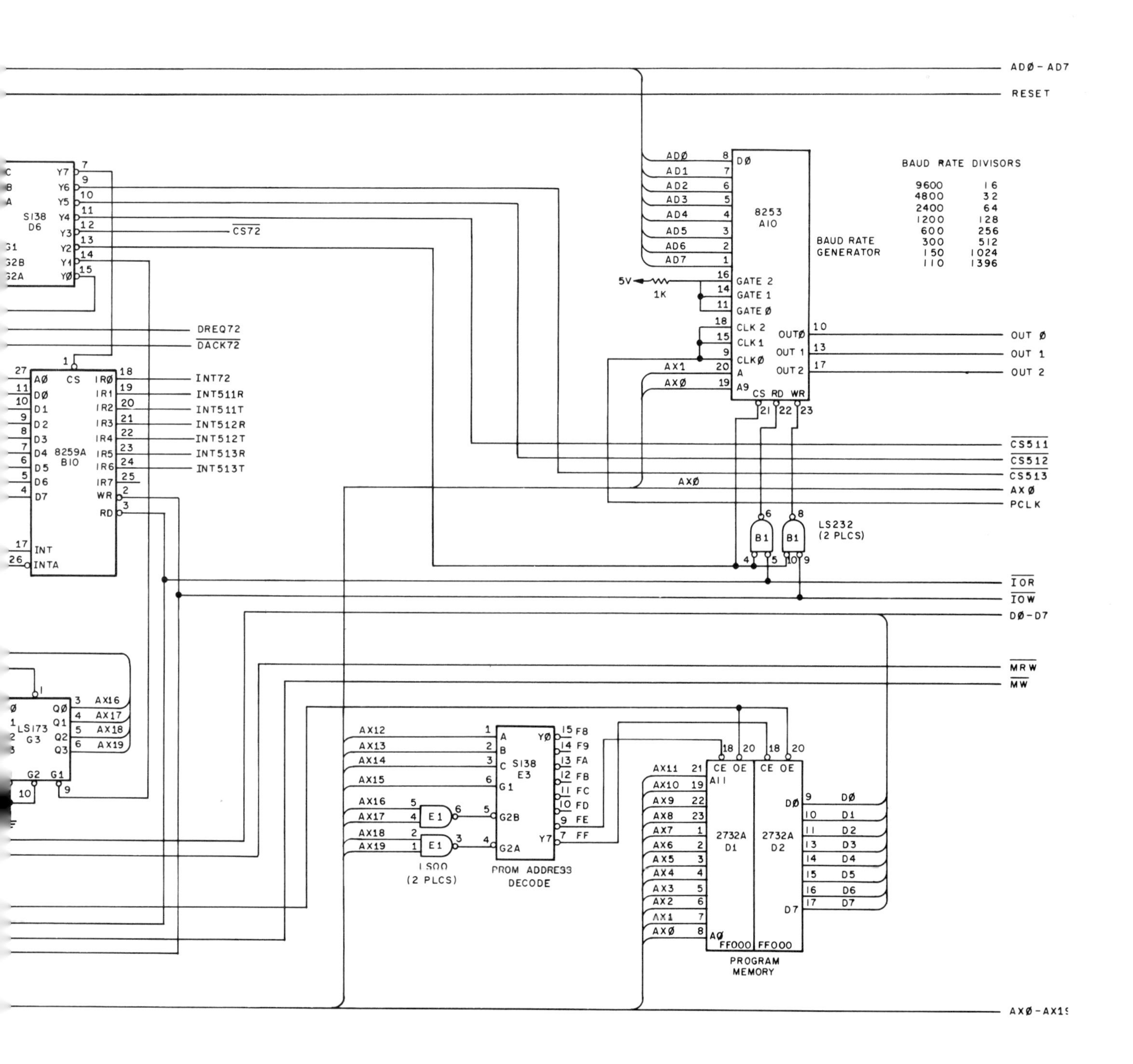

ADØ - AD7
RESET
S138
D6
8253
A10
BAUD RATE GENERATOR
BAUD RATE DIVISORS
9600 16
4800 32
2400 64
1200 128
600 256
300 512
150 1024
110 1396
CS72
DREQ72
DACK72
8259A
B10
INT72
INT511R
INT511T
INT512R
INT512T
INT513R
INT513T
OUT Ø
OUT 1
OUT 2
CS511
CS512
CS513
AXØ
PCLK
LS232
(2 PLCS)
IOR
IOW
DØ-D7
MRW
MW
LS173
G3
AX16
AX17
AX18
AX19
S138
E3
LS00
(2 PLCS)
PROM ADDRESS DECODE
2732A
D1
2732A
D2
FF000
PROGRAM MEMORY
AXØ-AX19

TMS 9909

Floppy Disk Controller

The TMS 9909, by Texas Instruments, is a Floppy Disk Controller designed to remove the complexity of floppy disk management and data transfer from the microprocessor. The controller is compatible with all practical recording formats and data encoding schemes, including the IBM 3740 and System 34 standards.

Requiring but a single 6-MHz crystal for all data rates, the TMS 9909 supports Direct Memory Access and creates a most efficient, flexible interface. In fact, this single chip will independently control up to four separate disk drives.

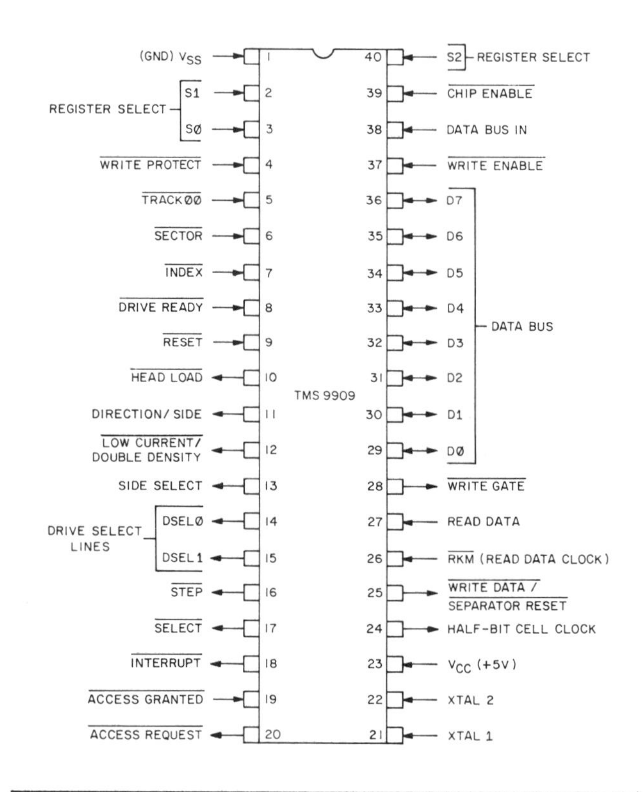

CPU Interface

The 9909 is designed as a memory-mapped CPU peripheral, with all communications between the CPU and 9909 carried out through eight memory-mapped registers. The registers are accessed by addressing the REGISTER SELECT (pins 3, 2, 40) inputs. These REGISTER SELECT lines are usually connected to the three LSBs of the host microprocessor so that the 9909 appears to consist of eight consecutive memory locations when used in conjunction with a decoded CHIP ENABLE (pin 39) input.

Communications is done through an 8-bit DATA BUS (pins 29-36) which, in the Texas Instrument tradition, are numbered in reverse order; D7 represents the LSB and D0 the MSB. The state of the DATA BUS IN (pin 38) line determines whether a command is being written into the chip or a status register is being read. To write from the CPU into the 9909, both DATA BUS IN and WRITE ENABLE (pin 37) inputs must be taken LOW; this is reversed for a read operation.

DMA Operations

This seemingly complicated bus control procedure is due to the fact that the 9909 has been designed primarily for DMA control, and these signals are required to properly sequence the TMS 9911 DMA controller chip normally used for DMS operations. During DMA access, the sense of the DATA BUS IN line is inverted with respect to the 9909 controller. DMA transfers are initiated when the ACCESS REQUEST output (pin 20) goes LOW and are performed when the ACCESS GRANTED input (pin 19) is enabled. Using this scheme, no action is required by the CPU until an error occurs or an operation is completed.

Disk Drive

The 9909 will control up to four disk drives. The binary-coded DRIVE SELECT LINES (pins 14, 15) set up the address for the diskette drive, with the select code valid when the SELECT output (pin 17) is LOW. The controller will wait up to 1.25 seconds after issuing the SELECT code to allow the disk drive to respond. If a DRIVE READY input (pin 8) is not received within this time, the drive is assumed inoperative, and operations to it are suspended. The chip will control both 8-inch and 5¼-inch drives. Each drive has its own set of parameters stored in the 9909 registers, and the two sizes may even be intermixed within the system.

The read/write head movement is also controlled by the 9909. When stepping the head, the DIRECTION output (pin 11) defines the direction of stepping. The STEP output (pin 16) generates one pulse per track. The pulse train has a nominal duty cycle of 50%, with the period equal to the applicable head step time. The head is brought into contact with the floppy disk when the HEAD LOAD (pin 10) command is executed.

When double-sided disks are used, the SIDE SELECT output (pin 13) dictates head selection. For those drives requiring a multiplexed direction and side control signal, both functions are supplied by the DIRECTION/SIDE (pin 11) pin.

Floppy Disk Signals

Both soft-sectored and hard-sectored floppy disks are acceptable to the 9909. Track location and positioning are controlled by the TRACK 00 (pin 5) and INDEX (pin 7) signals, respectively. For hard-sectored disks, the SECTOR input (pin 6) identifies sector locations.

Data is written onto a disk through the WRITE DATA (pin 25) output. The WRITE GATE output (pin 28) becomes active at least 1-us before data appears on the WRITE DATA output, and stays active until the last bit of data is transmitted. When writing on the inner tracks, the LOW WRITE CURRENT output (pin 12) reduces the recording current. Writing is prohibited if the WRITE PROTECT input (pin 4) is LOW.

Reading data from the disk is done through the READ DATA (pin 27) input. Although the chip contains an internal data separator, it requires an external read data clock ($\overline{RKM}$, pin 26) to extract the data pulses from the data stream. This clock, normally generated by a PLL, is synchronized to the raw data input using the HALF-BIT CELL CLOCK (pin 24) to initiate synchronization. When the chip is decoding MFM data, the DOUBLE DENSITY output (pin 12) is set HIGH.

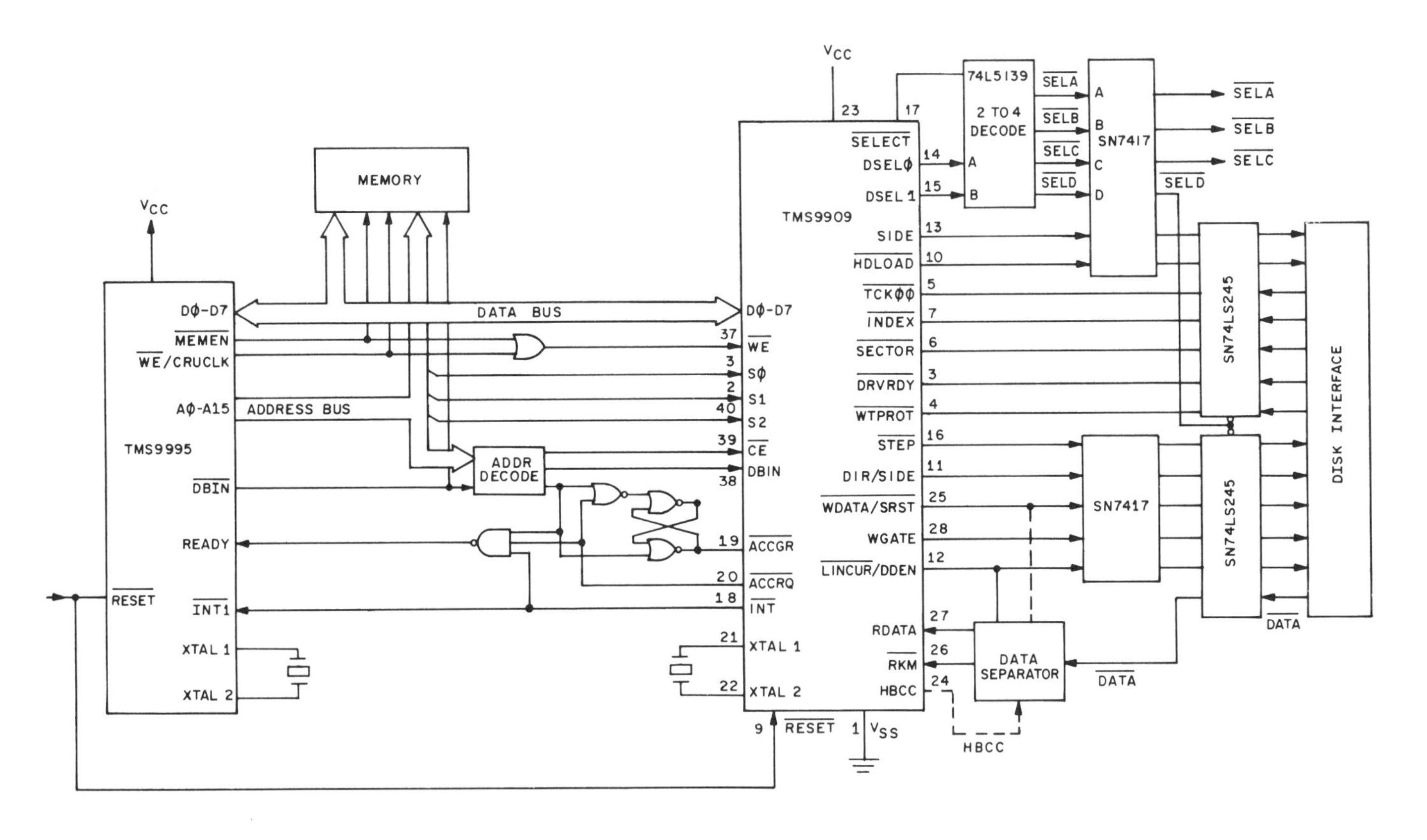

Figure 1

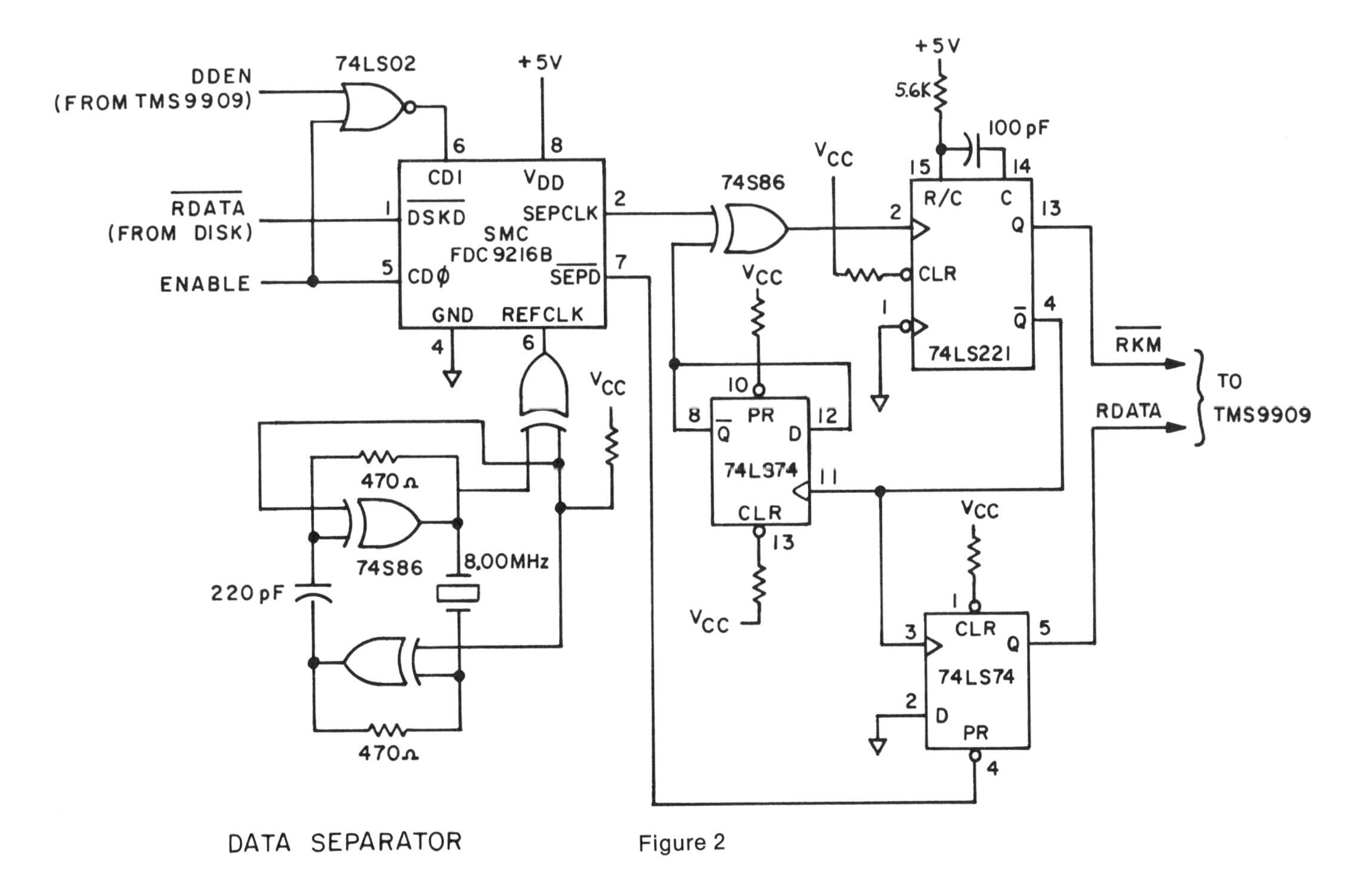

DATA SEPARATOR

Figure 2

FDC 9216

Floppy Disk Data Separator

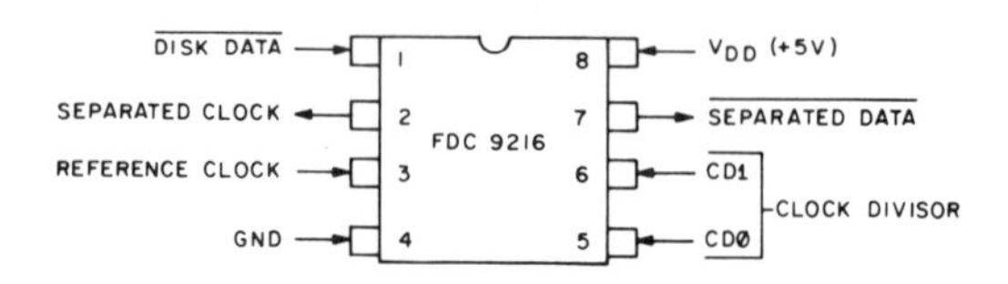

The FDC 9216, by Standard Microsystems Corp., is a Floppy Disk Data Separator supplied in an 8-pin Dual-In-Line package designed to save board space. The FDC 9216 replaces several SSI and MSI devices normally used for data separation, and provides a low-cost solution to the problem of converting a single stream of pulses from the floppy disk drive into separate clock and data inputs for a floppy disk controller.

The FDC 9216 is available in two versions; the FDC 9216, which is intended for 5¼-inch disks, and the FDC 9216B for 5¼-inch and 8-inch disks.

Operation

Data written on a floppy disk memory storage medium is typically a self-clocking encoded serial data stream which contains a mixture of clock pulses and digital information. Basically, the clock pulses form a bit cell that essentially frames the data bit. The presence or absence of a digital pulse between the clocking pulses determines the value ("0" or "1") of the data bit.

The function of a data separator is to extract the clock signal from the combined clock/data waveform as it comes from the floppy disk, and present it to the floppy disk controller chip for processing. Once the controller has a qualified clock reference signal, it can easily derive the original digital information from the raw data stream.

There are two methods presently used for performing this chore. One method is to synchronize the data stream to an analog phase-locked loop (PLL) oscillator, and use the oscillator's output to drive the data clock input. Another scheme is to analyze the raw data pulses digitally, logically comparing them to the established data format, to excerpt the clock signals. The 9216 uses the latter method.

Data Separation

The raw data is removed from the disk, conditioned, and presented to the 9216 Data Separator at pin 1, the DISK DATA input. The position of a pulse in the data waveform is taken to be its leading edge. Therefore, an edge detection logic circuit channels the data to the separation logic, where the clock pulse is recovered. The SEPARATED CLOCK output (pin 2) is a reconstruction of the original clock signal used when the information was stored on the disk.

The clock signal and data bits are then recombined, just as they are when they come off the disk, and output through the SEPARATED DATA (pin 7) pin. The 9216 can separate either Single Density (FM) or Double Density (MFM) encoded data from any magnetic media.

Clock

A REFERENCE CLOCK (pin 3) signal between 2- and 8-MHz is used to drive the internal separator logic. The internal clock frequency, the one that operates the logic circuits, is nominally 16 times the SEPARATED CLOCK frequency. To obtain this frequency from the REFERENCE CLOCK input, an internal dividing circuit reduces the input frequency to the required value. The division ratio is selected by configuring the CLOCK DIVISOR inputs, pins 5 and 6. With both inputs grounded, the ratio is 1:1.

Timing Errors

Due to the very nature of mechanical disk drives, errors are introduced into the pulse stream. In fact, the variation in the rotational speed of many floppy disk drives is as much as two percent. If data is recorded on a drive running two percent slow and retrieved on a drive running two percent fast, the data waveform presented to the data separator will be four percent faster than nominal.

Magnetic effects on the disk, mainly peak shift, also cause pulse positions to vary from the ideal. As a result, many pulses occur early or late within their half-bit slots. It is, therefore, important that the derived clock delineate the half-bit slots as accurately as possible so that the position of the pulse may vary with the greatest margin and still be associated with the correct half-bit slot.

The 9216 corrects for these variations by adjusting the frequency of its internal clock. In fact, it's not unusual for the nominal 16 to 1 frequency ratio to shift to as little as 12 to 1 and as great as 22 to 1. All this is, of course, done automatically inside the chip with no intervention necessary by the user. Such an arrangement eliminates critical adjustments sometimes encountered when working with phase-locked loops.

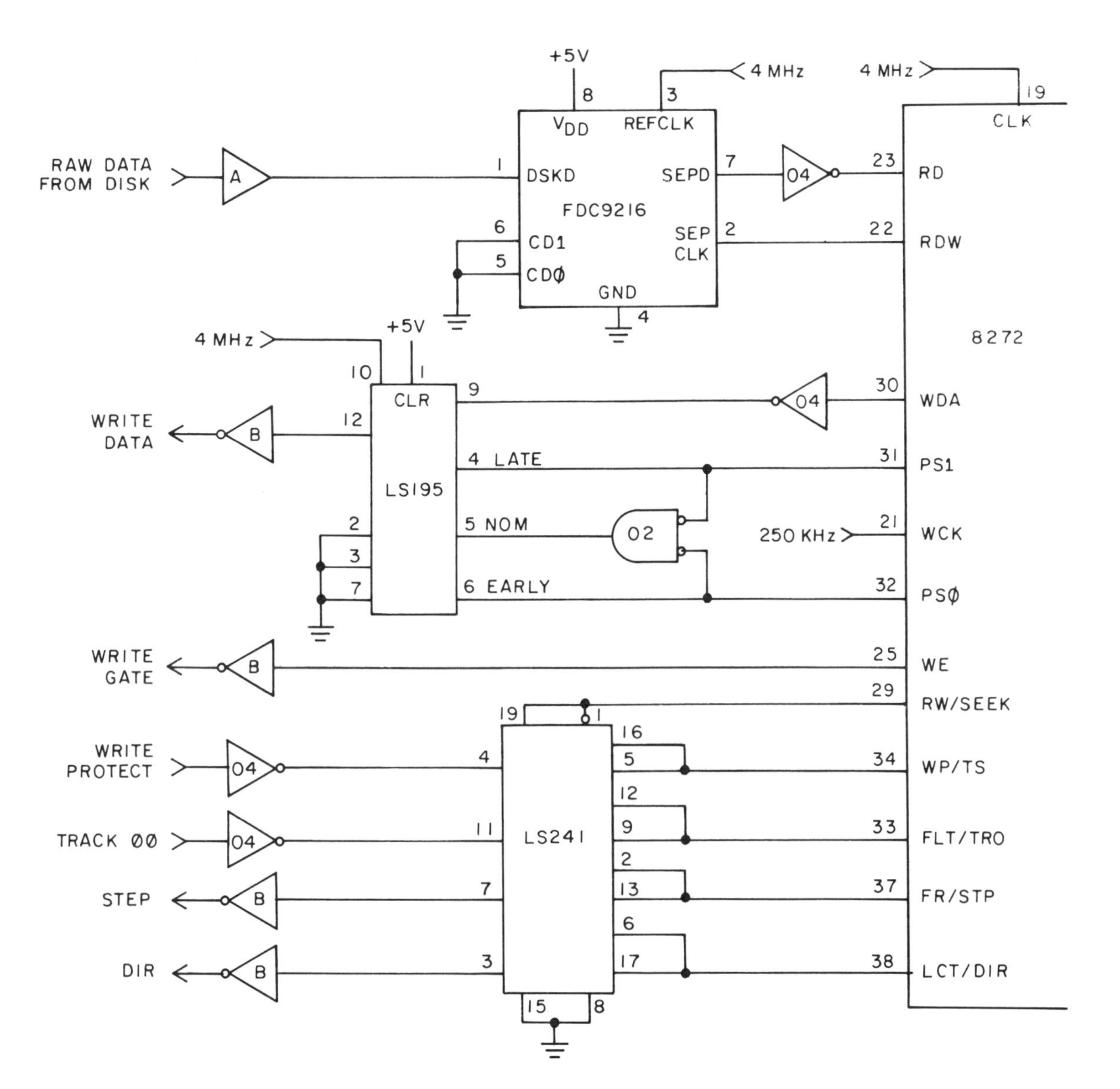

Figure 1

DIVISOR	CD1	CDØ	EXAMPLE
1	0	0	WHEN REFCLK = 8 MHz AND THE DATA SEPARATOR IS DECODING A 5 1/4-INCH DRIVE, USE THIS ← CODE.
2	0	0	
4	1	0	
8	1	1	
4	1	0	

Figure 2

WD1691

Floppy Support Logic

The WD1691, by Western Digital, is a Floppy Support Logic device that has been designed to minimize the external circuitry required to interface the 179X family of Floppy Disk Controllers to a disk drive. The chip includes write precompensation logic and playback data recovery circuits.

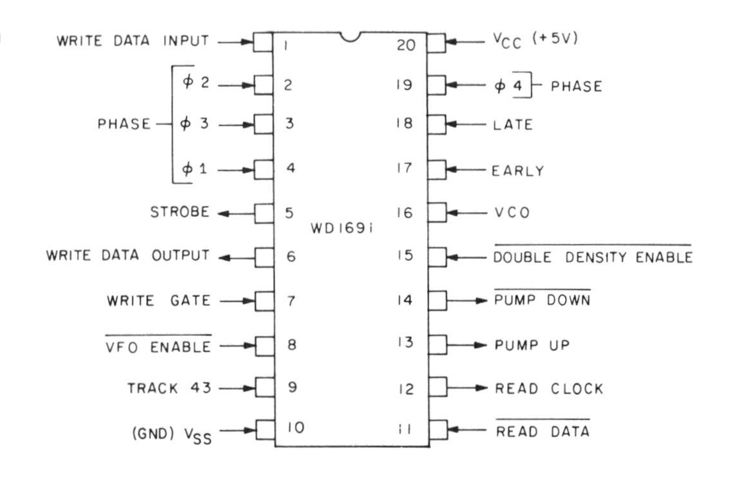

Data Recovery

The 1691 data separator uses a PLL arrangement for its method of data recovery. An external voltage-controlled oscillator (VCO), with a free-running frequency of 4-MHz, feeds its signal to pin 16 of the 1691, forming the input portion of the loop. The loop feedback signals consist of PUMP UP (pin 13) and PUMP DOWN (pin 14) outputs.

The two PUMP outputs are logically steered by the 1691 to track the relationship between the composite READ DATA (pin 11) signal from the disk drive and the VCO frequency. If the incoming frequency is higher than the VCO frequency, the PUMP UP output goes HIGH; when the input data rate is lower than the VCO, the PUMP DOWN output goes LOW. Should the VCO frequency equal the READ DATA rate, the PUMP outputs assume a high-impedence state.

By tying the PUMP UP and PUMP DOWN outputs together, and integrating them with a filter, an adjustment signal is created which forces the VCO control voltage higher for a frequency increase and lower for a decrease. Eventually, the PUMP signals will have corrected the VCO frequency to an exact multiple of the original writing clock signal, regardless of the timing errors present in all disk drives. The reconstructed clock pulses are directed to the READ CLOCK (pin 12) output and used by the floppy disk controller chip to extract the data bits from the raw data waveforms.

The data recovery system, which is enabled when the VFO ENABLE input (pin 8) is set LOW, has provisions for dealing with both FM and MFM encodings. When decoding single-density recordings (FM), the DOUBLE DENSITY ENABLE input (pin 15) activates internal logic which divides the VCO frequency in half to match the lower data rate.

Write Precompensation

Due to the mechanical nature of magnetic recording techniques, the data pulses placed on the memory disk may not always be exactly reproduced as they were entered. When two pulses are recorded in close proximity to each other, the read head—with its finite response time — has a tendency to integrate the high-frequency transitions when playing them back. The effect is a stretching of the time between the two transitions, known as peak shifting. This is especially prevalent in double-density recordings (or nearer the center of the disk), where the data bits are packed tightly together.

To overcome this problem, write precompensation is typically performed on the write data pulses to reduce the amount of peak shifting in the recovered data. Since peak shifts are data pattern sensitive, i.e. surrounding bits always create the same effect, their direction can be predicted. This pattern is used to write particular flux transitions (pulses) either earlier or later than normal, causing the recovered data to appear closer to its nominal bit position.

The 1691 write precompensation circuit has been designed to be used with the Western Digital 2143-01 clock generator. Four PHASE inputs, ø1, ø2, ø3, and ø4 (pins 4, 2, 3, and 19, respectively), interface to the clock generator to create the timing sequences necessary for the precompensation. A STROBE (pin 5) pulse keeps the two chips synchronized. When the 1691 is operated in the single-density mode only, write precompensation is not normally needed and the WD 2143-01 generator may be omitted. In this case, the PHASE inputs and STROBE output should be tied together.

A WRITE GATE (pin 7) signal forces the chip into the Write Mode and disables data recovery, regardless of the status on any other input. Precompensation is enabled when the DOUBLE DENSITY ENABLE is active LOW. In the Double Density Mode, the EARLY (pin 17) and LATE (pin 18) signals from the floppy disk controller select a PHASE input for pulse shifting. Precompensation may be specified from TRACK 43 on by tying pin 9 to the TG 43 output from the controller; or, as in most cases, all tracks are precompensated by simply leaving this input open.

The floppy disk write data is presented to the WRITE DATA INPUT (pin 1) and the precompensated write signal is delivered to the WRITE DATA OUTPUT (pin 6). When TRACK 43 is LOW or DOUBLE DENSITY ENABLE is HIGH, precompensation is disabled and any transitions occurring on the WRITE DATA INPUT pass through unaltered to the WRITE DATA OUTPUT.

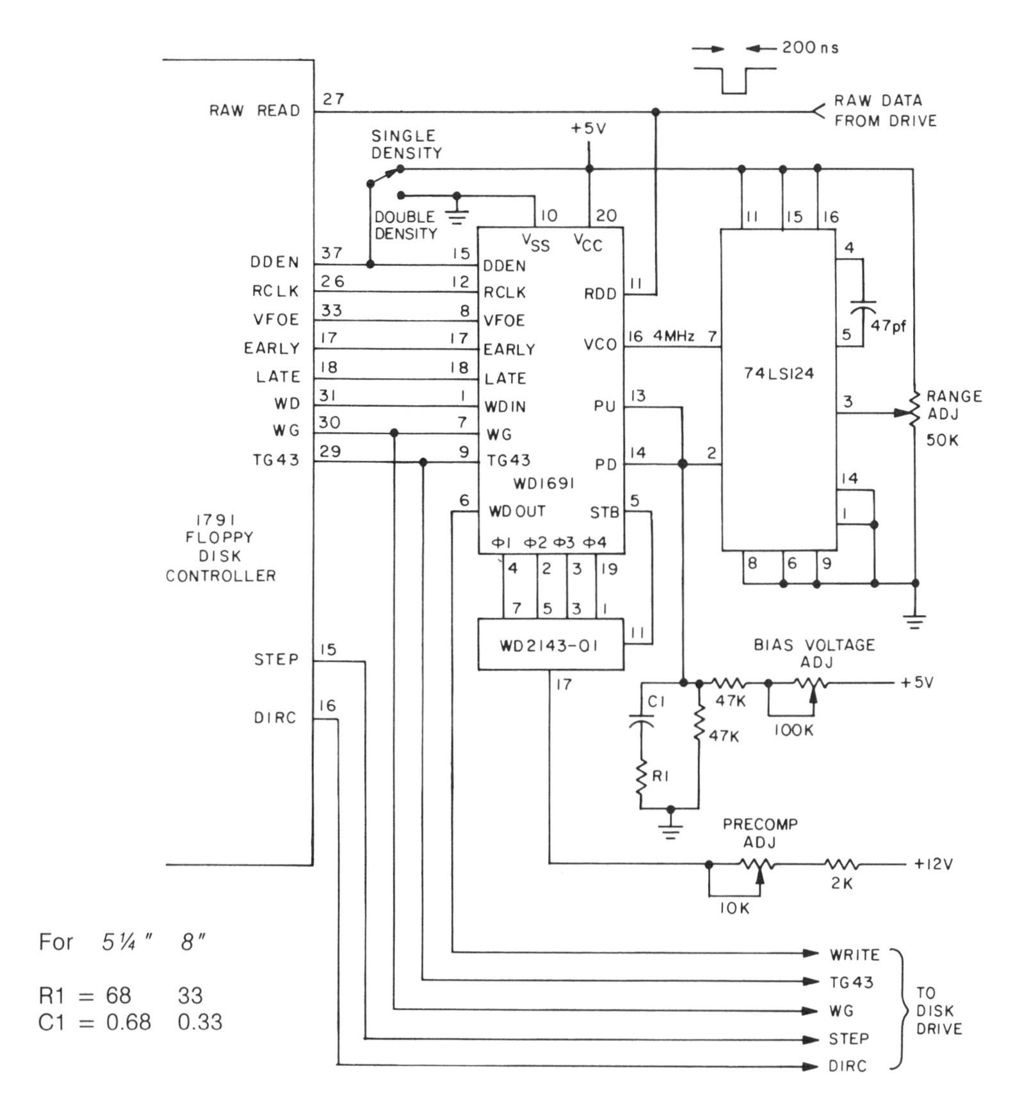

Figure 1

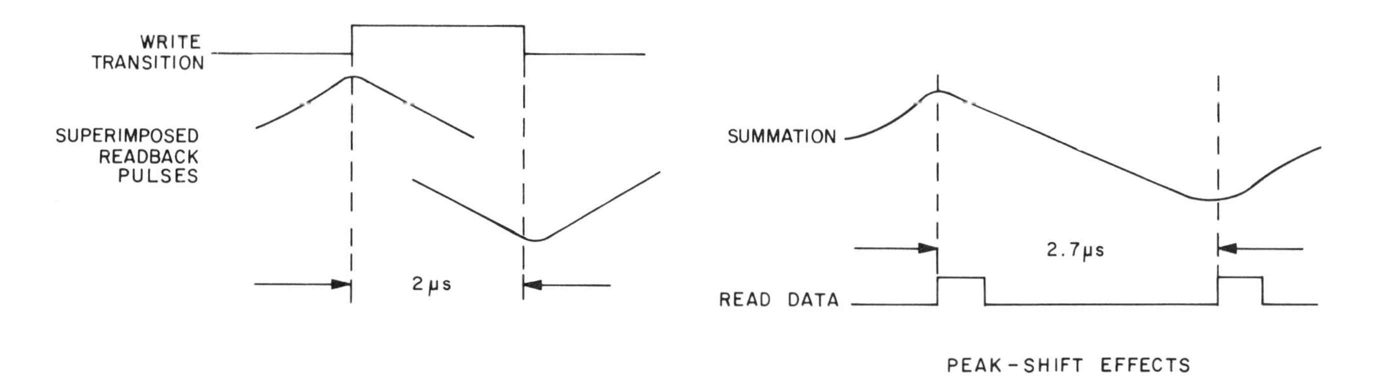

Figure 2

WD1100

Winchester Disk Controller Chip Set

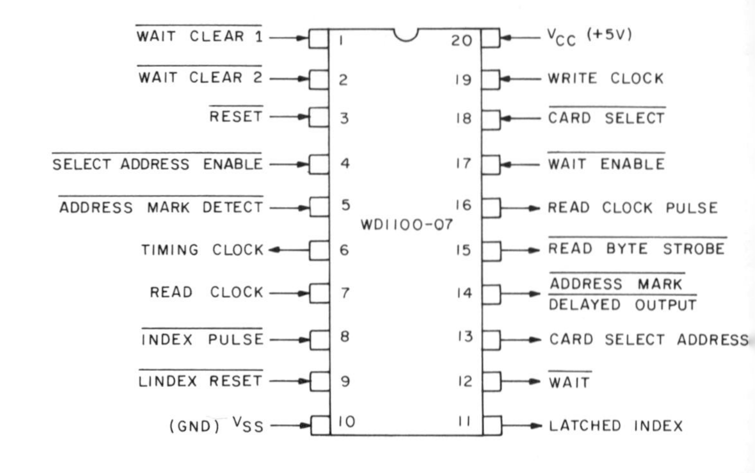

The WD1100 series of Winchester Controller Chips, by Western Digital, provides a low-cost alternative for developing a Winchester controller. These devices have been designed to read and convert MFM disk information into parallel data for CPU interfacing and vice versa for disk writing.

Altogether, seven chips will be discussed within this text. However, because their story is far reaching, it is spread over the next four pages. Accompanying the text is a complete controller schematic utilizing these chips. The controller circuit is the WD1001 Winchester Disk Controller developed by Western Digital; it will interface with most of the popular Winchester drives that use the Shugart Associates System Interface or Seagate Technology's ST506/406 "standard" interface.

Winchester Drives

The Winchester drive is a mass memory storage system which stores data on a disk coated with magnetic material. In many ways, the Winchester drive is similar to the floppy disk drives already discussed. It differs, though, in the disk itself and the drive mechanism.

The disk is a rigid platter, quite unlike that of the floppy disk, that rotates at a very fast 3600 RPM. By making the disk rigid and spinning it at high speeds, it is possible to store considerably more data per disk than is possible with any floppy disk in production today.

However, the Winchester owes its large storage capacity to a number of factors. First, the magnetic coating is extremely thin. The thinner the coating, the more data that can be stored on it. In addition, the magnetic head that reads and writes the pulses to and from the disk actually flies over the platter on a cushion of air. Due to the enormous speeds at which the disk spins, a "ground effect" is created between the head and the disk surface — much the same as an airplane experiences when flying low to the ground.

The flying height is directly related to the bit density of the disk. The lower the height, the greater the density. A typical flying level for a Winchester is about 1.5-*microns*—a distance considerably smaller than the diameter of a dust speck or smoke particle. Consequently, Winchester disks are sealed in an enclosed environment, where the elements can be controlled.

Positioning the head over the desired track is done in much the same way as with floppy disk controllers, using stepper motors. The disks range in size from 14-inch to 5¼-inch; 3-inch and 3¼-inch are currently being discussed.

An interesting, but little-known, fact about the Winchester disk is the origin of its name. In its infancy, a small, rigid disk drive was developed by IBM with one fixed and one removable disk pack. Each held about 30-megabytes of data, so the model was dubbed the "3030." As legend has it, this name soon evolved into the expression "Winchester," presumably borrowed from the 30-calibre Winchester rifle. IBM has long since abandoned the model "3030," but the nickname lingers.

Functional Description

In order to provide maximum data recording density and storage efficiency, data is recorded on the disk using a Modified Frequency Modulation (MFM) technique. This technique (see FD1791) requires clock bits to be recorded only when two successive data bits are missing in the serial data stream. This reduces the total number of bits required to record a given amount of information on the disk, resulting in an effective doubling of the amount of data capacity. Hence, MFM is often referred to as double-density recording.

Because clock bits are not recorded with every data bit cell, circuitry is required that can remain in sync with the data during the absence of the clock pulses, while generating the missing clock pulses at the same time.

Host Interface Logic

The WD1100-07 Host Interface Logic chip simplifies the design of a Winchester Hard Disk Controller using the WD1100 chip series by performing CPU interfacing logic functions that would otherwise require considerable discrete logic. Additionally, there are signals provided for ECC (Error Correction Code) implementation.

The chip is basically divided into four sections. The control logic provides logic signals which select the function of interest and interfaces it to the CPU. A timing clock generates the reference signals necessary for interfacing the chip set to the SA1000 type disk drive, while INDEX PULSE (pin 8) circuitry synchronizes the controller to the spinning disk. The last function supplies control logic for the various data processing signals.

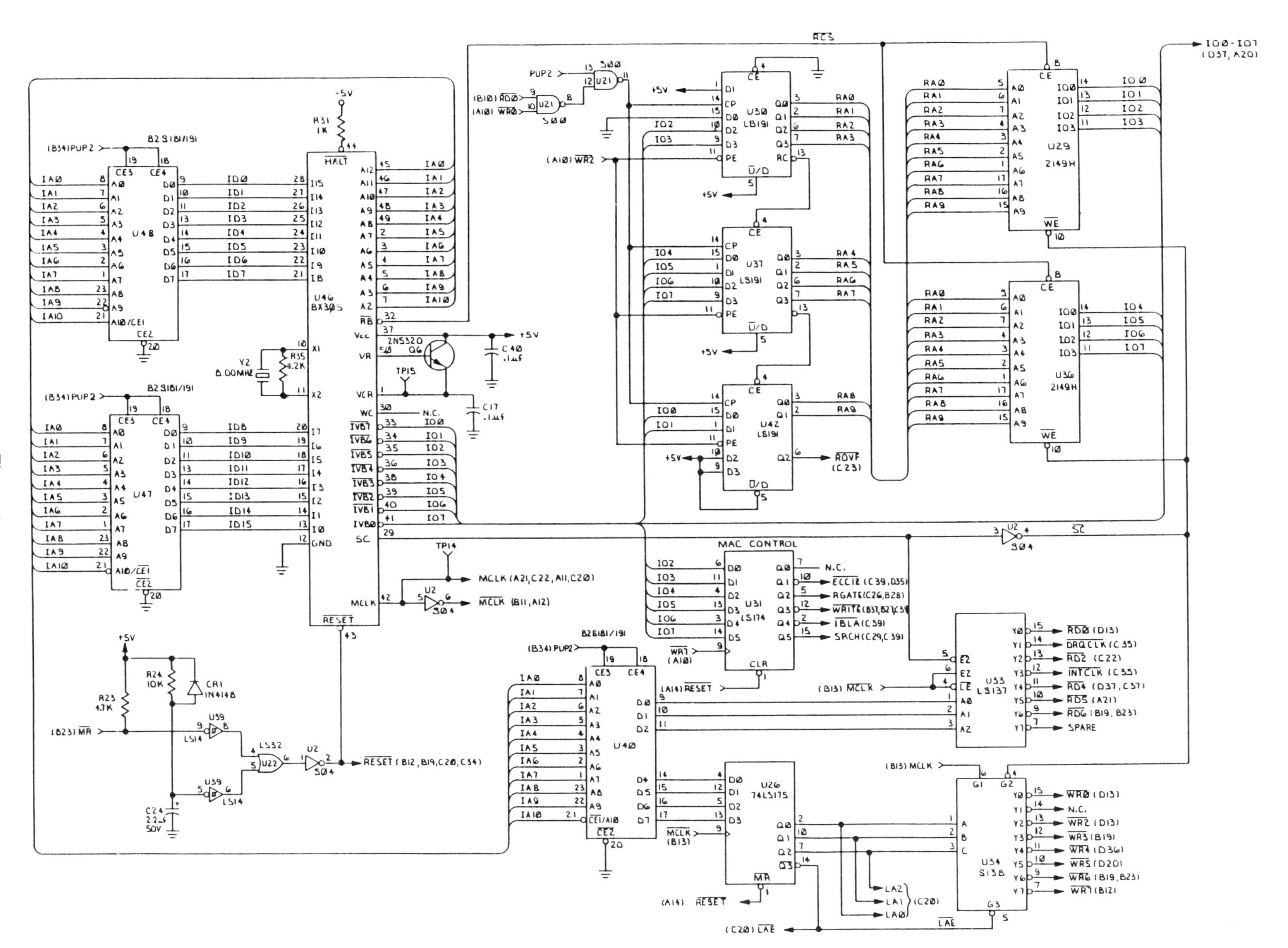

A complete WD1001 Winchester Controller schematic is presented on pages 135, 137, 139, and 141.

Figure 1

WD1100

Winchester Disk Controller Chip Set

The WD1100 series of Winchester Controller Chips, by Western Digital, provides a low-cost alternative for developing Winchester controllers. This is the second of four installments describing the operation of the WD1100 family.

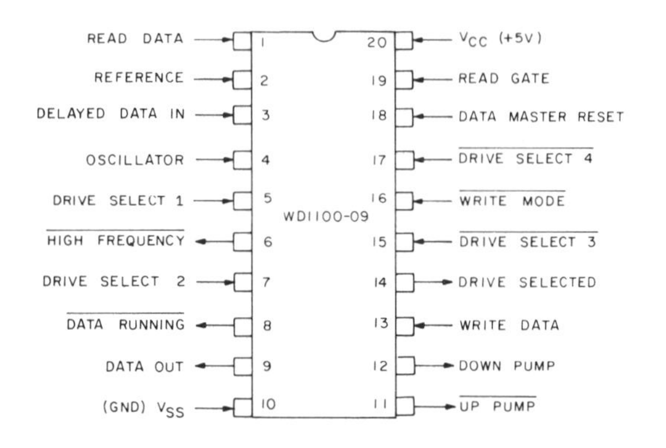

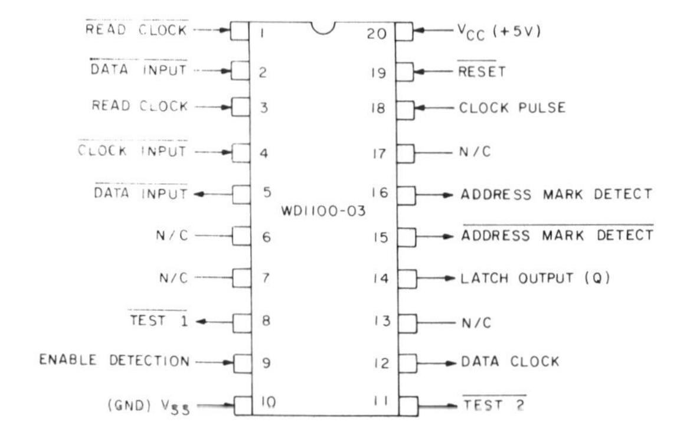

Data Separator Support Logic

The WD1100-09 Data Separate Support Logic chip is specifically designed to process incoming data from the Winchester drive using a procedure called data separation. The circuit employs an external voltage-controlled oscillator (VCO) to distinguish the data pulses from the clock.

The VCO frequency is determined by a voltage that is generated by internal logic. The VCO output is input to the OSCILLATOR (pin 4) input. The raw data from the disk is input to the READ DATA (pin 1) input; the data stream is also passed through an external delay line and fed back into the chip at the DELAYED DATA IN (pin 3). These two signals are compared to the VCO clock, and according to the signal arrival times, either the UP PUMP (pin 11) or DOWN PUMP (pin 12) output is activated.

These two PUMP outputs are combined, filtered, and presented to an error amplifier. The error amplifier drives the VCO, correcting for frequency drift errors between the raw data and the VCO clock.

To prevent the external VCO from locking onto a harmonic of its operating frequency, a REFERENCE (pin 2) pin is input with a clock signal that is twice the data rate. This signal is output to the VCO through the DATA OUT (pin 9) port whenever the separator is not actually reading data. Care is taken to switch in READ DATA to the VCO error detector only when it is known that the data stream frequency is in phase with the receiver clock. This can occur only when the data input is a solid stream of all ones or all zeros.

The switching function is initiated immediately after external logic detects the pattern and drives the READ GATE (pin 19) true. After sixteen consecutive pulses have been detected at the READ DATA, the HIGH FREQUENCY (pin 6) and DATA RUNNING (pin 8) outputs are asserted. At this point, the REFERENCE signal is switched out and the READ DATA stream is switched into the DATA OUT line, locking the VCO onto the raw data timing.

The WD1100-09 is also used for writing data onto the disk. When the WRITE MODE input (pin 16) goes active, the MFM-encoded data presented to the WRITE DATA (pin 13) input appears at the DATA OUT. During a write, the DELAYED DATA IN must be locked to a crystal-controlled oscillator clock that holds the VCO on frequency.

AM Detector

The WD1100-03 Address Mark Detector provides an efficient means of identifying clock pulses in an MFM (NRZ) data stream. Due to the very nature of MFM data encoding, it is impossible to know exactly which MFM bits are data and which are clock. This problem is solved by a uniquely recorded data/clock pattern called an Address Mark (AM).

The Address Mark consists of a data pattern of hexidecimal 'A1' with a missing clock pattern of hex '0A.' This pattern is used to identify the start of an information field within a sector. Preceding each AM on the disk is a long run of zeros; zeros have a clock bit for every bit cell.

Before the chip can begin sceening for an AM pattern, though, the internal logic must be reset by forcing the ENABLE DETECTION input (pin 9) LOW. This clears the registers and inhibits input or output of the chip.

When the ENABLE DETECTION is set HIGH, the search begins. NRZ data from the disk is entered into the DATA INPUT (pin 2), and clocked into an 8-bit synchronous serial shift register on the falling edge of the READ CLOCK – (pin 1). NRZ clock pulses, derived from the external Data Separator, are input to the CLOCK INPUT (pin 4), while the falling edge of READ CLOCK + (pin 3) clocks them into another 8-bit shift register.

As each bit is shifted through the registers, a 16-bit comparator constantly checks the contents for the AM pattern; AM detection occurs when both register detectors are true, thus enabling the ADDRESS MARK DETECT outputs (pins 15, 16). These two outputs are complementary latches that remain in effect until the device is reinitialized. At the instant AM occurs, the exact relationship between data and clock is known.

As soon as an AM is detected, the DATA CLOCK output (pin 12) begins to toggle. The MFM data is clocked out the end of the data register as it ripples through, and presented to the DATA OUTPUT (pin 5) line eight bits after it enters the chip.

An uncommitted edge-triggered flip-flop has been provided to facilitate the detection of HIGH FREQUENCY by the data separator, but may be used for other purposes. These pins are located at CLOCK PULSE (pin 18), LATCH OUTPUT (Q) (pin 14), and RESET (pin 19).

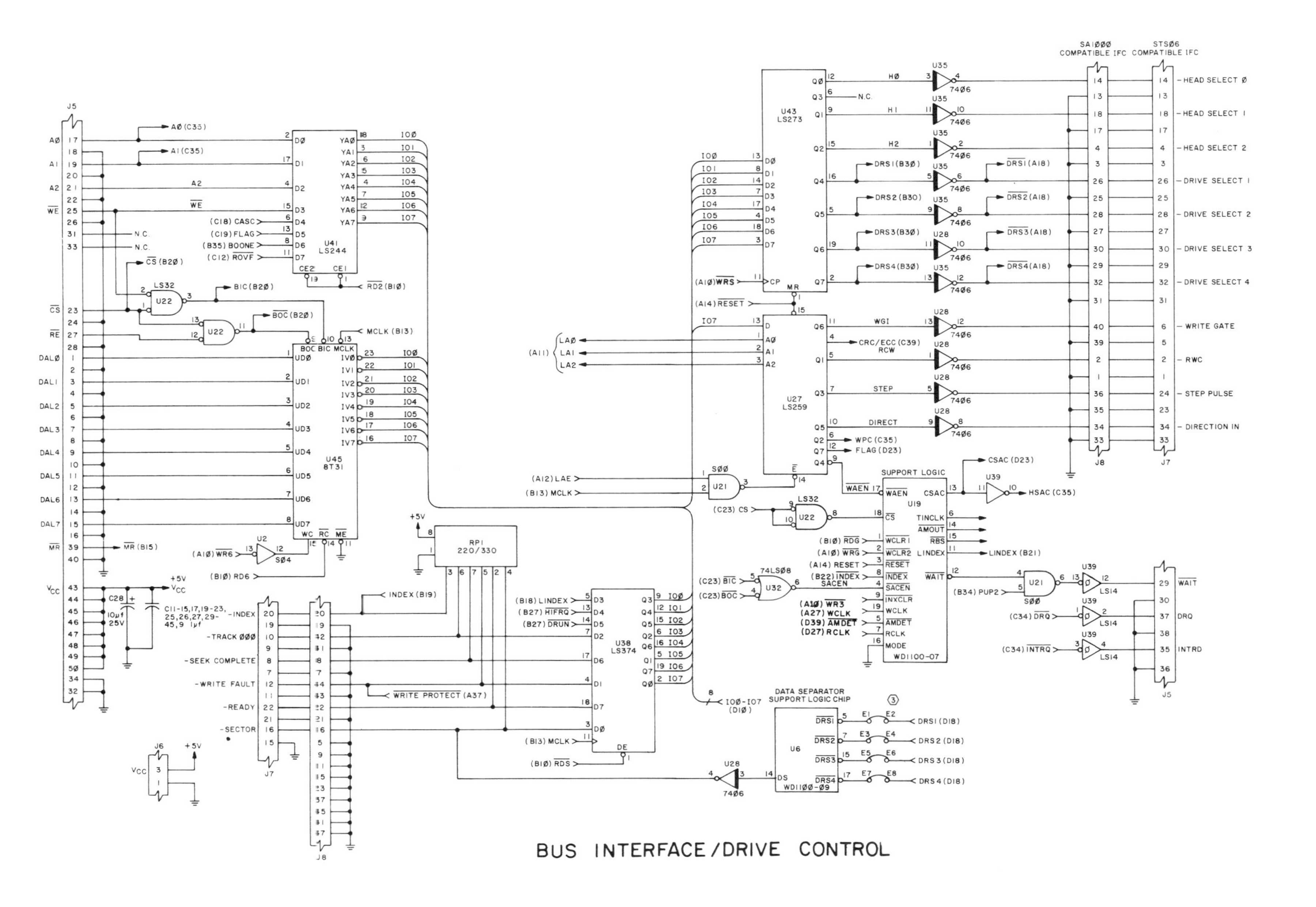

BUS INTERFACE/DRIVE CONTROL

A complete WD1001 Winchester Controller schematic is presented on pages 135, 137, 139, and 141.

WD1100

Winchester Disk Controller Chip Set

The WD1100 series of Winchester Controller Chips, by Western Digital, provides a low-cost alternative for developing Winchester controllers. This is the third of four installments describing the operation of the WD1100 family.

ECC/CRC Logic

The WD1100-06 ECC/CRC Logic chip gives the user of the WD1100 series easy ECC or CRC implementation. With proper software, it will provide single burst correction up to 8-bits and double burst detection.

Data recorded on magnetic media is prone to several types of errors which could render the data unusable if some form of error detection were not employed. The WD1100-06 performs both Cyclic Redundancy Check (CRC) error detection *and* Error Correction Code (ECC) data correction using a sophisticated polynomial. The user has the option of selecting either one of these procedures by properly setting the SELECT (pin 6) input.

To ensure proper operation of the device, it must first be reset to a predetermined state by strobing the ECC INITIALIZE (pin 7) input. The DATA/CHECK SYNDROME SELECT input (pin 17) is then raised HIGH and the input lines set for operation.

The chip has two sets of data inputs: One for read signals and one for write. Selection of the Read or Write mode is determined by the status of the READ/WRITE (pin 1) input. In the Read Mode, data is serially clocked through the READ DATA input (pin 4) by the READ CLOCK PULSE (pin 2). Write data, on the other hand, is input to WRITE DATA (pin 5) by the WRITE CLOCK PULSE (pin 3). The status of pin 17 can be latched in the appropriate mode by the READ BYTE (pin 18) and WRITE BYTE (pin 19) inputs, respectively.

Since most disk drives use an Address Mark, the WD1100-06 uses this feature to start off the ECC/CRC calculation. The first active LOW-going edge of the input data releases the internal Set flip-flop and drives the ECC ENABLE output (pin 9) LOW, indicating the internal circuitry is ready to begin the computation. Immediately following the Address Mark, data is supplied in serial fashion. This data passes through the chip and is output at the DATA OUTPUT (pin 15) port.

Sometime after the next to last byte of data, but before the last byte, the DATA/CHECK SYNDROME SELECT is taken LOW. This signal can be derived from those external devices marking the byte boundaries. The WD1100-06 now outputs the calculated polynomial remainder. A one-bit delay through a D flip-flop has been added to the DATA OUTPUT to prevent glitches during the transition from data to ECC/CRC. An unbuffered output is also available at the EARLY DATA OUTPUT (pin 16) port.

During the Write operation, the input data stream is divided by the polynomial and the 32-bit remainder used as the four check-syndrome bytes. If the syndrome is zero, no errors occurred. Otherwise, the non-zero syndrome is used by a software algorithm to compute the displacement and the error vector within the bad sector.

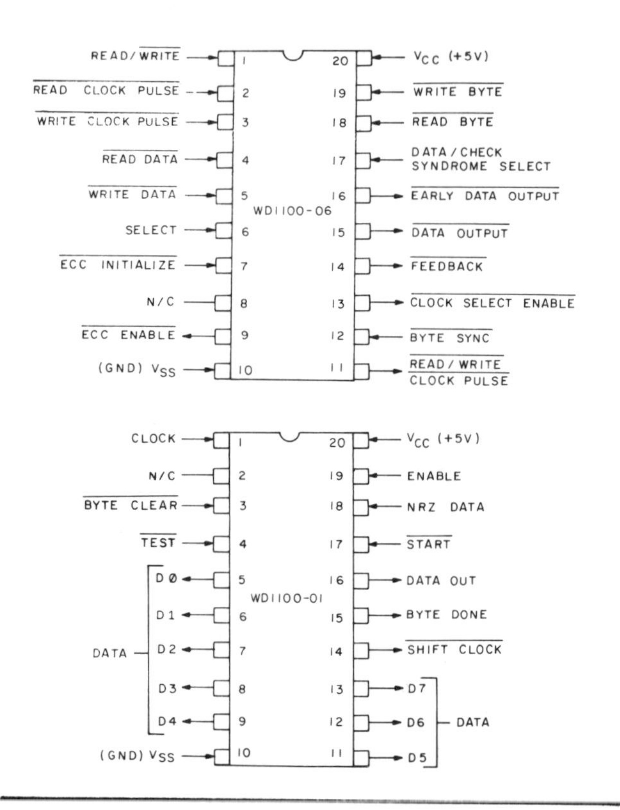

Serial/Parallel Converter

The WD1100-01 Serial/Parallel Converter allows the user to convert NRZ data from a Winchester disk drive into 8-bit parallel form. Additional inputs are provided to signal the start of the parallel process, as well as Byte Strobes to signify the end of the conversion.

After data has been processed by the WD1100-06 ECC/CRC device, the WD1100-01 takes over. Prior to shifting data through the device, the converter must synchronize to the data stream. The START line (pin 17) is raised HIGH to hold the internal bit counter in a cleared state until valid NRZ DATA (pin 18) and CLOCK (pin 1) pulses are entered. After one full cycle of the CLOCK occurs, the device is ready to perform conversions.

Data is entered on the NRZ DATA line and clocked into an 8-bit shift register. After eight data bits have been entered, the final HIGH-to-LOWtransition of the CLOCK sets an internal latch tied to the BYTE DONE (pin 15) output, signifying 8-bits of data have been assembled. At the same time, the contents of the shift register are parallel loaded into an 8-bit output register, making the parallel data available on the D0 to D7 (pins 5-9, 11-13) DATA outputs.

When interfacing to a microprocessor, BYTE DONE is used to indicate a parallel byte is ready to read. While the CPU reads the DATA lines, the BYTE CLEAR (pin 3) input should be strobed to reset the BYTE DONE line in anticipation of a next assembled byte. The BYTE CLEAR is a level triggered input that must be set to a logic HIGH before the next 8-bits can be shifted into the register. An address decoded signal generated by the host CPU may be used for this purpose.

The serial data bits are shifted out the last stage of the shift register into the DATA OUT (pin 16) output. An inverted copy of the CLOCK is available at the SHIFT CLOCK (pinl 4) output. Together, these signals can be used to drive another shift register external to the device. The ENABLE input (pin 19) sets these lines, and the BYTE DONE output, in a high-impedence state when it is LOW.

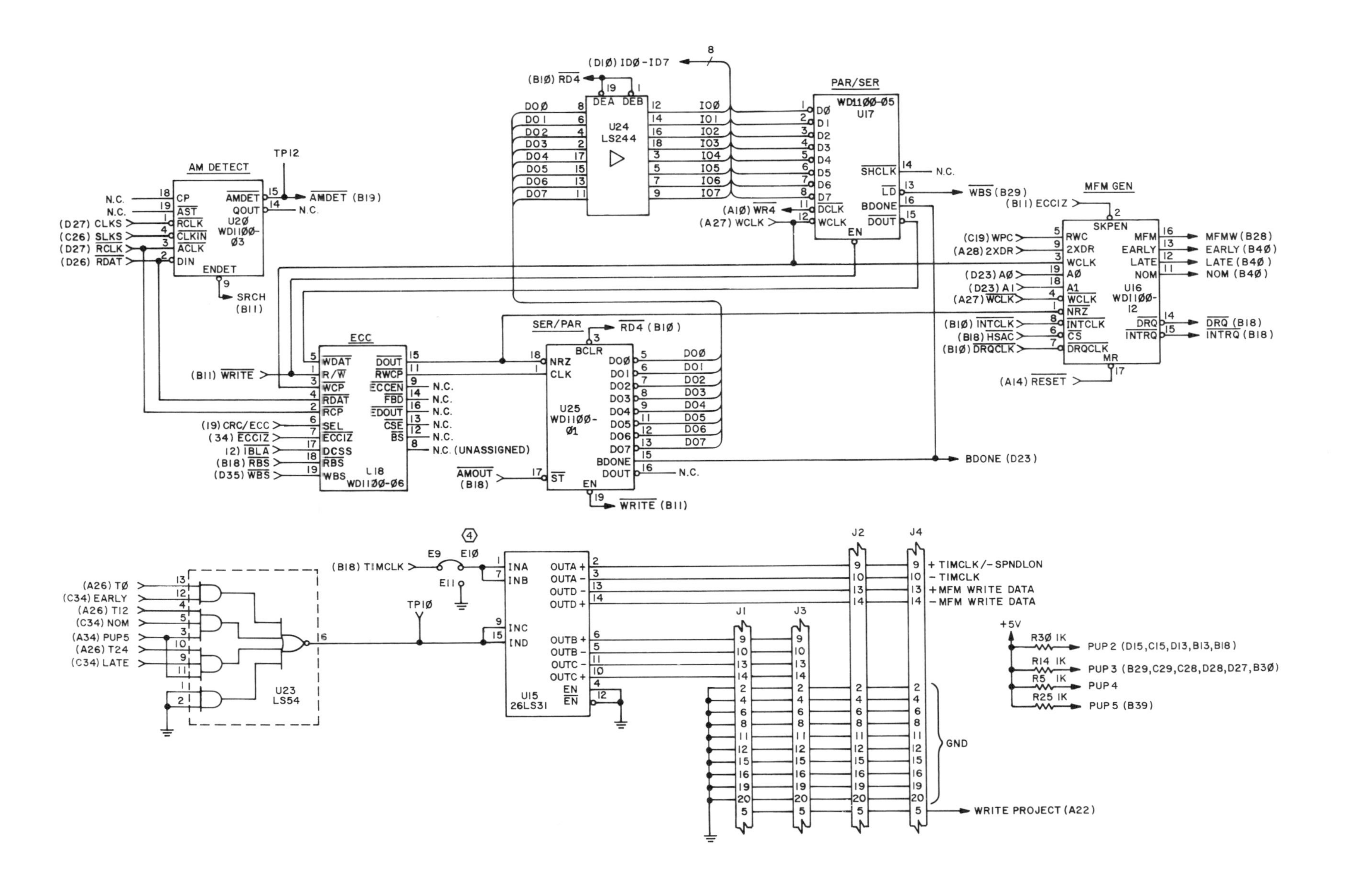

A comp ete WD1001 Winchester Controller schematic is presented on pages 135, 137, 139, and 141.

WD1100

Winchester Disk Controller Chip Set

The WD1100 series of Winchester Controller Chips, by Western Digital, provides a low-cost alternative for developing Winchester controllers. This is the last of four installments describing the operation of the WD1100 family.

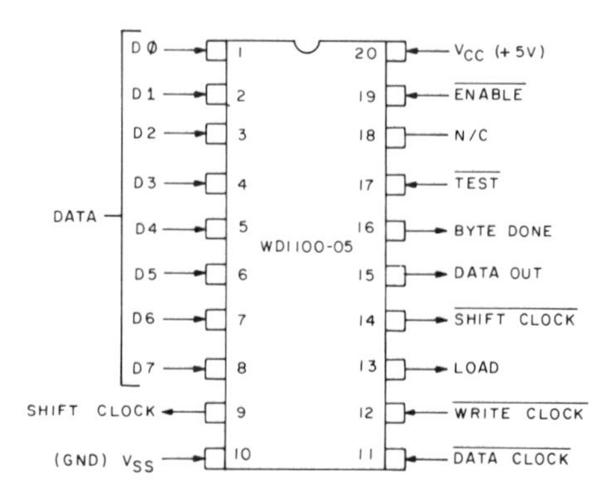

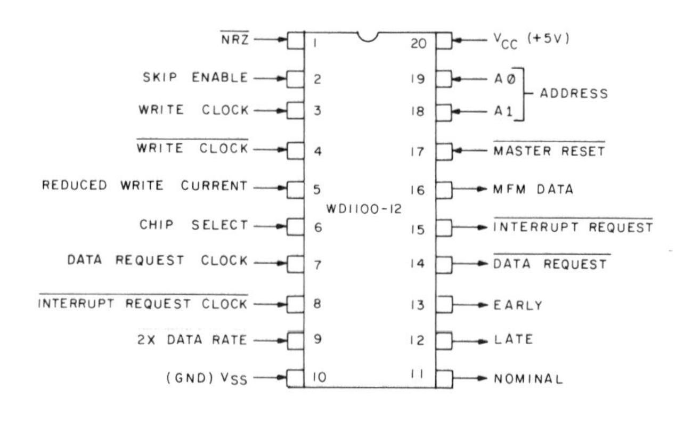

Parallel/Serial Converter

The WD1100-05 Parallel/Serial Converter allows the user to convert an 8-bit data byte into a serial data stream for writing to a disk or other serial device.

Prior to loading the chip, it is recommended that 00H or FF first be loaded into the input buffers to initialize the chip. The ENABLE (pin 19) line is then set LOW to enable the device outputs.

Parallel data is entered on the DATA (pins 1-8) input lines, where it is strobed into the data latches on the rising edge of the DATA CLOCK (pin 11). From here, the data is serially shifted out the DATA OUT (pin 15) output on the HIGH-to-LOW transition of the WRITE CLOCK (pin 12). The LOW-to-HIGH edge of the WRITE CLOCK increments a byte counter, which in turn sets the BYTE DONE output (pin 16) HIGH after 8-bits of data have been shifted out. The data buffer is now ready to be reloaded with the next byte.

Entry of the next byte automatically clears the BYTE DONE signal. BYTE DONE always needs servicing within eight WRITE CLOCK cycles — unless the next byte to be transmitted is the same as the previous byte. In which case, the converter will loop around and retransmit the byte in the register. If the BYTE DONE request is serviced prior to every eighth WRITE CLOCK, the output data will represent a contiguous block of the bytes entered. Due to the asynchronous nature of the converter, however, the input data may appear in serial form at the DATA OUT anywhere from 8 to 16 WRITE CLOCK cycles later.

Improved MFM Generator

The WD1100-12 Improved MFM Generator converts NRZ data into an MFM (Modified Frequency Modulated) data stream. The derived MFM signal, which contains both clock and data pulses, can then be used to record information on a Winchester disk. In addition to an MFM output, the device generates first level write precompensation signals for use with inner track densities. A unique feature of the WD1100-12 is its ability to delete a clock pulse from the outgoing MFM data stream in order to record Address Marks.

The chip is divided into two sections: MFM Generator and Interrupt Logic. The MFM generator converts NRZ data into MFM data for disk writing, while providing write precompensation signals. The interrupt logic, however, is used specifically for the WD1001 Winchester Controller Board (schematic opposite, courtesy Western Digital), but may be used in similar designs to generate interrupt signals. The two sections of the device are isolated and have no input or output signals in common.

Prior to entering data for encoding, the SKIP ENABLE input (pin 2) must be set LOW to enable clock pulses into the data stream. Data is entered on the NRZ (pin 1) input during the HIGH-to-LOW transition of the WRITE CLOCK (pin 3). Normally encoded NRZ data then appears on the MFM DATA (pin 16) output, lagging by one clock cycle.

The SKIP ENABLE signal is used to record a unique data/clock pattern, such as an Address Mark, on the Winchester disk. When the SKIP ENABLE input is set HIGH, the internal skip logic is enabled. As long as zeros are being shifted into the NRZ data input, the device generates normal MFM data. On the receipt of the first non-zero bit, typically the MSB of the Nex A1 code, the skip logic begins counting the WRITE CLOCK cycles. When the MFM generator tries to produce a clock pulse between bits 2 and 3, the skip logic prohibits it. The result is a unique Mark Address. After the skip logic has performed, it disables itself and normal MFM encoding resumes.

Write precompensation is performed on the data stream by the EARLY (pin 13), LATE (pin 12), and NOMINAL (pin 11) outputs when the REDUCED WRITE CURRENT input (pin 12) is enabled. These outputs are logically steered and latched by the 2X DATA RATE (pin 9) clock.

The Interrupt Logic is used to clear the DATA REQUEST (pin 14) and INTERRUPT REQUEST (pin 15) outputs by enabling CHIP SELECT (pin 6) in combination with A0 and A1 (pins 19, 18). The MASTER RESET (pin 17) input clears both request lines simultaneously. DATA REQUEST AND INTERRUPT REQUEST can be set to logic "0" only by a LOW level of DATA REQUEST CLOCK (pin 7) or INTERRUPT REQUEST CLOCK (pin 8), respectively. The signals remain LOW until cleared as per above.

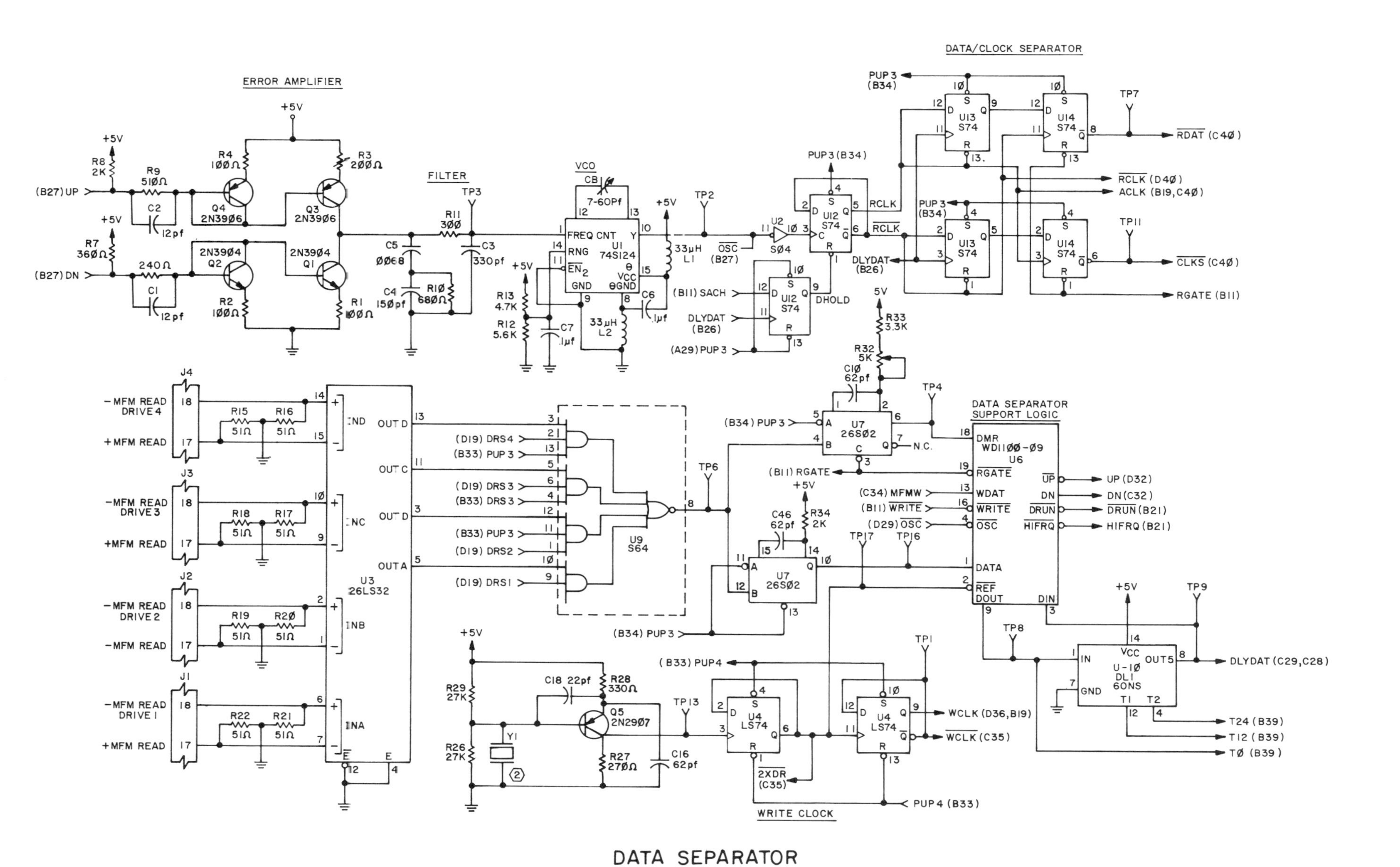

DATA SEPARATOR

A complete WD1001 Winchester Controller schematic is presented on pages 135, 137, 139, and 141.

DP8460

Data Separator

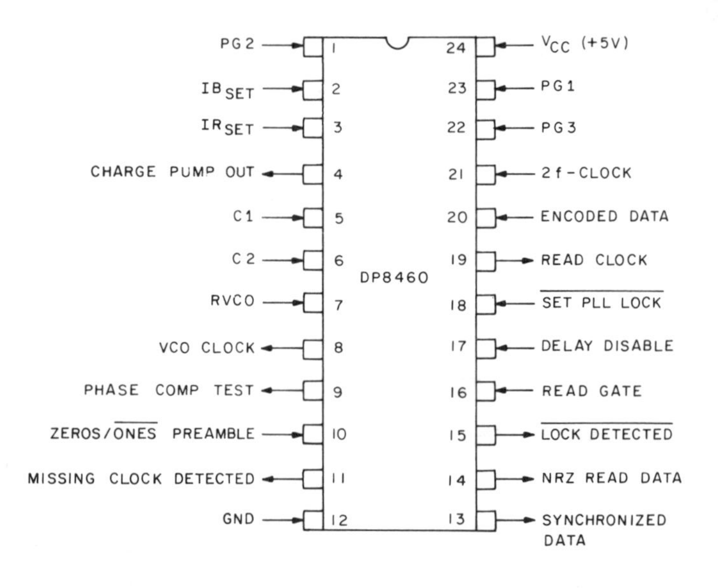

The DP8460 Data Separator, by National Semiconductor, is designed for application in disk driver memory systems. Primarily targeted for use with Winchester drives, the separator decodes MFM serial data into NRZ, and even provides a synchronized raw data output that allows external circuitry to decode other run-length-limited codes (such as FM or '2, 7').

The circuit has been designed with an on-chip Phase-Locked Loop for MFM decoding. The few external resistors and capacitors necessary to complete this function permits the chip to be included within the drive electronics.

Operation

The raw MFM data from the disk drive is delivered to the ENCODED DATA (pin 20) input. The chip is placed in the Read Mode by enabling the READ GATE (pin 16) input.

Once activated, the chip attempts to lock onto the Preamble data pattern recorded on the disk. The DELAY DISABLE input (pin 17) determines whether attempting lock-on will begin immediately following a READ GATE, or wait for two bytes before starting. This allows the user a selection of disk drive types for use with the chip.

In a typical hard-sectored drive, the READ GATE is set active as soon as the sector pulse is detected, indicating a new sector is about to pass under the head. Normally, the Preamble pattern does not begin immediately after this signal because gap bits from the preceding sector often overrun their sector. Therefore, it is advantageous to allow two bytes to pass before enabling the PLL to ensure it will begin locking onto the Preamble, and not start chasing non-symmetrical gap bits. This is achieved by setting the DELAY DISABLE input LOW.

For soft-sectored drives, however, the controller doesn't usually wait for the index pulse before attempting a lock-on, and the READ GATE is likely to go active at any time. Chances are the head will not be over the Preamble when it happens, and waiting the delay period is just a waste of time. Tying the DELAY DISABLE input HIGH engages the PLL within one clock cycle.

As soon as the bytes of the Preamble are recognized, the PLL is firmly locked into the incoming data stream and the LOCK DETECTED output (pin 15) drops LOW. Since there are two commonly acceptable ways to record a Preamble, strings of ones or zeros, the ZEROS/ONES PREAMBLE input (pin 10) allows the user to select between them. Any deviation from the all one or all zero Preamble pattern during lock acquisition resets and recycles the search. In many cases, an Address Mark is used to indicate the Preamble field. The detection of an Address Mark sends the MISSING CLOCK DETECTED output (pin 11) HIGH.

PLL Data Separator

The 8460 features as PLL data separator consisting of a pulse gate, phase comparator, charge pump, buffering amplifier, voltage-controlled oscillator (VCO), and decoder circuitry.

The pulse gate multiplexes the VCO signal with the data input and conditions the pulses, using an external filter, to reduce jitter. There are four external filter elements connected to the pulse gate, the values of which are dependent upon the data rate. Fig. 3 shows typical values. The signals are passed through the phase comparator and input to the charge pump.

The charge pump is a switchable current source that steers the control voltage of the VCO. The direction of the charging current is determined by the phase comparator, which in turn, is determined by the phase relationship between the VCO frequency and the incoming data rate.

The charge pump has two rates. During a lock-on attempt, the faster rate is used to quickly gain a lock on the signal. Resistors to IR_{SET} (pin 3) and IB_{SET} (pin 2) establishes the two charging rates. Bypass capacitors are required for each resistor; a 1000-pf value is suitable.

After the lock-on has been achieved, the charge rate is reduced by enabling the SET PLL LOCK (pin 18) input. This prevents the VCO from hunting, and possibly locking onto noise that may be picked up by the data stream. In most cases, the LOCK DETECTED output is connected back to the SET PLL LOCK input; as the PLL achieves lock-on, the charge pump automatically switches to the slower tracking rate.

The control voltage from the charge pump is filtered, at CHARGE PUMP OUT (pin 4), passed through a buffer amplifier, and input to the VCO. The VCO center frequency is set by a capacitor across C1 (pin 5) and C2 (pin 6) VCO center with a resistor from RVCO (pin 7) to ground. The recovered clock pulse is available at the VCO CLOCK (pin 8) output.

Using this clock, the incoming data stream is decoded and the resulting NRZ data output at the NRZ READ DATA (pin 14) pin. Alternatively, unprocessed data is passed through the chip and output at SYNCHRONIZED DATA (pin 13). The VCO generates a READ CLOCK (pin 19) that is in step with the SYNCHRONIZED DATA output, which can be used for external data processing.

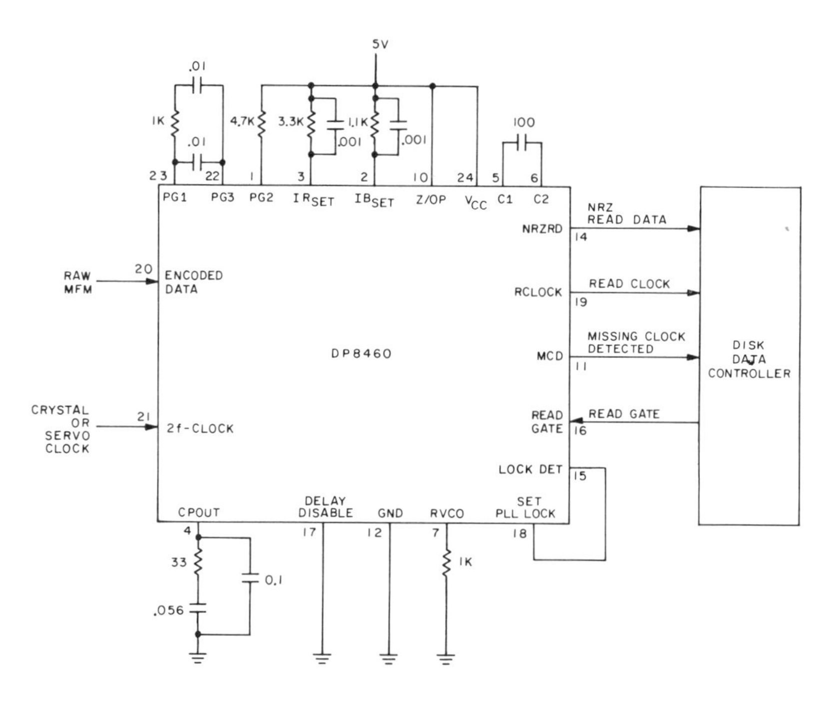

Figure 1

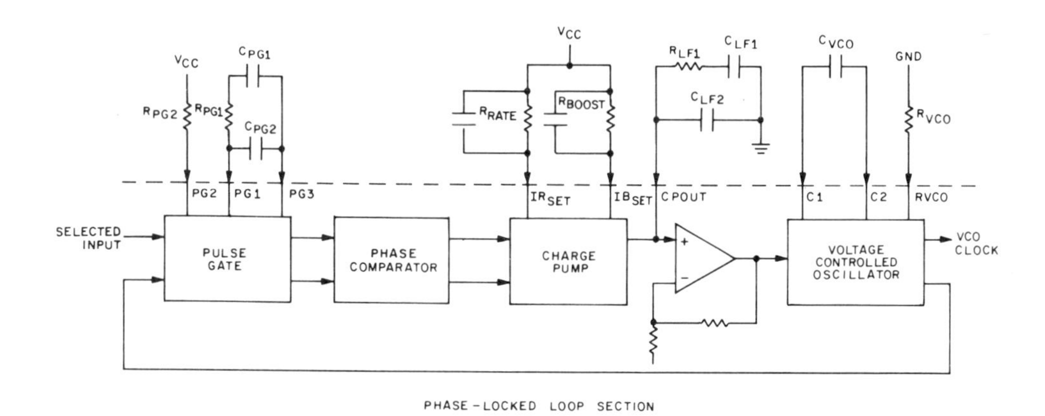

Figure 2

DATA RATE	R_{PG2}	R_{PG1}	C_{PG2}	C_{PG1}
2 M bit/sec	16K	1K	0.01μF	0.01μF
5 M bit/sec	4.7K			
10 M bit/sec	1.9K			
15 M bit/sec	750			
20 M bit/sec	300			
25 M bit/sec	0			

Figure 3

The 2f-CLOCK is a system clock that runs at twice the VCO frequency.

CHAPTER FIVE
VIDEO

CRT 5027

Video Timer and Controller

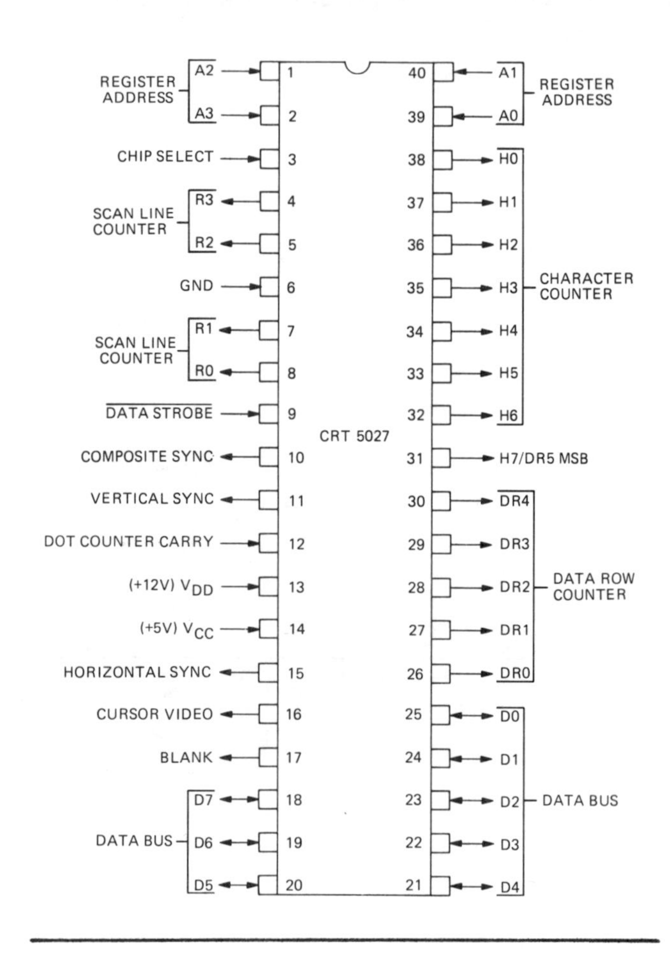

The CRT 5027, by Standard Microsystems Corporation, is a user programmable device containing the logic required to generate the timing signals for the presentation and formatting of video data on a CRT monitor. With the exception of the dot counter, all functions necessary for total frame formatting are contained in this single chip.

Display Formatting

Altogether, there are 9 programmable control registers in the 5027. Seven of the registers control the horizontal and vertical format for the screen.

The first three registers, R0 through R2, define the horizontal parameters, the next three define the vertical format. The seventh register (R6) is a line counter (not to be confused with the programmed number of lines to be displayed) which can be used for scrolling.

There are also two registers which control the movements of the cursor. All registers are accessible through four REGISTER ADDRESS inputs, A0 through A3 (pins 39, 40, 1 and 2, respectively).

Programming

Once the desired register has been selected by the REGISTER ADDRESS, data is loaded into the register through eight bidirectional DATA BUS lines (pins 25-18). To initiate loading, the CHIP SELECT input (pin 3) must be HIGH. Data is subsequently transferred on the DATA STROBE (pin 9, LOW) signal.

The 5027 has no read or write select function, and the registers are strictly on a write-only basis. This means data can only be entered into the formatting registers, not reviewed. Once programmed, the screen format remains fixed until reset. This usually means that the normal Read/Write signal from the processor must be NORed externally before input to the DATA STROBE input.

Cursor

The cursor, however, requires a read output. This is accomplished by providing two registers, instead of the usual one, for cursor operations. Two separate addresses are used to access each of the *cursor registers*: one address is used for write operations and the other is for reading.

When the screen memory address is the same as that of the *cursor register* address, the CURSOR VIDEO output (pin 16) goes HIGH. This produces a continuous stream of dots at that character position, generating a cursor that always appears as a block symbol. However, external logic may be used to generate other cursor symbols using the CURSOR VIDEO output as a condition signal.

Initialization

When the system is first powered up, a succession of write operations must be executed to establish initial values in the control registers. Once complete, the 5027 will display the format.

The chip can be reset under microprocessor or program control by presenting a 1010 address on the REGISTER ADDRESS inputs A3 through A0 (pins 2, 1, 40, and 39, in that order). The controller chip will remain reset at the top of the even field page until a Start Command is presented by advancing the address to 1110.

Self-Loading

Another method for establishing initial values in the control registers is to use the self-loading capacity of the device. The self-loading sequence is initiated by presenting and holding the 1111 address on the REGISTER ADDRESS inputs. Provided the CHIP SELECT input is HIGH, initialization begins on the receipt of a DATA STROBE pulse.

The strobe command also starts the SCAN LINE COUNTER, R0 through R3 (pins 8, 7, 5, and 4), counting. These outputs are used to increment the program memory chip storing the register instructions. The REGISTER ADDRESS code must remain valid during the entire loading procedure (under one millisecond in most applications). Once loading is complete, the address code is removed and the timing sequence begins one line scan later.

Self-loading is also available for processor-based systems. However, the initialization code becomes 0111, and the timing chain is *not* initiated until a Start Command is received from the processor.

Other Versions

The 5027 CRT Controller comes in three other versions: 5037, 5047, and 5057. Each of these chips has a particular variation on the basic theme, as outlined in the CRT 5047 data sheet in this book. You will find additional operational information relating to the CRT 5027 in that section.

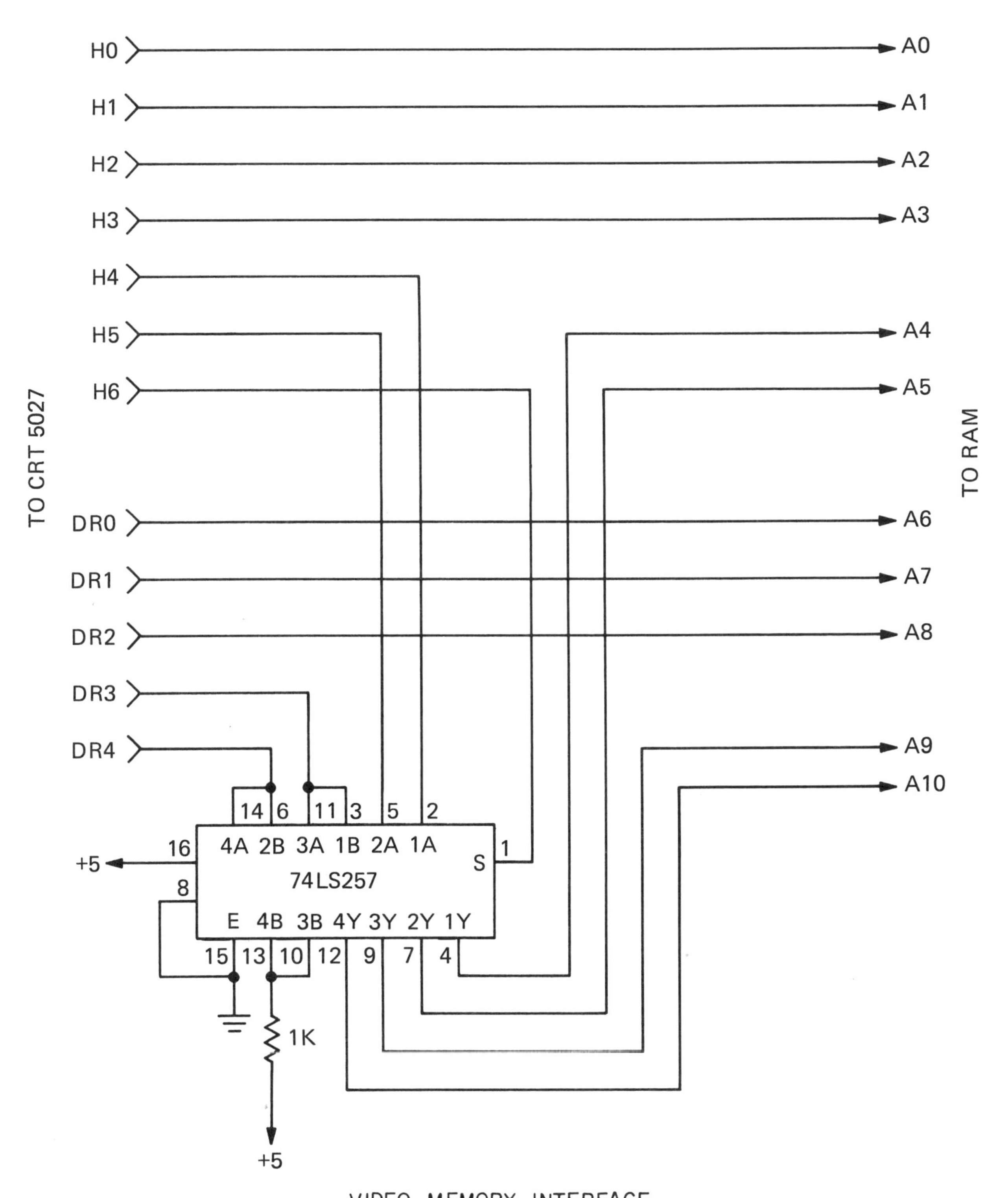

VIDEO MEMORY INTERFACE

Figure 1

A complete CRT terminal is assembled from the diagrams on this page and pages 148 and 149.

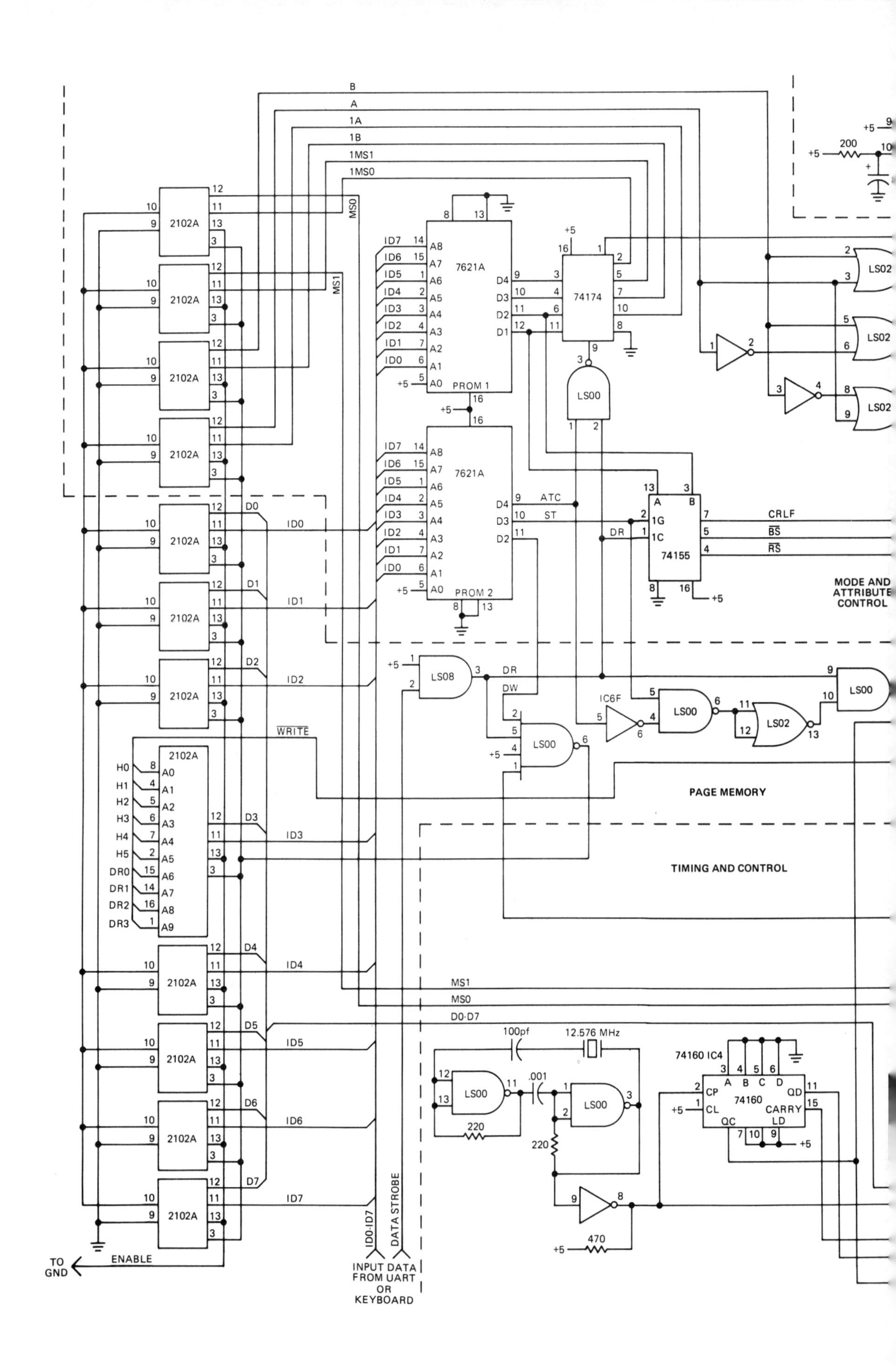
MODE AND ATTRIBUTE CONTROL
PAGE MEMORY
TIMING AND CONTROL
2102A
7621A
PROM 1
PROM 2
74174
74155
74160 IC4
LS00
LS02
LS08
IC6F
CRLF
BS
RS
ATC
ST
DR
DW
WRITE
MS1
MS0
D0-D7
100pf
12.576 MHz
.001
220
470
200
ID0-ID7
DATA STROBE
INPUT DATA FROM UART OR KEYBOARD
TO GND
ENABLE

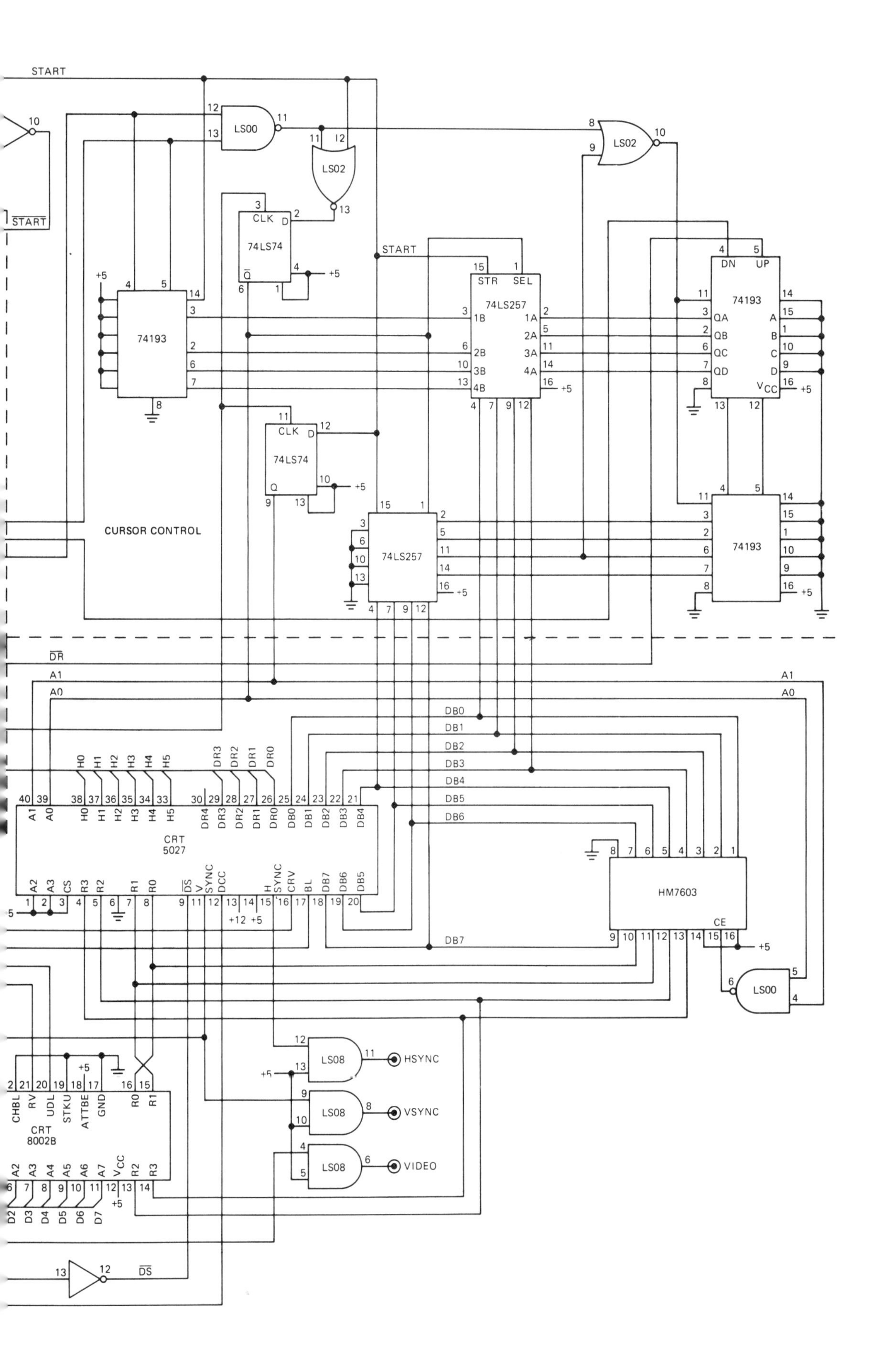
START
LS00
LS02
74LS74
CLK
74193
74LS257
STR
SEL
DN
UP
CURSOR CONTROL
CRT
5027
HM7603
LS08
HSYNC
VSYNC
VIDEO
CRT
8002B
DR
A1
A0
DB0
DB1
DB2
DB3
DB4
DB5
DB6
DB7
DS

CRT 5047

Video Timer and Controller

The CRT 5047 Video Timer and Controller, made by Standard Microsystems Corporation, is a special version of the CRT 5027 which has been ROM programmed with a fixed format. The use of a fixed program eliminates the software overhead normally required to specify the display parameters and simplifies terminal design.

Included in this section is a discussion on two other versions of the CRT 5027 controller chip: the CRT 5037 and CRT 5057.

Screen Formats

Unlike the totally programmable 5027, the parameters of the 5047 have been selected to be compatible with most CRT monitors. The device uses an 80 × 24 display format with a 5 × 7 character matrix in a 7 × 9 block. It uses a non-interlaced scan mode and is compatible with the 8002B Video Attributes Controller.

The 5037 is the more general mask-programmable version of the 5047, and is its base component. The 5037 may be programmed at the time of manufacture for an odd or even number of scan lines per data row in both interlaced and non-interlaced modes. Programming for an odd number of scan lines eliminates character distortion caused by uneven beam currents normally associated with odd field/even field interlacing of alphanumeric displays.

Clock and Power Supply

The entire family of 5027 controllers requires two power supplies for operation. The first is a standard + 5-volt voltage on pin 14. For the second, a + 12-volt supply is connected to the V_{DD} input, pin 13.

Internal circuit operation is clocked by an input signal to the DOT COUNTER CARRY (pin 12). This signal is derived from the character generator and is easily obtained from a standard 11.004-MHz crystal that is often used to drive the character generator.

Memory Address

There are two sets of signals associated with screen addressing logic, simply because the 5027 family addresses the memory on a column/row basis. The horizontal component, the CHARACTER COUNTER outputs, is H0 through H7 (pins 38-31). Vertical, or DATA ROW COUNTER, outputs are DR0 through DR5 (pins 26-31). Notice that pin 31 does a double duty in this respect. When required, it will function as the most significant bit of the CHARACTER COUNTER (H7), otherwise it is the most significant bit of the DATA ROW COUNTER output (DR5). Other than this variation, the memory addressing is rigid and unalterable.

The character generator interfacing consists of four raster address, or SCAN LINE COUNTER, outputs (pins 8, 7, 5, 4). These outputs comprise one set of inputs to the character generator, while the screen memory supplies the second set of character codes for input to the character generator logic. If you are using the 5027 or 5057 in the interlace mode, R0 (pin 8) will also serve as the odd or even field indicator.

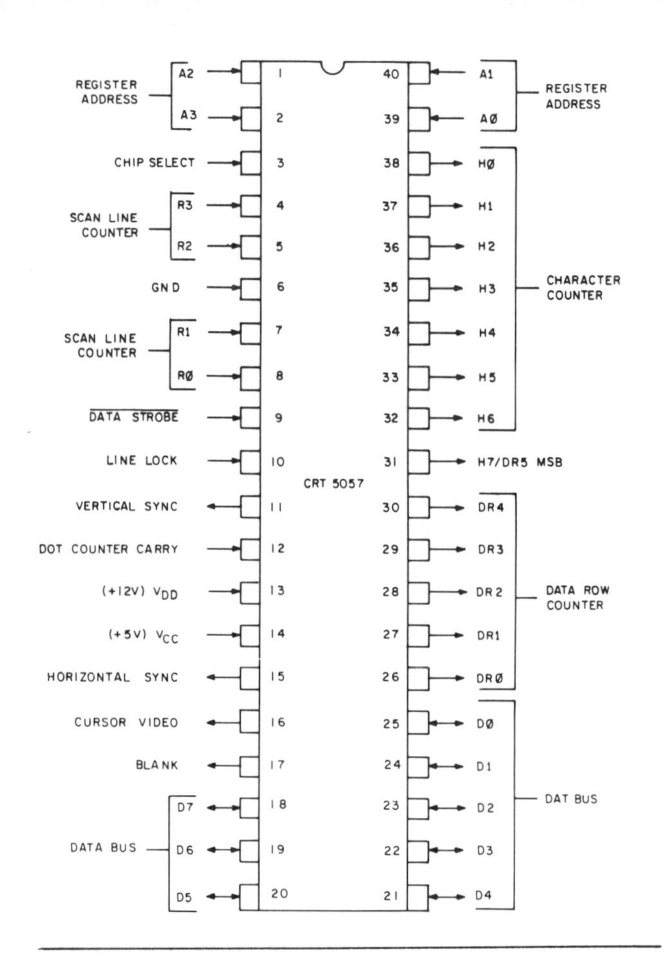

Video Outputs

The controller has four video output functions, plus a cursor, that interface with the CRT monitor. The HORIZONTAL SYNC (pin 11) pulses are standard video signals compatible with most three-terminal monitors. They initiate their respective retrace functions as specified by the values located in the control registers. The duration of the HORIZONTAL SYNC is also program-specified.

The VERTICAL SYNC pulse, on the other hand, is fixed equal to three horizontal scan lines, and external circuitry may be necessary to obtain proper pulse widths for any given CRT monitor. This timing sequence, however, is appropriate for most monitors.

The COMPOSITE SYNC output (pin 10) combines both the horizontal and vertical synchronization pulses. It can be used for CRT monitors so equipped, or fed to an RF modulator to operate a standard television receiver.

The 5057 version of the controller has the ability to lock a CRT's vertical refresh rate to the 50-Hz or 60-Hz line frequency. This feature eliminates the "swim" phenomenon so common with many displays. To accomplish this, though, the COMPOSITE SYNC output is eliminated from the chip and replaced with a LINE LOCK (pin 10) input.

The BLANKing output (pin 17) is set HIGH during horizontal and vertical retrace times, and is used to turn off the video during these intervals. However, the blanking signal is activated at all times except during those intervals in which it is specified that a character is to appear on the screen, and actually bears no relationship to either of these retrace signals other than by default.

Additional information concerning the operations of the controller chips in this section can be found by referencing to the CRT 5027 device.

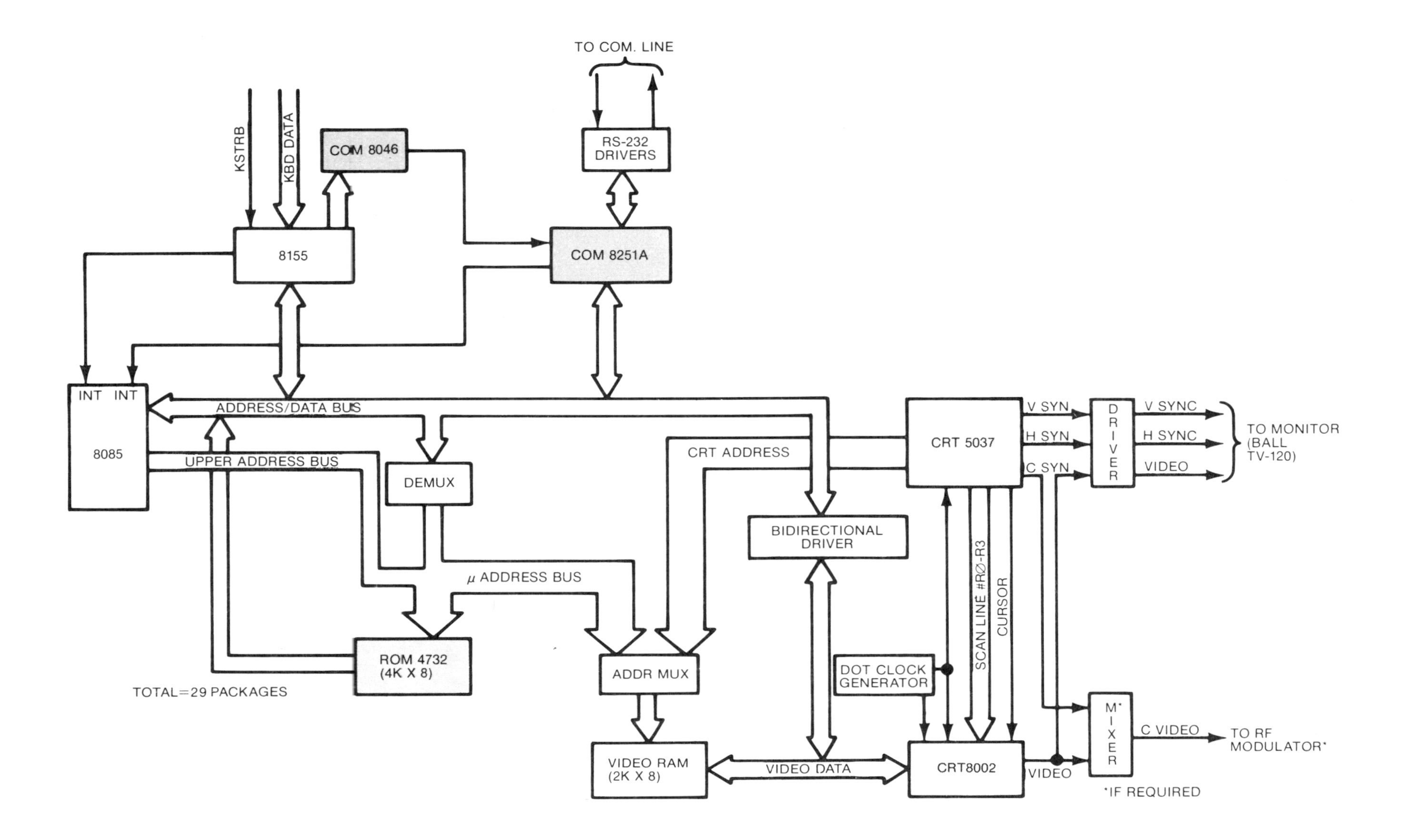
TO COM. LINE
KSTRB
KBD DATA
COM 8046
RS-232 DRIVERS
8155
COM 8251A
INT
INT
8085
ADDRESS/DATA BUS
UPPER ADDRESS BUS
DEMUX
CRT ADDRESS
CRT 5037
V SYN
H SYN
C SYN
DRIVER
V SYNC
H SYNC
VIDEO
TO MONITOR (BALL TV-120)
BIDIRECTIONAL DRIVER
μ ADDRESS BUS
SCAN LINE #RØ-R3
CURSOR
ROM 4732 (4K X 8)
ADDR MUX
DOT CLOCK GENERATOR
TOTAL=29 PACKAGES
MIXER*
C VIDEO
TO RF MODULATOR*
VIDEO RAM (2K X 8)
VIDEO DATA
CRT8002
VIDEO
*IF REQUIRED

CRT 8002

CRT Video Display Attributes Controller

The CRT 8002, by Standard Microsystems Corp., is a Video Display and Attributes Controller containing an on-chip ROM character generator in addition to its video control functions. The chip controls five video attributes, all of which can be implemented in character and/or field modes. The CRT 8002 is a companion chip to the CRT 5027, and together these two chips comprise all the circuitry required for the display portion of a CRT video terminal.

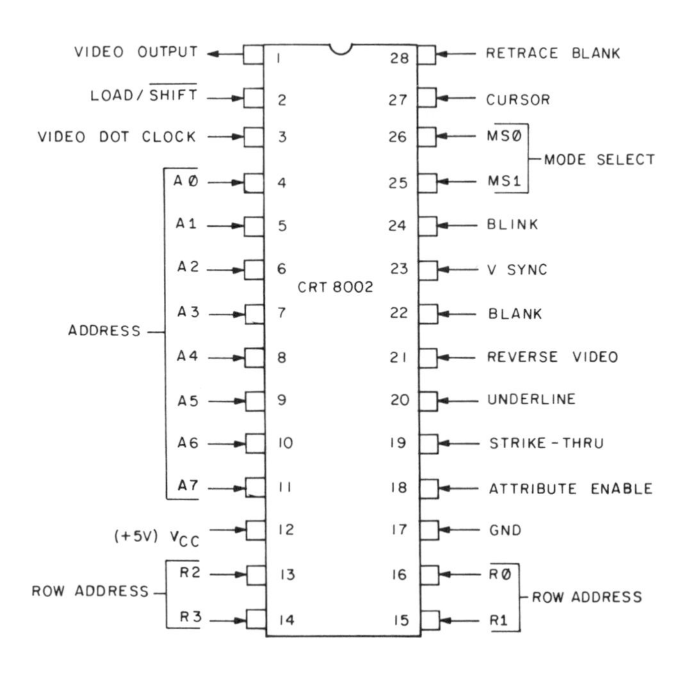

Operation

The 8002 has been designed to directly interface to most CRT monitor video inputs using a single VIDEO OUTPUT (pin 1) signal. This signal also contains the horizontal and vertical blanking pulses when the chip is modulated by the RETRACE BLANK (pin 28) input. These timing pulses can be supplied by the 5027's BLANKING OUTPUT.

There are four display modes available with the 8002, including Alphanumeric, Wide Graphics, Thin Graphics, and External Input. Selection of a mode is dictated by the configuration of the MODE SELECT (pins 26, 25) inputs, and the four modes can be intermixed on a per character basis. Since the chip contains a character generator and all the logic necessary to produce the graphic patterns, creating the desired video image is a simple matter of selecting the appropriate mode and addressing the proper video data location.

The video data locations are accessed by the ADDRESS inputs (pins 4-11), and are latched into the chip on the HIGH input of the LOAD/SHIFT (pin 2) control line. When this input is returned LOW, the chip automatically shifts the video information out the VIDEO OUTPUT port at the rate set by the VIDEO DOT CLOCK (pin 3).

To completely assemble a character, each character row must be scanned sixteen times. Each scan line contains the dot pattern necessary to construct that portion of the character. The scan line of current interest is selected by the ROW ADDRESS (pins 16, 15, 13, 14) inputs; sequencing these pins in binary fashion writes an entire character row.

Alphanumeric Mode

The 8002 contains a mask-programmable 7 × 11 character generator ROM capable of storing 128 symbols and figures. Although several standard formats are available in the 8002 family, including 5 × 7 ASCII generated by the CRT 8002-003 chip, the user may opt to encode the ROM with a custom pattern. Each ROM character is selected using seven of the eight ADDRESS inputs, A0 to A6.

Graphics Modes

There are two graphics modes available for use with the 8002. The Wide Graphics Mode defines 256 graphic patterns by addressing all eight ADDRESS inputs (A0 through A7), while the Thin Graphics Mode has a set of thin-lined figures located at ADDRESSES A0 through A2. These entities can butt up against each other to form a contiguous pattern.

External Mode

When the 8002 lacks the necessary symbols or designs, the user can supply them externally using the External Mode. In this mode, the ADDRESS inputs become data inputs, and any word presented to these inputs goes directly into the *output shift register* (with no decoding) via the attributes logic, and is so displayed.

Video Attributes

In addition to generating the video information for display, the 8002 also manages five video attributes. They include: underline, strike-thru, reverse video, character blink, and blank.

The UNDERLINE and STRIKE-THRU functions are directed by pins 20 and 19, respectively. Internal logic will BLINK (pin 24) a character at 1.875-Hz by blanking the character during 25% of the display period. Should the CHARACTER BLANK input (pin 22) be active HIGH, however, it takes precedence over the blinking attribute and masks the character altogether. All attributes, including CHARACTER BLANK, are effectively reversed when REVERSE VIDEO (pin 21) is applied.

The attributes are latched into the chip, along with the MODE SELECT status, on the falling edge of the LOAD/SHIFT strobe when the ATTRIBUTE ENABLE input (pin 18) is HIGH. These attributes will remain in effect until the ATTRIBUTE ENABLE input becomes HIGH once again. When changing attributes character by character, this input should be tied HIGH to facilitate latching.

The CURSOR input (pin 27) enables a 3.75-Hz blinking reverse video block to indicate the cursor position on the screen. Both the CURSOR and BLINK functions extract their blinking rates and duty cycle from the V SYNC (pin 23) clock, a clock pulse that is normally derived from the 60-Hz vertical sync generator.

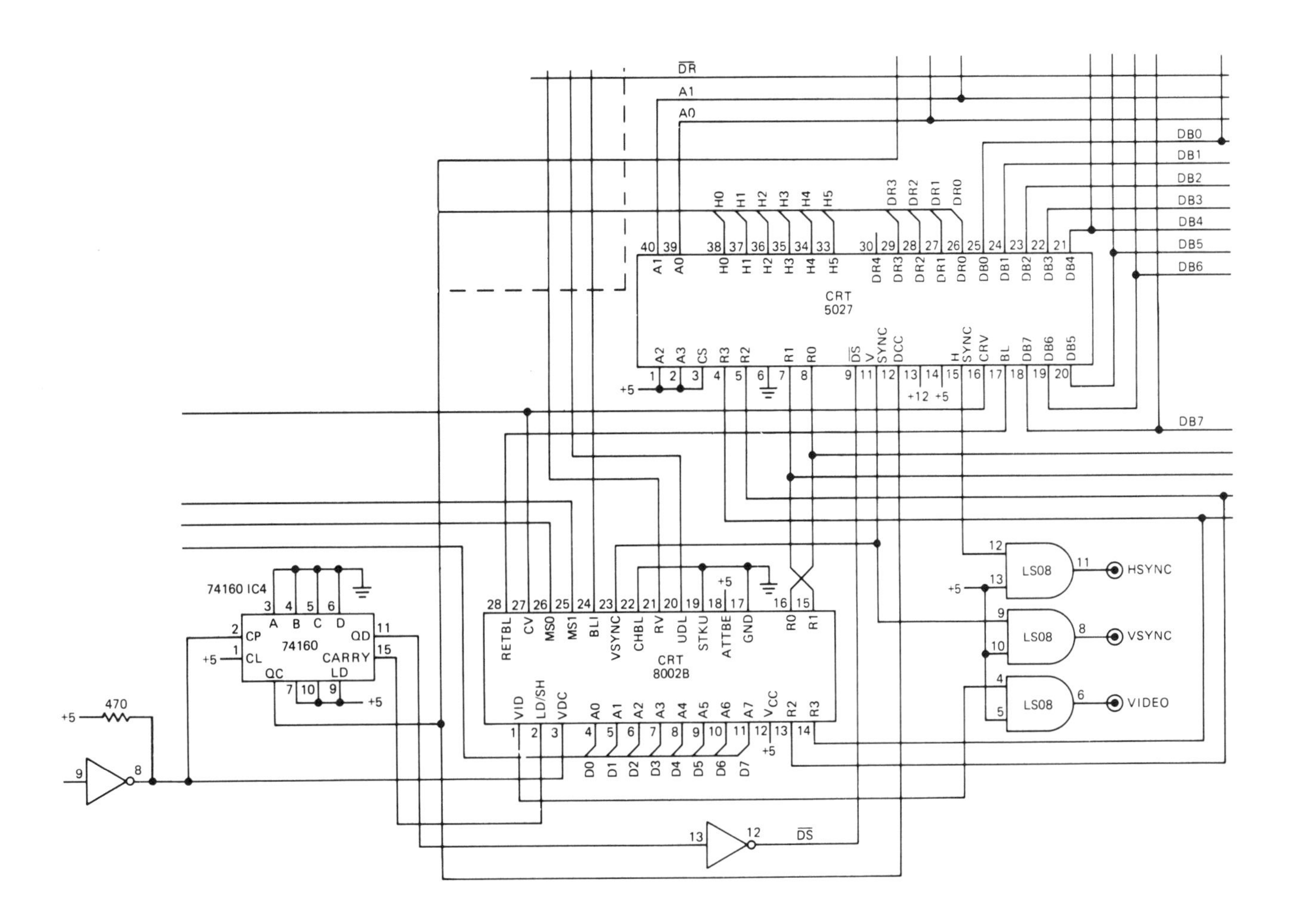

Figure 1

MC6845

CRT Controller

The MC6845, made by Motorola, is a CRT Controller chip which generates the signals necessary to interface a digital system to a raster scan CRT display. Once programmed, the chip has total control over the video display with intervention by the CPU only when new information is put into the display memory. The chip contains a programmable horizontal and vertical timing generator, programmable Linear Address Register, cursor logic, Light Pen Strobe Register, and control circuitry for interface to a processor bus.

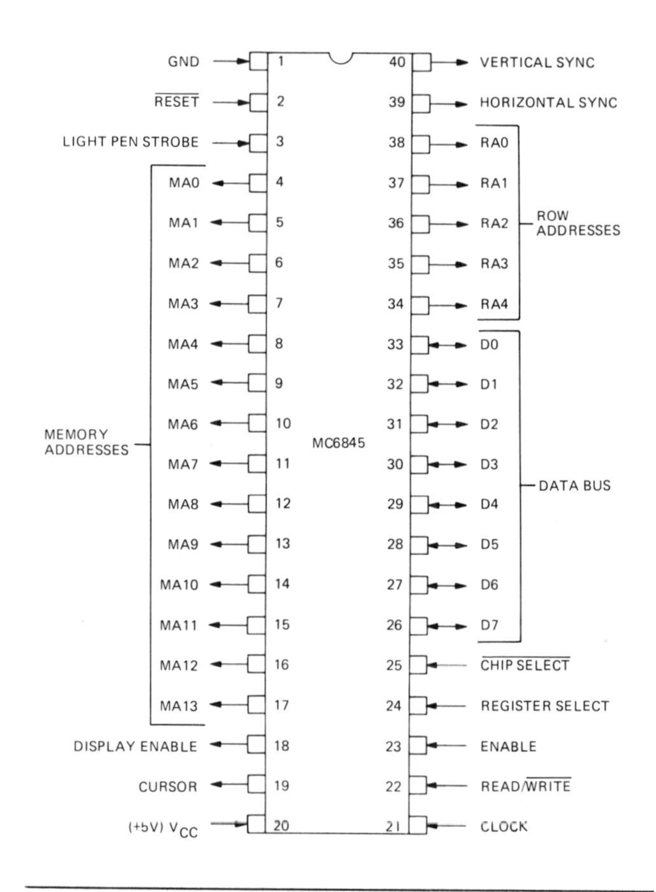

System Operation

A total CRT display system can be built around the MC6845 with only 25 support chips, plus the memory. The processor communicates with the CRT controller through an 8-bit bidirectional DATA BUS (pins 33-26) using ASCII code. The data to be displayed is stored in an external refresh (screen) memory accessed directly by the CPU, and once logged needs no further attention until updated. A character generating ROM is used to convert the ASCII codes into a 5×7, or 7×9, dot matrix for display.

Device Implementation

The MC6845 has 18 programmable registers, R0 through R17, that control the horizontal and vertical sync, number of characters per row, number of scan lines per row, number of rows per screen, the portion of the memory to be displayed (multiple pages), cursor format and position, and the choice of one of three scanning modes. These *data registers* are accessed by the *linear address register* using the REGISTER SELECT (pin 24).

Display Sizing

The first four registers, R0 through R3, are concerned with the horizontal format. Programming considerations are based on the sweep, retrace time, etc. The next four registers, R4 to R7, are used to set up the vertical format.

Loading the registers begins with valid data on the DATA input lines (D0 through D7). The ENABLE pin (pin 23) is a high-impedence input which allows the data bus buffers to clock data to and from the MC6845. This signal is usually derived from the processor clock, and the high-to-low transition is the active edge.

To enter a program, the READ/WRITE input (pin 22) must be LOW, as must be the CHIP SELECT (pin 25). However, the latter signal should only be present when there is a valid address being decoded. The REGISTER SELECT input selects either the *linear address register* (LOW) or the *data register* (HIGH) defined in the *address register.*

Scanning Format

Three operating modes are available to the user which have to do with whether the screen field format is even or odd. In the Normal Sync Mode, information is written into one field only. When writing into one field, each dot will be refreshed 60 times every second.

The second mode, Interlace Sync, writes in both fields, in which case the odd field is an exact duplicate of the even field. Essentially, the same information is written twice, giving a solid appearance to the vertical lines. However, each dot is scanned only 30 times per second, which may produce objectionable flicker in some applications.

The third mode, Interlace Sync and Video, also writes in both fields. However, only one-half the character is written in each field, producing a symbol that is half as tall as the other two formats. This is often used when high screen intensity is desired.

Display Memory

The CRT controller provides MEMORY ADDRESSES (pins 4-17) to scan the display RAM. These 14 outputs are used to refresh the CRT screen with pages of data located within the 16-K of memory. In addition, five ROW ADDRESSES (pins 38-34) output an address to the character generator ROM to clock out the proper display symbol.

Both the MEMORY ADDRESSES and ROW ADDRESSES continue to run during vertical retrace, thus allowing the MC6845 to provide refresh addresses required by dynamic RAMs.

Light Pen

The MC6845 is designed to interface with a light pen using standard logic circuits. A low-to-high transition on the LIGHT PEN STROBE input (pin 3) causes the current contents of the *address counter* to be latched into the *light pen register.* The contents of this register can then be read by the processor for subsequent processing.

The MC6845 is virtually identical to the MC6835, a mask programmable version of the controller, and is often used to emulate it in prototyping. Further information concerning the chip's operation can be found under the MC6835 listing.

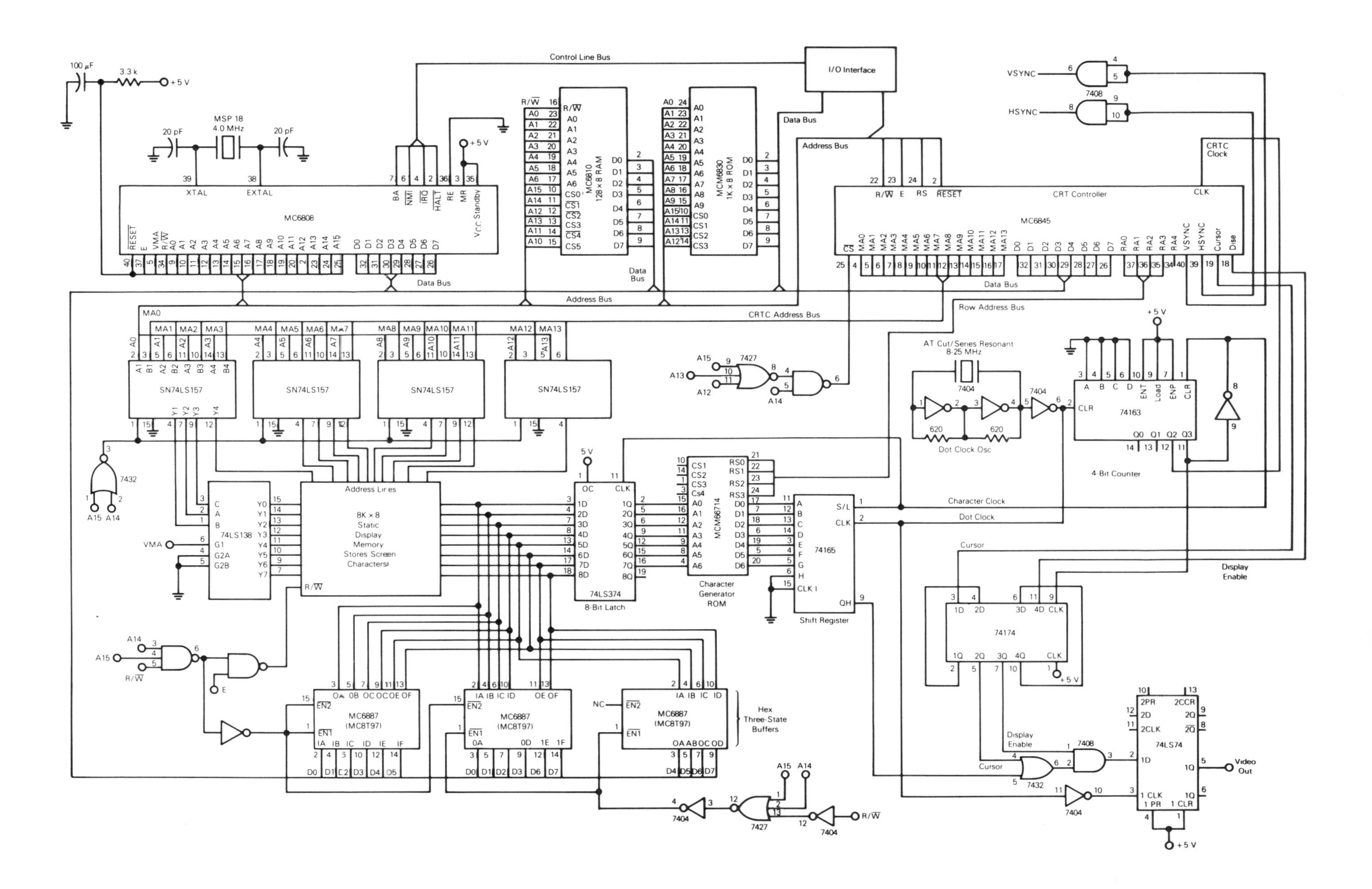

Control Line Bus
I/O Interface
Data Bus
Address Bus
MC6808
MSP 18
4.0 MHz
MC6810
128 × 8 RAM
MCM6830
1K × 8 ROM
CRT Controller
MC6845
CRTC Clock
VSYNC
HSYNC
CRTC Address Bus
Row Address Bus
SN74LS157
74LS138
Address Lines
8K × 8 Static Display Memory Stores Screen Characters
74LS374
8-Bit Latch
MCM66714
Character Generator ROM
74165
Shift Register
AT Cut/Series Resonant 8.25 MHz
Dot Clock Osc
74163
4-Bit Counter
Character Clock
Dot Clock
Cursor
74174
Display Enable
74LS74
Video Out
MC6887 (MC8T97)
Hex Three-State Buffers

MC6835

CRT Controller

The MC6835, by Motorola, is a ROM-based CRT Controller which interfaces a CPU to a raster scan CRT display. The chip, which is virtually identical to the MC6845, contains a mask programmable horizontal and vertical timing generator, software programmable Linear Address Register, mask programmable cursor logic, and control circuitry for 8-bit data bus interfacing.

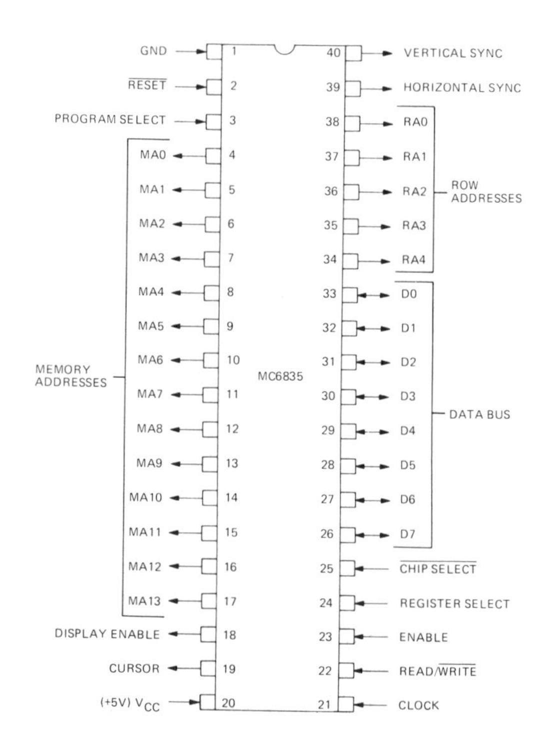

Screen Format

The MC6835 supports two mask programmed screen formats, as determined by the programming of the *horizontal* and *vertical registers*; the scanning formats are outlined in the MC6845 discussion. The PROGRAM SELECT input (pin 3), which replaces the light pen function, allows selection of one of the two programs: Set zero is selected when the pin is LOW, set one is implemented when the input is HIGH.

Cursor

The cursor is controlled by four internal registers—two under mask programming (MC6835) and two software programmable—which determine cursor format, position, and blink rate. These mask programmed registers allow a block cursor of up to 32 scan lines in height to be started, then stopped on any scan line of the character block. An active HIGH cursor (pin 19) output indicates a valid cursor address for external video processing logic.

Sync Outputs

The VERTICAL SYNC (pin 40) and the HORIZONTAL SYNC (pin 39) outputs are active HIGH signals which either drive the monitor directly or can be fed to a video processor to generate a composite video signal. In addition, a DISPLAY ENABLE output (pin 18) indicates when the CRT controller is addressing in the active screen area, and is equivalent to a video blanking signal.

Care should be exercised when interfacing the chip to a CRT monitor, as many monitors claiming to be "TTL compatible" have transistor inputs which may exceed the maximum rated drive currents. It is best to avoid the problem altogether by interfacing with a buffer.

Clock

An external CLOCK (pin 21) input signal synchronizes all CRT functions except for the microprocessor interface. An external dot counter is used to derive this signal, which is normally the character clock in an alphanumeric display. The active transition is high-to-low.

The ENABLE clock (pin 23) is used to synchronize the I/O buffers on the controller with the CPU for data transfer. This signal is received from the CPU, and in a 6800-based system it connects to the ø 2 signal.

Power Up and Reset

Registers R12 and R15 must be initialized after the system is powered up. The processor will normally load these registers from a firmware table. Of course, the MC6845, in which the registers are all software programmed, requires registers R0 through R15 to be initialized after a power off condition.

The RESET input (pin 2) is used to reset the controller during operation. An input to this line must remain LOW for at least one cycle of the character clock, and forces the chip into the following states: All counters are cleared and the device stops the display operation. In the MC6845 chip, all the outputs are driven LOW. With the MC6835, on the other hand, the outputs go LOW with the exception of the MEMORY ADDRESSES (pins 4 through 17) which assume the current value of the *start address register*. The *control registers* are not affected by a reset and remain unchanged. The MC6835 CRT controller resumes normal display operation immediately after the release of the RESET input.

Additional Applications

Additional services can be provided by the MC6835, most notably the refreshing of dynamic RAMs. This is easily implemented since the refresh addresses are continually running. In addition, the VERTICAL SYNC and HORIZONTAL SYNC outputs can be used as a real-time clock.

Characteristic to the MC6845, the LIGHT PEN STROBE input (pin 3) will support system functions beyond that of a light pen. For instance, a D/A converter can be added to the system by referencing the refresh address to a resistive divider connected to a comparator. When the refresh address analog value and unknown voltage are equal, the comparator inputs a signal to the LIGHT PEN STROBE input, which can then be read by the processor.

The LIGHT PEN STROBE input could also be used as a character strobe to allow the refresh address to decode a keyboard matrix. Debouncing is usually a software function in this application.

The MC6835 is pin compatible with the MC6845, which may be used as a prototype part to emulated the MC6835. Therefore a software related discussion for both parts can be found under the MC6845 listing.

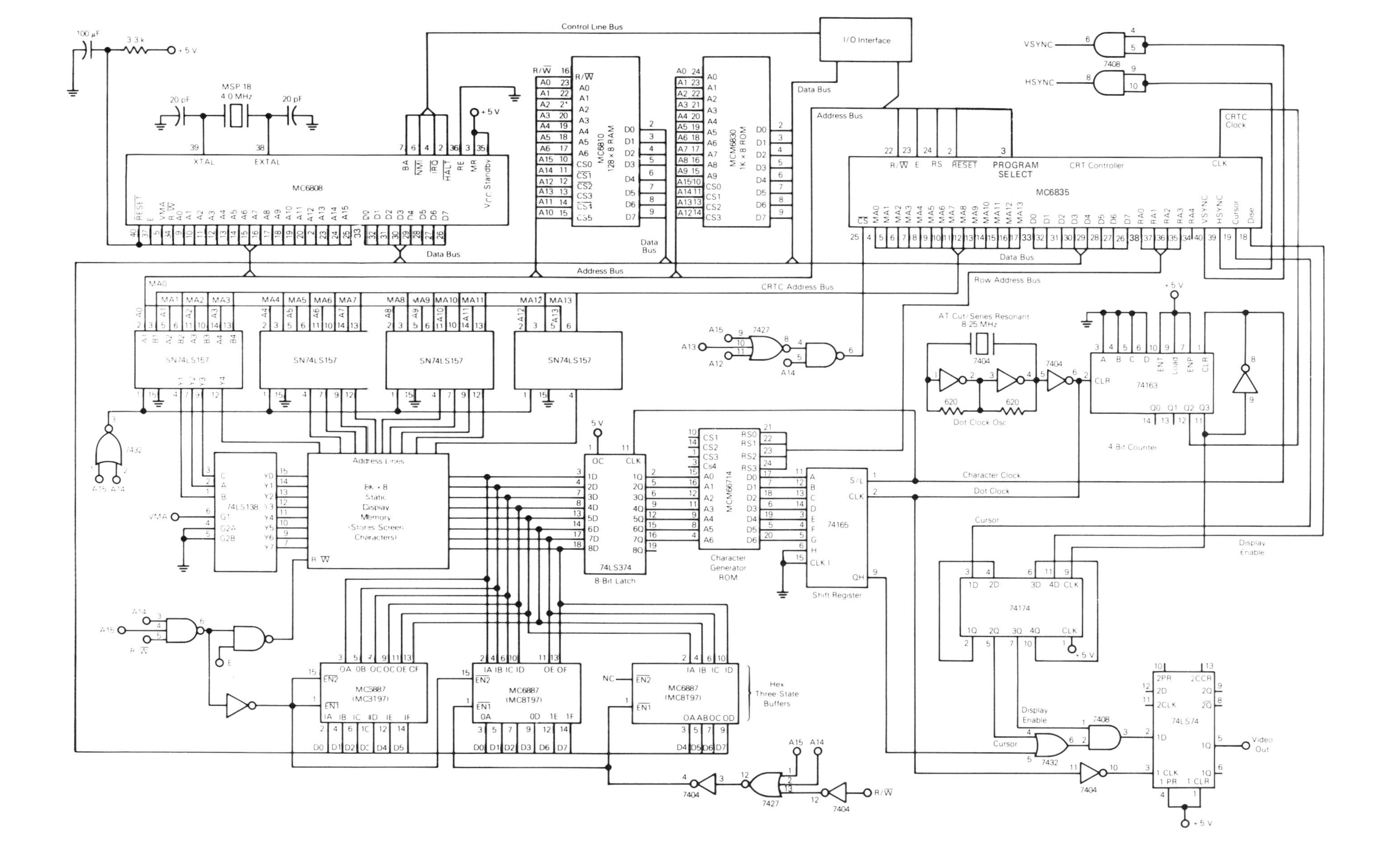
Control Line Bus
I/O Interface
Data Bus
Address Bus
VSYNC
HSYNC
7408
CRTC Clock
MSP 18
4.0 MHz
20 pF
100 μF
3.3 k
+5 V
MC6808
XTAL
EXTAL
Vcc Standby
MC6810
128 × 8 RAM
MCM6830
1K × 8 ROM
PROGRAM SELECT
CRT Controller
MC6835
CRTC Address Bus
Row Address Bus
SN74LS157
7432
74LS138
VMA
Address Lines
8K × 8 Static Display Memory (Stores Screen Characters)
74LS374
8 Bit Latch
MCM66714
Character Generator ROM
74165
Shift Register
7427
AT Cut/Series Resonant 8.25 MHz
7404
Dot Clock Osc
74163
4 Bit Counter
Character Clock
Dot Clock
Cursor
Display Enable
74174
MC6887 (MC8T97)
Hex Three State Buffers
NC
74LS74
Video Out

SY6545

CRT Controller

The SY6545, made by Synertek, Inc., is a CRT Controller designed to interface the 6500 microprocessor to a raster scan CRT monitor. The chip is basically of the same design as the MC6845 (made by Motorola) and is pin compatible with it. However, some rather significant improvements exist in the SY6545 which aren't available in the MC6845. Since the operation of the family has been established in the sections on MC6845 and MC6835, this discussion will center only on the differences between the devices.

Video Memory

Like its CRT controller relative, the 6545 has 14 lines of MEMORY ADDRESS (pins 4-17) output, designated MA0 through MA13, respectively. These outputs address the video memory in straight binary sequence.

With the 6545, however, the MEMORY ADDRESS outputs have a second set of control signals associated with them. These outputs can be programmed to address the video memory in a row/column sequence rather than in straight binary. The first eight addresses (pins 4-11) are defined as the CHARACTER COLUMN (CC0 to CC7) outputs, while the remaining six addresses (pins 12-17) serve as the CHARACTER ROW (CR0-CR5) outputs.

Program Registers

Basically, the programmable registers in the 6545 are identical to the MC6845. However, some modifications have had to be made in them to accommodate the added instructions. For instance, the *mode control register* (R8) defines the new memory contingencies, and there are three new registers.

The data registers are accessed by the *address register*, just as before. This time, though, the *address register* has a slight twist to it. To load the *address register*, the REGISTER SELECT input (pin 24) is first taken LOW. When a write operation is initiated, using the CHIP SELECT (pin 25) and READ/WRITE (pin 22) operators, the address of the desired data register is placed into the *address register*.

Now, by forcing the REGISTER SELECT input HIGH (and maintaining READ/WRITE at LOW), the data register whose address is contained in the *address register* is accessed. Because the 6545 is tailored to the 6500 CPU, the ø2 CLOCK enable (pin 23) is asynchronous.

Transparent Memory Addressing

A significant difference between the MC6845 and the 6545 is in the memory contention logic function. In the M6845 no provisions have been incorporated for the allocation of memory time, and external circuitry is necessary. With the 6545, on the other hand, all memory addressing is done by the controller and fed to the memory on the MEMORY ADDRESS lines, allowing no direct access by the CPU.

When new data is to be entered into memory, the CPU places the address of the location to be changed in the *update registers*, R18 and R19. The 6545 then decides when access to the memory

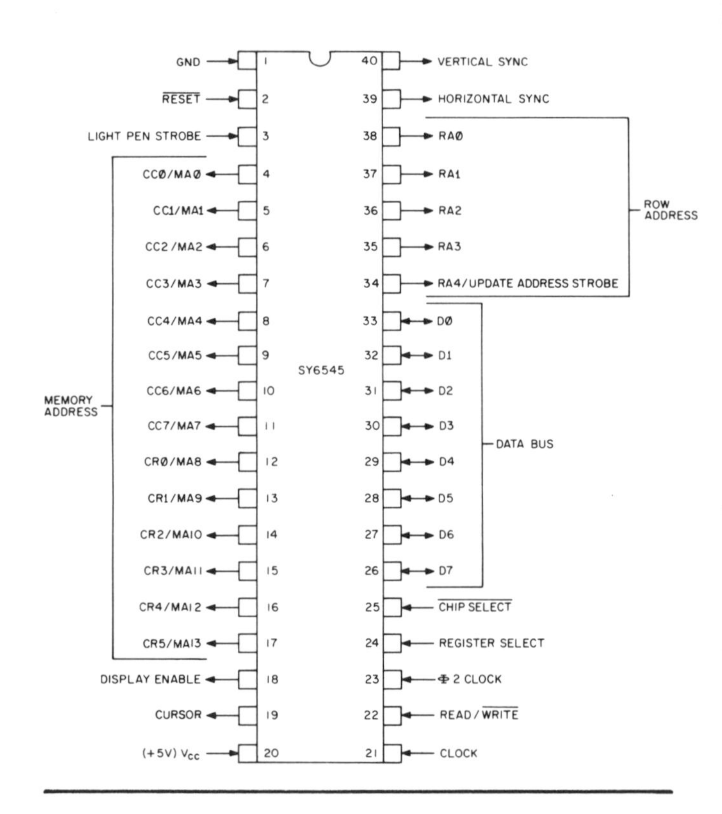

is available to the CPU by addressing that location at the appropriate time. Two schemes are used for CPU intervention.

Interleaved Memory Addressing

One method for CPU intervention is to split the time between the screen refresh and CPU access by dividing the microprocessor cycle in half. This is accomplished by using the ø2 CLOCK pulse to establish an access window.

During the ø1 portion of the clock cycle, the CRT controller will scan the memory and refresh the CRT screen. During the ø2 portion of the block cycle, access to the memory is available to the CPU, with the address of the desired location for updating being supplied by the controller chip via the *update registers*.

Retrace Memory Addressing

The second method of memory allocation is to allow the CPU access during the horizontal and vertical retrace periods. As before, the memory location to be modified is placed in the *update registers*; memory updating will take place during the retrace portion of the screen.

Since the CPU will not always have immediate access to the memory during every retrace request, an intermediary data latch is used to store the memory contents. To clock data to and from the latch, an UPDATE ADDRESS STROBE (pin 34) signal is available.

The microprocessor has no direct control over the CRT controller or the video memory; consequently it must poll the 6545 chip periodically to check the state of the last memory transfer. Exclusive to the 6545 is the presence of a *status register* at the *address register* location. The CPU can access the *status register* at any time by simply performing a read function on the REGISTER SELECT while the input is LOW. This operation will output a three bit signal ascertaining the progress of memory operation, revealing whether or not addressing space is available.

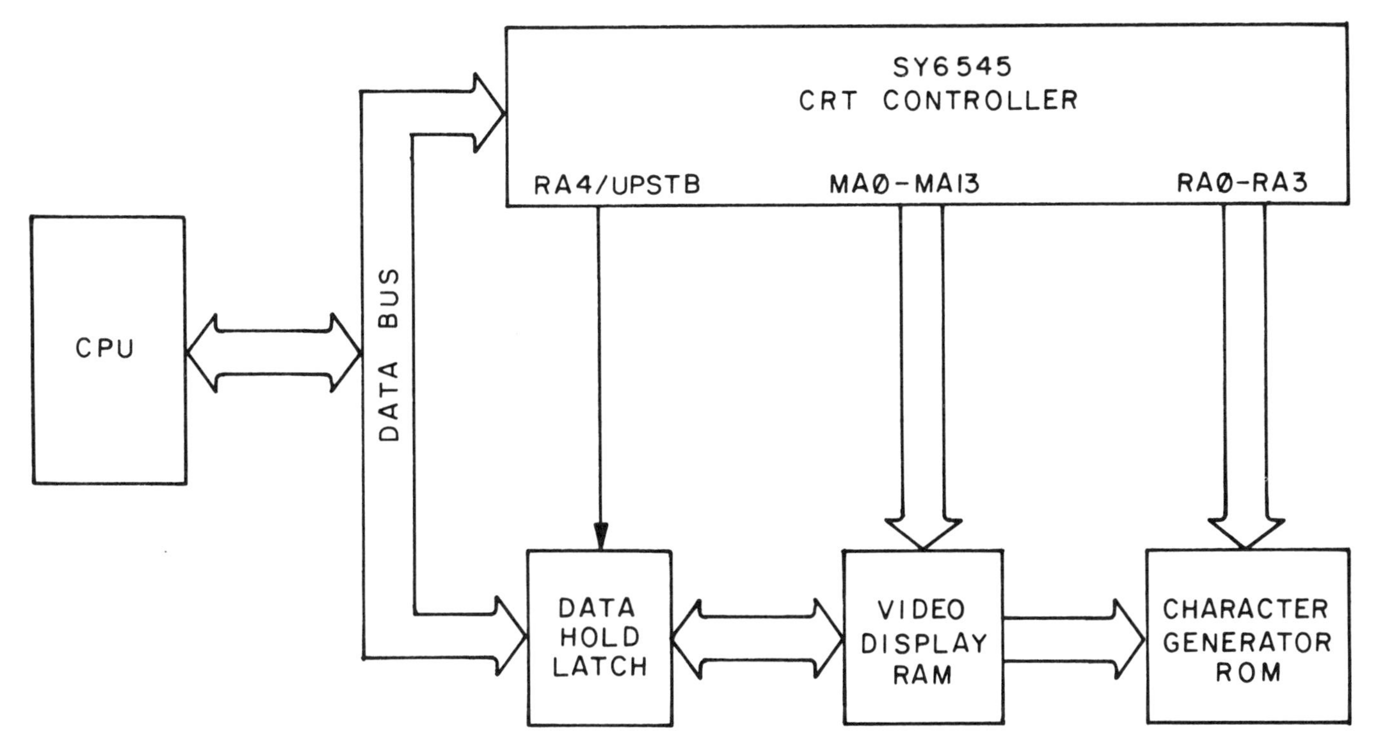

Figure 1

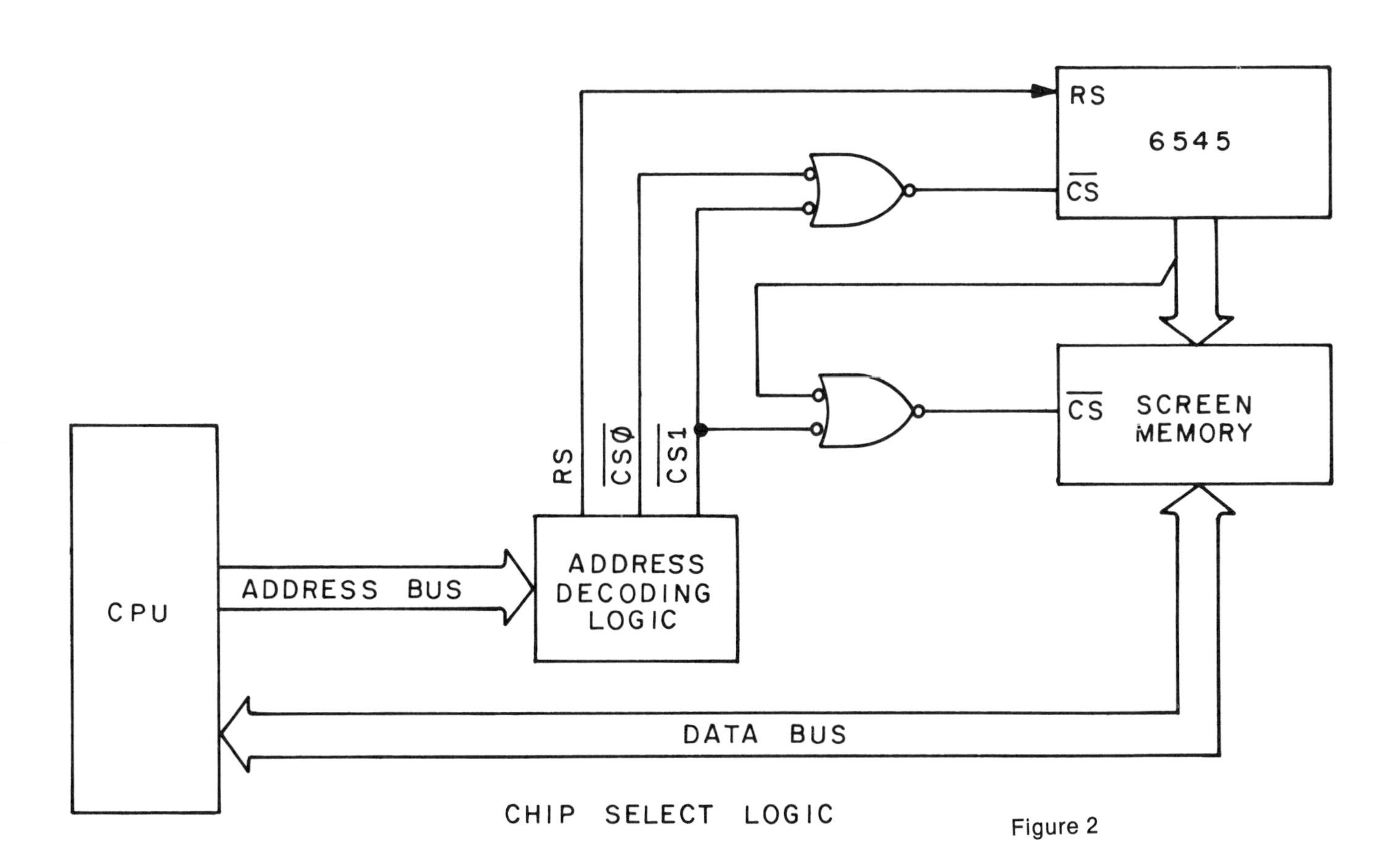

Figure 2

MC6847

Video Display Generator

The MC6847 Video Display Generator, by Motorola, provides a simple interface for display of digital information on a standard color or black and white television receiver. Since this chip also provides color graphics, applications including video games and controller displays are well within its realm, as well as computer terminal display.

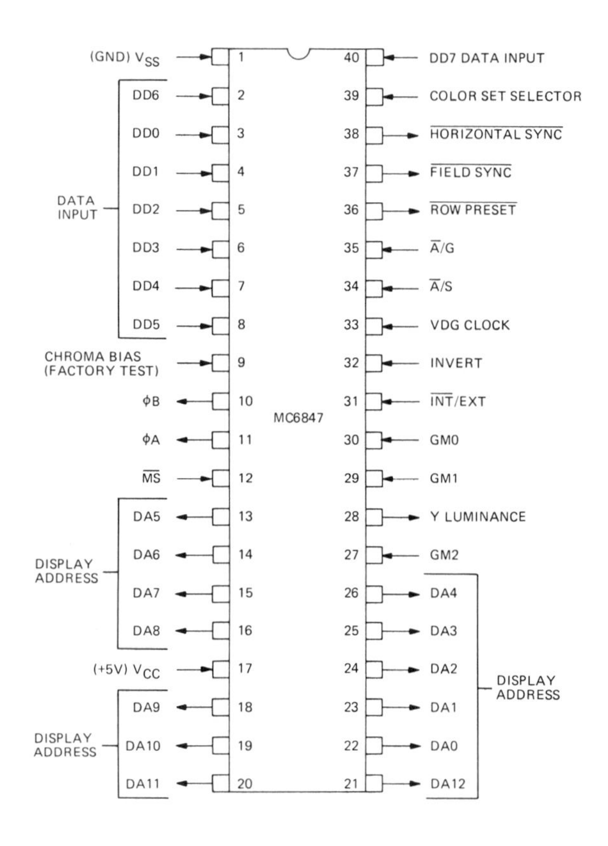

Operation

The MC6847 contains all the necessary timing and control circuits to directly interface with standard NTSC television receivers. The chip comes in two versions — the MC6847 and MC6847Y. The "Y" version uses interlace scanning and thereby increases the screen resolution.

The display generator reads data from a RAM memory, and translates it into a video signal which is displayed as alphanumeric symbols or graphics. The chip includes an internal character generator ROM which contains 64 5 × 7 dot matrix characters in ASCII code. This ROM, however, is mask programmed at the time of manufacture, and can be custom programmed, allowing you to modify the contents to display Chinese, Japanese, special graphics, etc.

Memory

Display data is stored in a RAM memory using ASCII code. Thirteen DISPLAY ADDRESS lines, DA0 through DA12 (pins 22-26, 13-16, 18-21), scan the display memory and read the contents through eight DATA INPUT lines (pins 2-8, 40). These address lines are under tri-state control, and can be forced into a high impedence state whenever the $\overline{MS}$ input (pin 12) goes LOW. This is done to allow direct access to the RAM by the CPU for information updating. The FIELD SYNC (pin 37) output coincides with the active display area, and when this pin is LOW, the CPU can access RAM without causing undesirable flicker on the screen.

Video Outputs

Three digitally synthesized analog outputs are used to transfer video and color information to the video monitor. The Y LUMINANCE output (pin 28) is a six level analog signal containing composite sync, blanking, and four levels of video luminance. The øA and øB outputs (pins 11 and 10, respectively) are three or four level signals (in that order) used in combination with Y to specify one of eight screen colors. The additional voltage level present on the øB pin is used to declare a time burst reference signal.

These outputs may be used to drive a video monitor directly, or be passed along to an RF modulator, such as the MC1372, for reception by a standard television receiver on channels 3 or 4.

Operating Mode

There are two major display modes in the MC6847 generator. These modes are determined by the $\overline{A}$/G input (pin 35). Major Mode One, which is input LOW, contains four alphanumeric and two limited graphic display formats. Input $\overline{A}$/S (pin 34) chooses between the Alphanumeric Mode (LOW) and the Graphic Display (HIGH). The alphanumeric internal mode uses an internal character generator to decode and display 64 ASCII symbols. The characters may be either green on a dark green background (virtually black) or orange on dark orange, depending on the status of the COLOR SET SELECTOR (pin 39) input. In addition, the INVERT pin (pin 32) can be used to display dark characters on a bright background.

For those who wish to display lower case letters, special characters, or even limited graphics, an external ROM may be used by forcing the INT/EXT input (pin 31) HIGH. But first, a four bit counter must be added to supply ROM row addresses. The counter is clocked by a HORIZONTAL SYNC signal (pin 38) and is cleared by the ROW PRESET output (pin 36). All display elements within Major Mode One, whether internal or external, are similar; therefore, mode switching between Alphanumeric Modes and Semigraphic Modes can be carried on freely on a character by character basis.

Major Mode Two (pin 35, HIGH) contains eight graphic modes. The specific mode selected is determined by a combination of the GM0 (pin 30), GM1 (pin 29), and GM2 (pin 27) inputs. These modes differ in the size and colors available for the graphic building blocks. Each element of the display (size determined by the graphic mode) is under individual control, and can be used to produce a variety of special effects. ROW PRESET pulses only occur in Major Mode One, making the use of an external ROM for Major Mode Two an impossibility.

Clock

The VDG CLOCK input (pin 33) requires an external 3.58-MHz (standard color burst frequency) square wave. The duty cycle of the clock input must fall between 45% and 55% for proper picture proportioning, and can usually be derived from the RF modulator or video monitor.

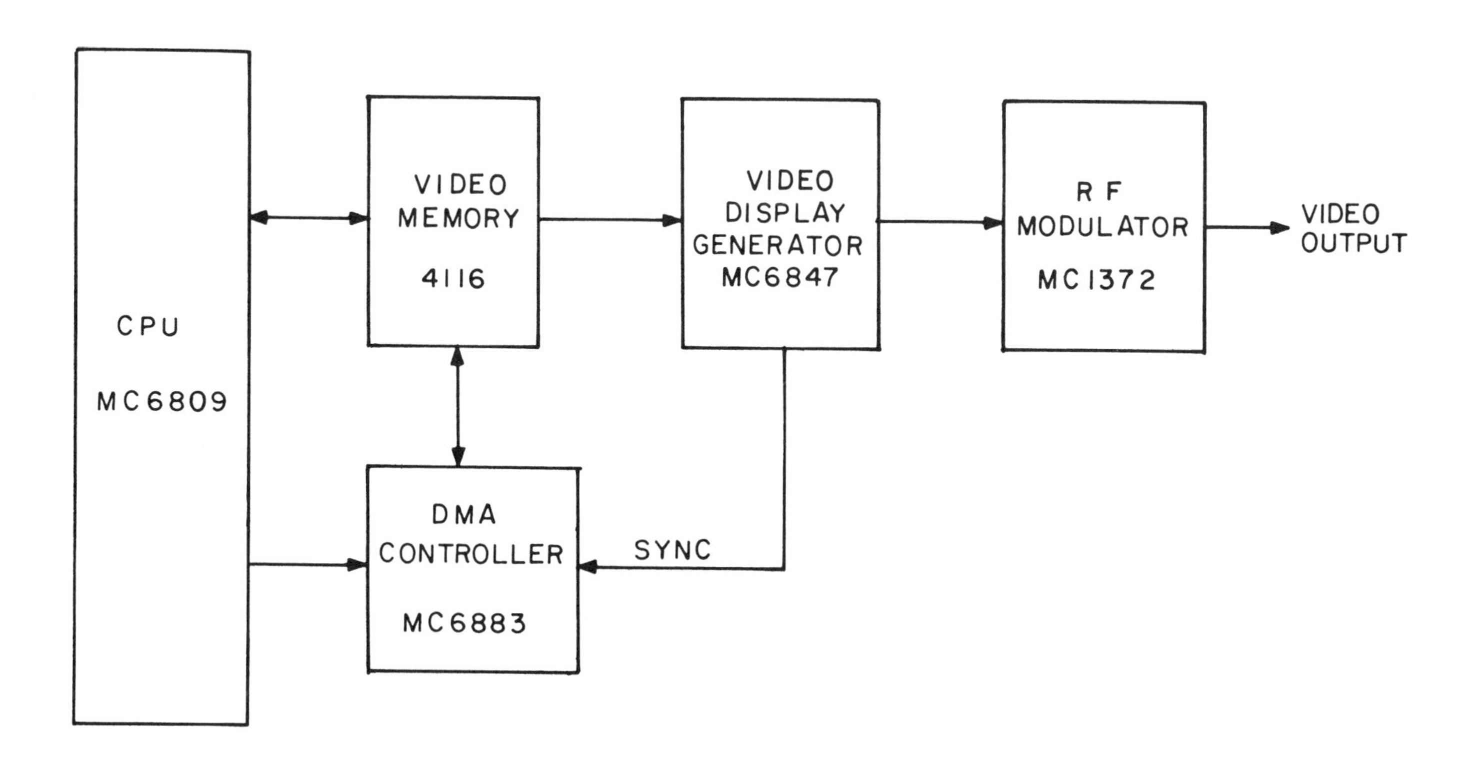

A system with the MC6847 and the MC1372 video modulator forms a video transmitter operating at 61.2-MHz (channel 3) or 67.25-MHz (channel 4), depending upon the values selected. Care must be taken to meet FCC requirements Part 15, Subpart H when the system is in operation.

Figure 1

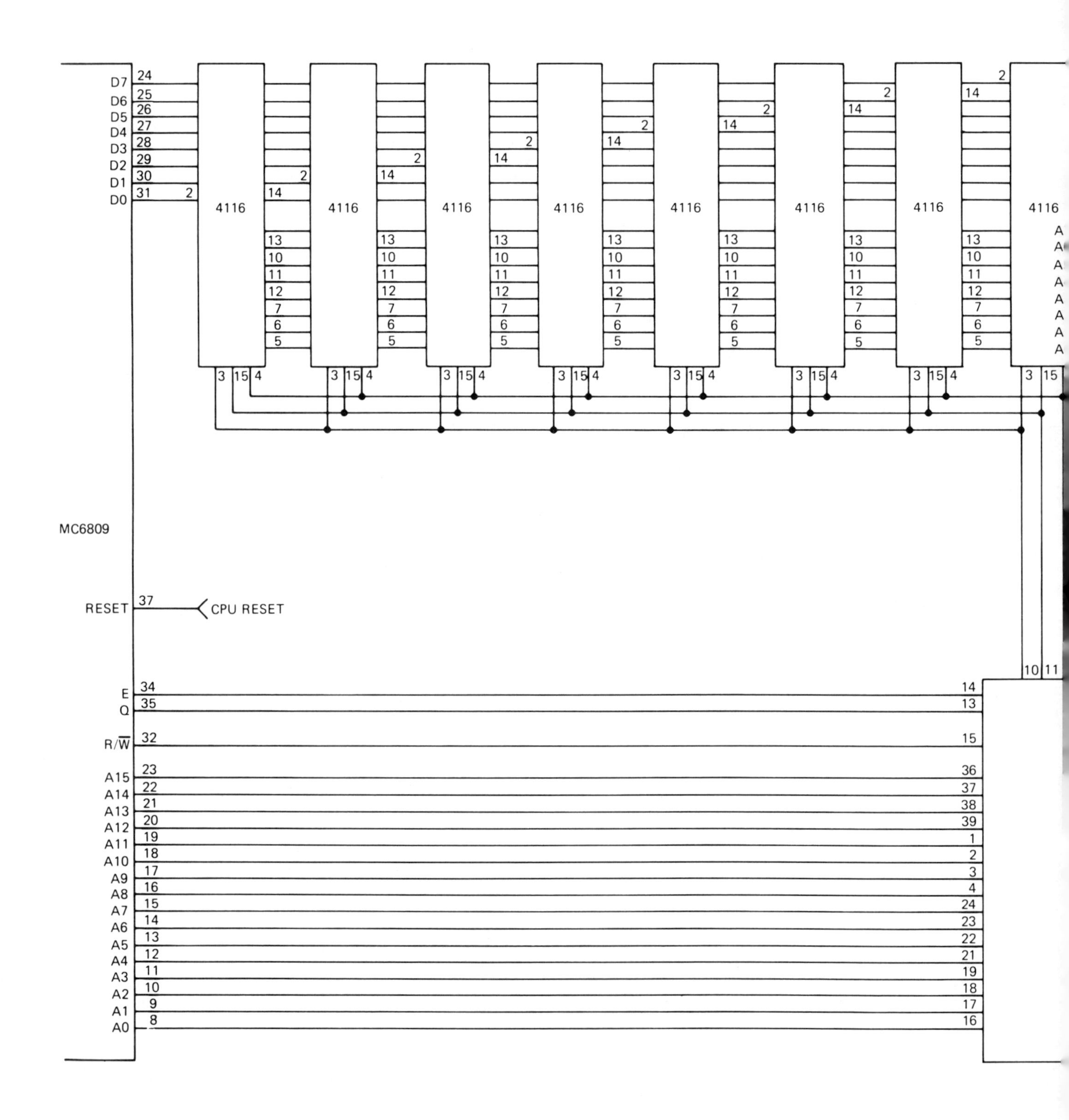
D7
D6
D5
D4
D3
D2
D1
D0
4116
4116
4116
4116
4116
4116
4116
4116
MC6809
RESET
CPU RESET
E
Q
R/W̄
A15
A14
A13
A12
A11
A10
A9
A8
A7
A6
A5
A4
A3
A2
A1
A0

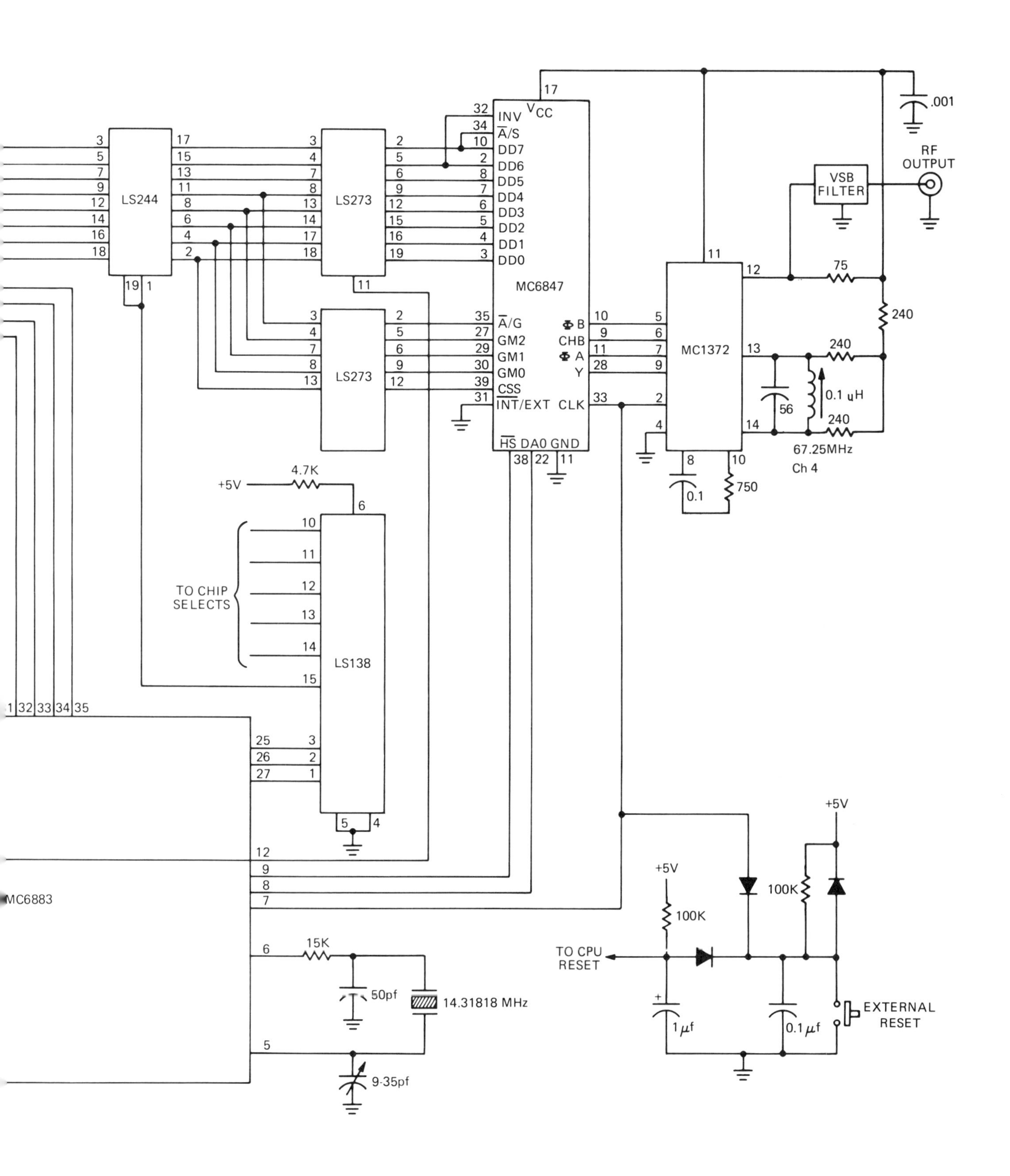
LS244
LS273
LS273
LS138
MC6847
MC1372
MC6883
INV
VCC
A/S
DD7
DD6
DD5
DD4
DD3
DD2
DD1
DD0
A/G
GM2
GM1
GM0
CSS
INT/EXT
CLK
HS DA0 GND
B
CHB
A
Y
.001
RF
OUTPUT
VSB
FILTER
75
240
240
240
56
0.1 uH
67.25MHz
Ch 4
0.1
750
+5V
4.7K
TO CHIP
SELECTS
15K
50pf
14.31818 MHz
9-35pf
+5V
100K
100K
TO CPU
RESET
1 μf
0.1 μf
EXTERNAL
RESET

2672

Programmable Video Timing Controller

The 2672 Programmable Video Timing Controller, by Signetics, is a programmable device designed for use in CRT terminals and display systems that employ raster scan techniques. The function of this chip is to generate the horizontal and vertical timing signals required to produce the TV raster on a CRT monitor, for both interlaced and non-interlaced video display, along with directing video memory management.

Operation

The user defines all video parameters in software. The programmable registers are first loaded with the instruction for the screen raster and display configuration. Once the video pattern has been established, the 2762 directs the loading of the video display memory with the desired characters and symbols, then manages the CRT display with no further intervention by the CPU until memory updating is needed. All events are sequenced using the CHARACTER CLOCK (pin 16).

Register Programming

There are 11 video control registers which must be initialized before operation can begin. The registers are identified by the ADDRESS LINES (pins 37-39). The data to be written is placed on the tri-state DATA BUS lines (pins 8-15), and the CHIP ENABLE (pin 2) input pulled LOW. A LOW input to the WRITE STROBE (pin 3) will transfer the data in to the register addressed.

Data is read from the 2672 by addressing the appropriate register and strobing the READ STROBE (pin 1) input.

Video Memory

The DISPLAY ADDRESS (pins 34-21) outputs generate linear addressing for up to 16K bytes of video memory. Altogether, there are four memory configurations possible with the 2672 chip: Independent, Transparent, Shared, and Buffer Row.

In the Independent Mode, data is transferred between the CPU and video memory using a latching buffer as an intermediary. The memory address is supplied by the CRT controller at the cursor position, or as specified by the *pointer address register*. Pin 4, the HANDSHAKE CONTROL #1, strobes the data from the data latch into the memory, while pin 5, the HANDSHAKE CONTROL #2, strobes data the opposite direction; i.e., from memory into latch. Pin 6, the HANDSHAKE CONTROL #3, provides the memory with the Chip Enable command.

When in either the Transparent or Shared Memory Mode, the video memory becomes part of the CPU memory system and the 2672 is the interloper. To direct traffic, the 2672 recognizes a Memory Access Request from the CPU on pin 4. The controller responds to a request by first giving a Get Ready warning on pin 6, followed by a Permission To Access signal on pin 5 when memory is available. In the Transparent Mode, memory is available during a horizontal or vertical retrace interval; in the Shared Mode, the controller will grant the CPU immediate access and blank the video.

The Row Buffer Mode allows the maximum video memory access time to the CPU. An entire row of data is transferred to a buffer under DMA control. The CRT controller then proceeds to

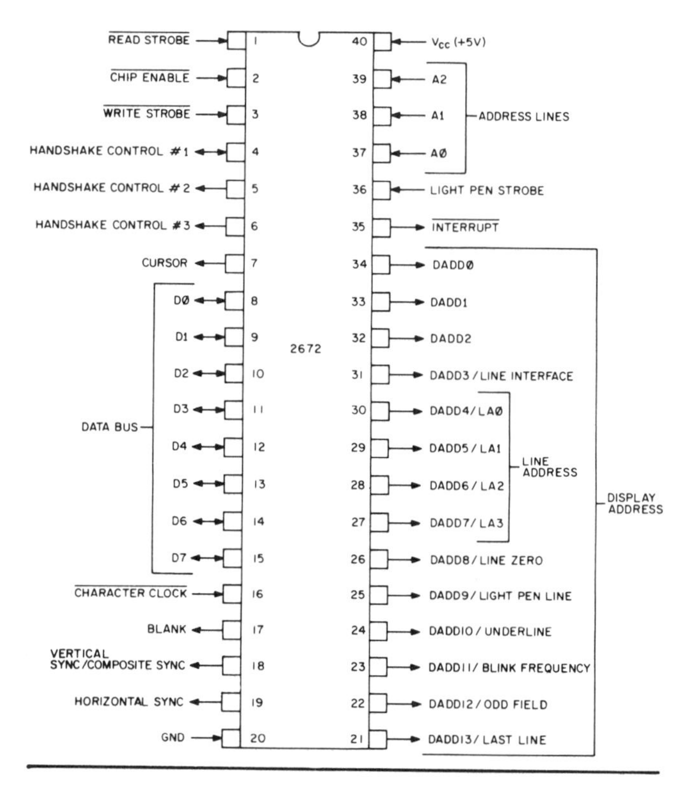

scan the row data from the buffer, leaving the CPU free to access the memory until the next row of characters is required. Pin 6 configures the memory for DMA data transfer; pin 5 halts CPU intervention during row data acquisition.

Video Signals

The 2672 provides a good complement of video output signals to direct the CRT display. A COMPOSITE SYNC is available at pin 18. When COMPOSITE SYNC is not desired, the HORIZONTAL SYNC (pin 19) and VERTICAL SYNC (pin 18) can be extracted separately using software instructions. A programmable CURSOR (pin 7) output goes active HIGH when the *cursor register* matches the DISPLAY ADDRESS, and a LIGHT PEN STROBE input (pin 36) loads the *light pen register* with the current value of the DISPLAY ADDRESS when strobed. A BLANKing output is available at pin 17.

Alternate Functions

Besides providing the addressing lines for the video memory, the DISPLAY ADDRESS outputs perform an alternate function. These are multiplexed output directives that indicate the status of the various functions inside the controller chip.

Pin 31, the DADD3 output, alternates with a LINE INTERFACE output to indicate whether there is an even or odd row being scanned in an even or odd field. The ODD FIELD (pin 22) output signals all odd lines in interlace scan. The LINE ZERO output (pin 26) asserts itself before the first scan line in each character row and the LINE ADDRESS (pins 30-27) enumerates the number of scan lines per row. The last scan line of each row is indicated by the LAST LINE (pin 21) output.

The cursor has an UNDERLINE (pin 24) signal every time the scan line matches the programmed cursor position and the BLINK FREQUENCY (pin 23) indicates cursor blink rate. The LIGHT PEN LINE (pin 25) shows the light pen scan line (line 3, 5, 7, or 9).

Although these functions are available in software, it is often more convenient to have them accessible in hardware form. These outputs are always valid at the trailing edge of BLANK.

INDEPENDENT BUFFER MODE CONFIGURATION

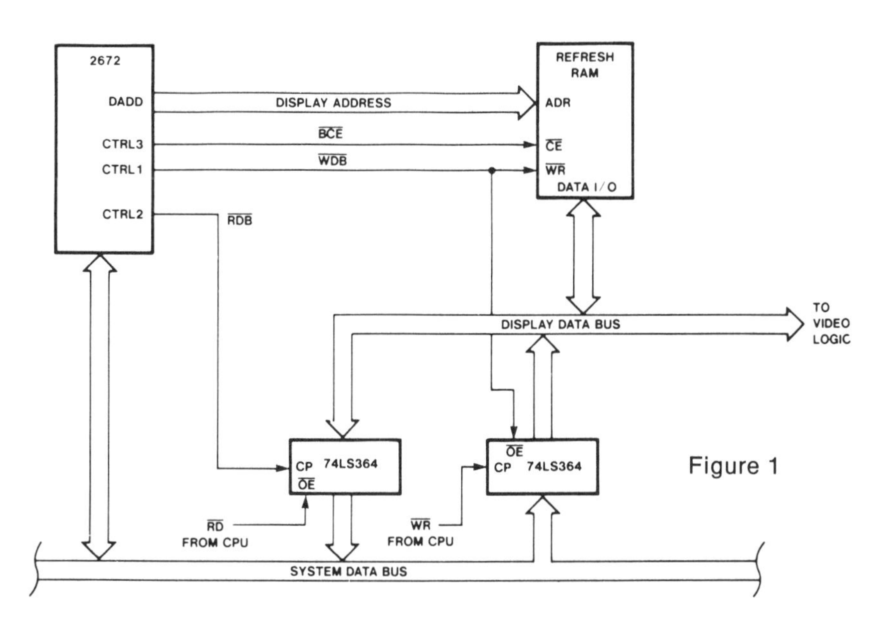

Figure 1

PVTC SHARED OR TRANSPARENT BUFFER MODES

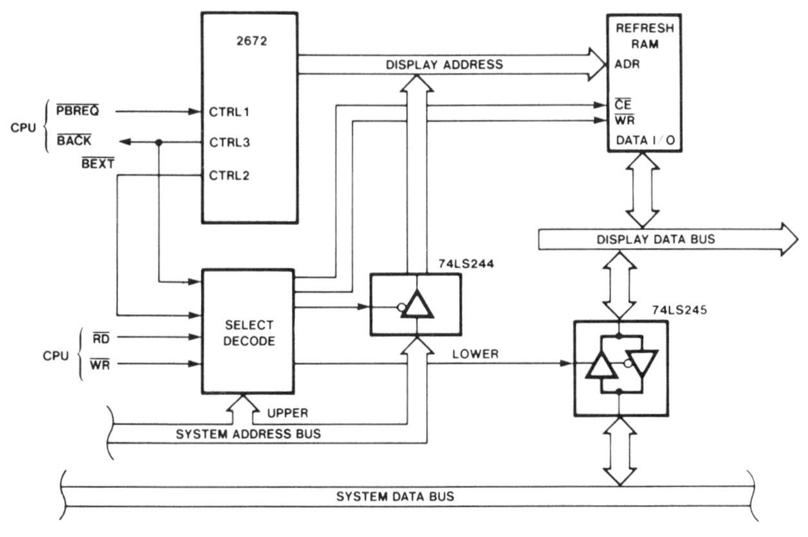

Figure 2

ROW BUFFER MODE CONFIGURATION

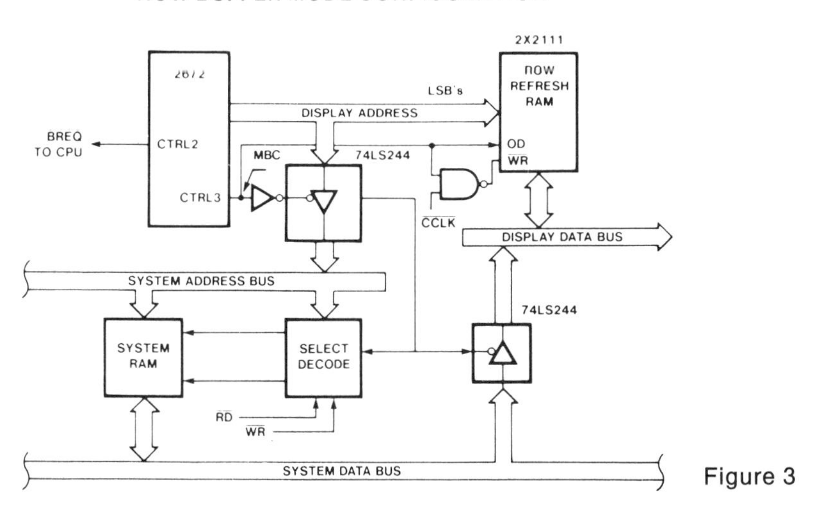

Figure 3

SC2673

Video Attributes Controller

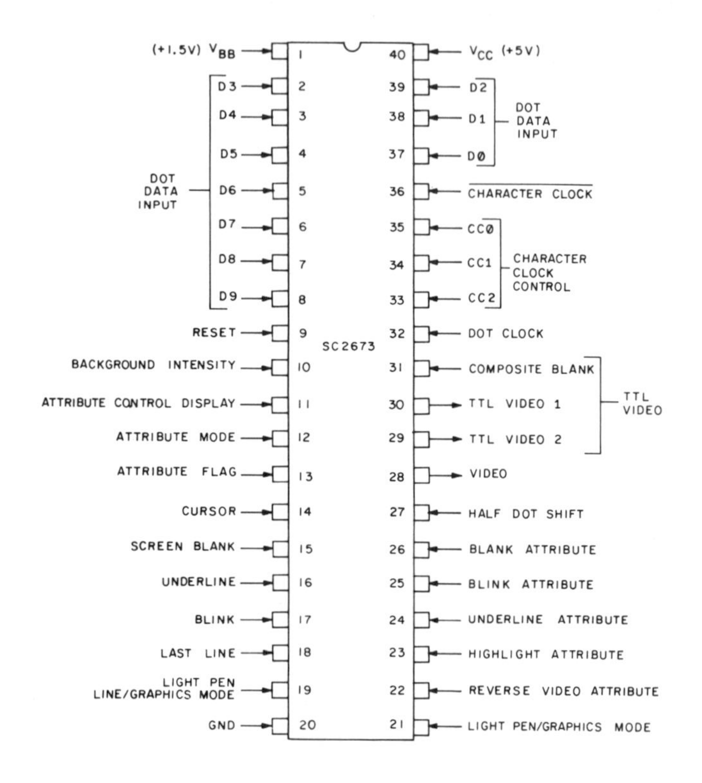

The SC2673 Video Attributes Controller, by Signetics, provides control of visual attributes for CRT terminals and display systems that employ raster scan techniques. While the CRT controller chip, such as the SC2672, provides the timing signals necessary to format the screen raster, it is the function of the video attributes controller to modulate the electron beam for the actual display of the video information.

Dot Clock

The DOT CLOCK (pin 32) is the one control signal that synchronizes all video display functions. The frequency of the DOT CLOCK determines the number of dots (pixels) that can be displayed by a single raster scan line.

The display of the characters, however, is governed by the CHARACTER CLOCK (pin 36). The CHARACTER CLOCK rate determines the number of horizontal dots that each character will contain, and is programmable from 6 through 12 dots per character. The CHARACTER CLOCK is derived from the DOT CLOCK frequency by dividing it according to the ratio specified by the CHARACTER CLOCK CONTROL (pins 35-33) inputs.

Character Display

The character to be displayed is entered at the DOT DATA INPUTs (pins 37-39, 2-8). In an average video display row, 16 scan lines are required to construct the character. In other words, 16 consecutive input words need be presented to the DOT DATA INPUTs, with each word representing the character's dot pattern for that scan line.

The characters are stored row by row in the video display memory, with each row accessed 16 times during character development. The actual dot pattern, however, is usually produced by a character generating ROM.

To further enhance the legibility of the characters, the 2673 includes a HALF DOT SHIFT control (pin 27) that delays the video output by one-half dot time when raised HIGH. This lends a pleasant rounding appearance to the symbol, and gets away from the harsh "computer" look.

Video Outputs

On each character boundary, the parallel data (D0-D9) is loaded into a *video shift register.* The data is then serially shifted out of the register at the DOT CLOCK frequency into the video logic, where it is encoded into three levels of video luminescence. The composite video signal is available at the VIDEO (pin 28) output. The three intensities are also binary encoded on two TTL compatible video outputs, TTL VIDEO 1 (pin 30) and TTL VIDEO 2 (pin 29).

Either light or dark screen backgrounds may be selected by manipulating the BACKGROUND INTENSITY (pin 10) input. Since the symbols are normally gray (middle-level intensity), they will appear as white on black when this input is HIGH, or as black over white with it LOW. A separate SCREEN BLANK input (pin 15) defines the active screen area by forcing the video outputs to the level specified by the BACKGROUND INTENSITY control, either high or low intensity, when HIGH. Overriding the SCREEN BLANK command is a retrace COMPOSITE BLANK (pin 31) which produces an all-black screen for the TTL outputs (only).

Video Attributes

The principle application of the 2673, though, is for video attributes control. Altogether, there are five visual attributes available: reverse video, character blink, highlight, underline, and non-display (blank).

The HIGHLIGHT ATTRIBUTE (pin 23) emphasizes the character by driving the dot intensity to the extreme opposite of the background intensity. Attention can also be focused on a symbol by underscoring it with the UNDERLINE ATTRIBUTE; the UNDERLINE control (pin 16) indicates which scan line(s) is to be used for underlining.

The BLINK ATTRIBUTE (pin 25) flashes the character at a rate determined by the BLINK input (pin 17), and the BLANK ATTRIBUTE (pin 26) washes it out altogether. When a CURSOR input (pin 14) is specified, it effectively reverses the video intensities. All video attributes are reversed by enabling the REVERSE VIDEO ATTRIBUTE (pin 22) input.

A sixth attribute is available, the nature of which depends upon the chip type selected. With the "A" version of the 2673, this attribute is a LIGHT PEN strike-through (pin 21) that is used in conjunction with the LIGHT PEN LINE (pin 19) input. Otherwise, the chip contains a GRAPHICS MODE input (pin 21) and GRAPHICS MODE output (pin 19) to synchronize graphic displays.

The attributes can be administered in two ways: as a field, or character by character. The Character Mode is specified by pulling the ATTRIBUTE MODE (pin 12) LOW and the ATTRIBUTE FLAG (pin 13) HIGH. This enables the attributes one character at a time.

When the ATTRIBUTE MODE input is HIGH, the attributes are permanently established in a Field Mode. The LAST LINE (pin 18) timing signal extends the attributes across character row boundaries. Using this method, all attributes are effective until another set is entered by strobing the ATTRIBUTE FLAG anew.

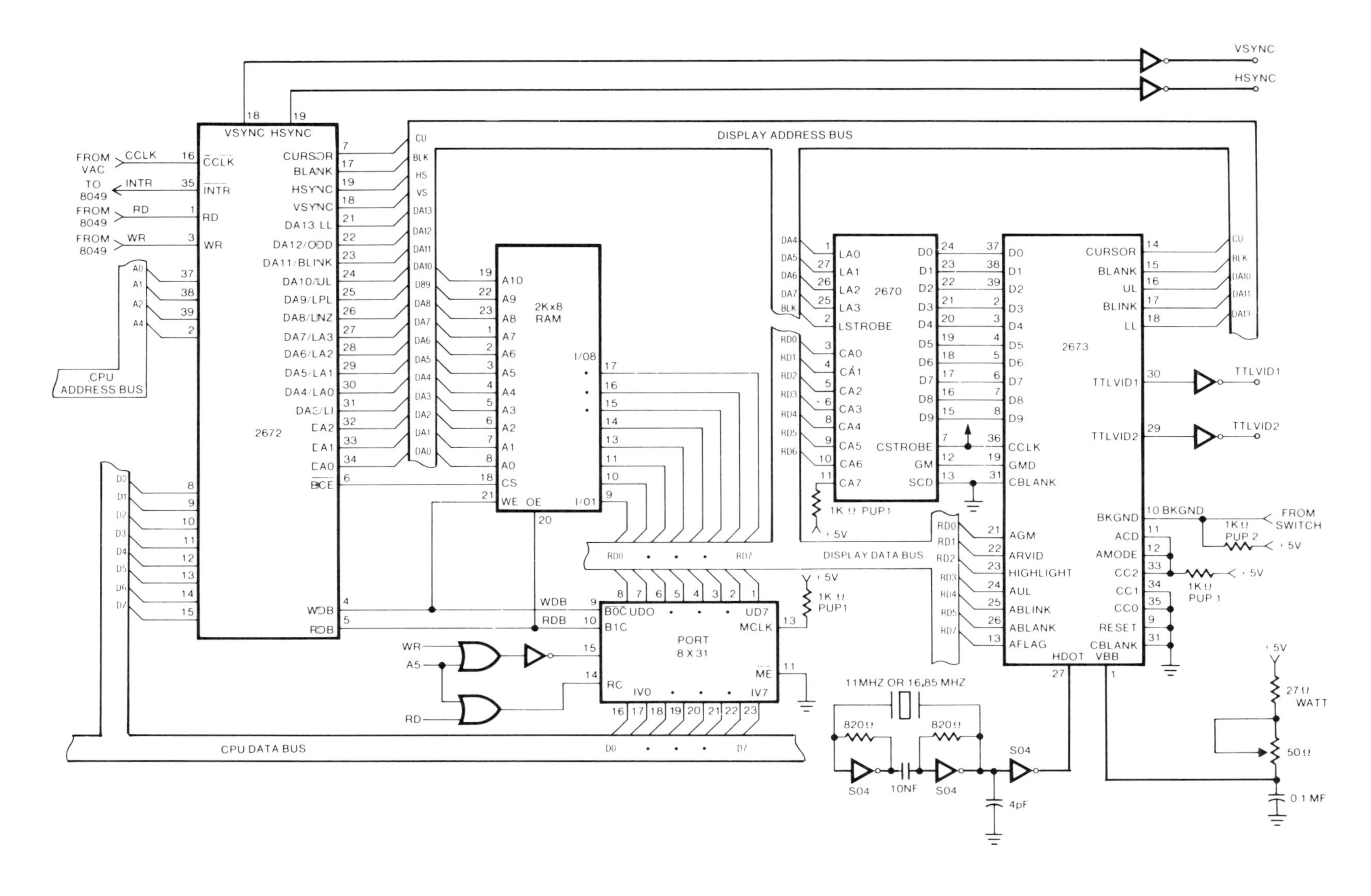

Complete Video Terminal (Sheet 1 of 2). Continues on page 171.

SC2670

Display Character & Graphics Generator

The SC2670, by Signetics, is a mask-programmable Display Character and Graphics Generator ROM which stores the symbols and graphics necessary for video display applications. This chip has been specifically designed to work with the Signetics 2673 Video Attributes Controller, yet be flexible enough to comply with other designs.

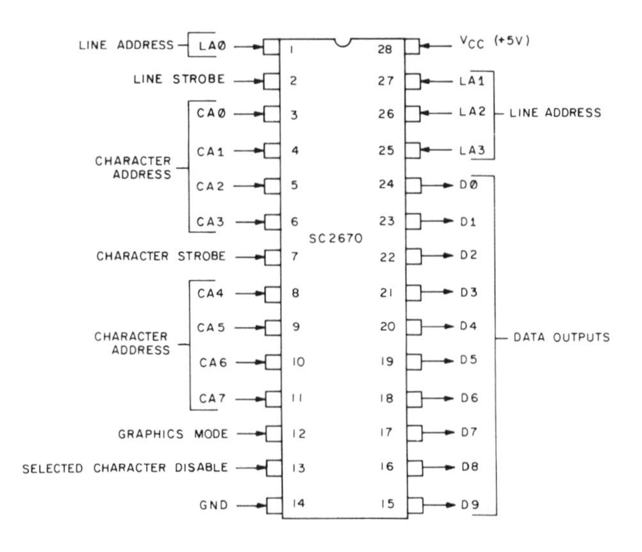

Characters

The chip contains 128 10 × 9 alphanumeric characters placed in a 10 × 16 matrix block. Since the character height is 7 scan lines less than the character row, the generator has the capability of shifting certain characters—such as the j, g, etc.—below the baseline by simply lowering the 9 active lines within the matrix structure. Consequently, there is no mistaking a lower-case symbol for its capital.

The character of interest is specified by a CHARACTER ADDRESS (pins 3-6, 8-11) code. This address is latched into the chip on the negative-going edge of the CHARACTER STROBE (pin 7). Choosing one of 128 characters, however, only requires seven bits of address. Therefore, the eighth bit, CA7 (pin 11), is used as a chip enable input. When this pin is HIGH, it enables the DATA OUTPUTS (pins 24-15).

The information present on the DATA OUTPUTS defines the dot pattern for one scan line of the requested character; the particular line involved is determined by a 4-bit word to the LINE ADDRESS (pins 1, 27-25) inputs. When the LINE ADDRESS inputs are binarily sequenced, the device will automatically place a 10 × 9 character on the 16-line matrix, with the vertical position defined within the ROM generator. To synchronize the character generator with the screen raster, the LINE ADDRESS word is entered into the *line address latch* on the falling edge of LINE STROBE (pin 2).

Basically, there are only 96 characters of general interest. With 128 characters available, 32 of them are classified as special purpose symbols, and are stored within the first two rows of ROM memory. The SELECTED CHARACTER DISABLE input (pin 13) precludes the use of these special characters, when set HIGH, by forcing the DATA OUTPUTS LOW whenever they are addressed.

Graphics

The 2670 also includes a graphics capability, wherein the 8-bit CHARACTER ADDRESS code is translated directly into 256 graphic patterns. For the Graphics Mode, the GRAPHICS MODE input (pin 12) is selected HIGH, and the output data is generated by the graphics logic instead of the ROM. Since it requires 8 bits of input to address the 256 graphic patterns, the CA7 bit no longer controls the DATA OUTPUTS, and they remain enabled at all times.

The graphics pattern is divided into eight blocks, each measuring 5 pixels wide by 4 pixels high. The blocks are organized 4 blocks tall by 2 blocks wide in order to fill the 10 × 16 character space they must occupy.

Due to the graphic arrangement, each scan line must generate two contiguous block patterns per character space. The presence or absence of any block is determined by decoding the CHARACTER ADDRESS inputs. The first five DATA OUTPUTS, D0 to D4, represent the first column of graphic blocks, the next five D5 to D9, reflect the next column. Notice that the DATA OUTPUTS are divided into two groups, with all five members of each group assuming the same value to create a solid line. By arranging these blocks in different sequences, the unique graphic patterns are formed.

The Graphics Mode also has provisions for Thin Graphics, or line drawings. These 16 thin-line figures are only one pixel wide, and are located at hex addresses H'80' to H'8F'.

CRT Terminal Chip Set

The 2670 is a member of a four-chip family designed for CRT terminal applications. Three of these chips alone—the 2672 Programmable Video Timing Controller, 2673 Video Attributes Controller, and 2670 Display character and Graphics Generator—make up a major portion of a CRT display system.

When these three chips are combined with the SC2671 Programmable Keyboard and Communications Controller, they form a complete CRT terminal. Only 15 IC packages are required altogether for a minimum count system when using the Signetics chip set. Despite this low chip count, the terminal provides many important features not found in more complex designs.

A complete schematic of a minimal count CRT terminal, which has a display format containing 24 rows of 80 characters each and graphic capabilities, is found by combining the schematics for the 2671 with those for the 2673.

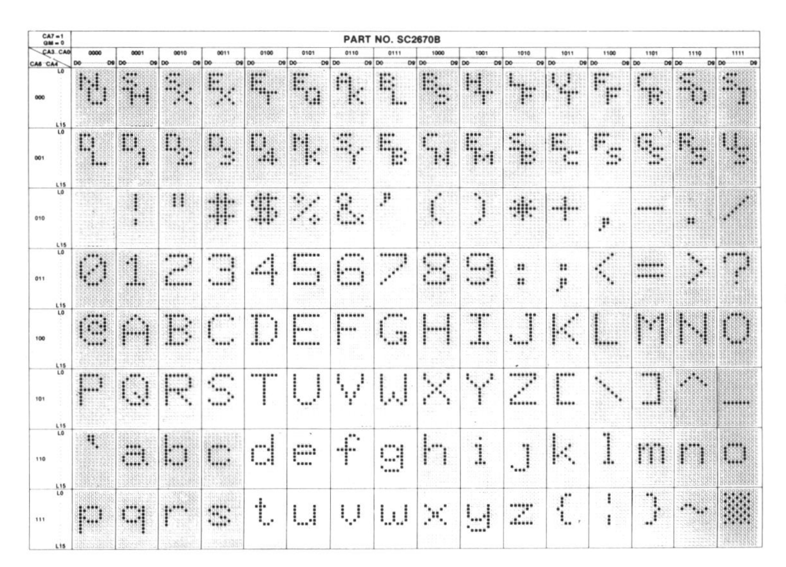

Figure 1

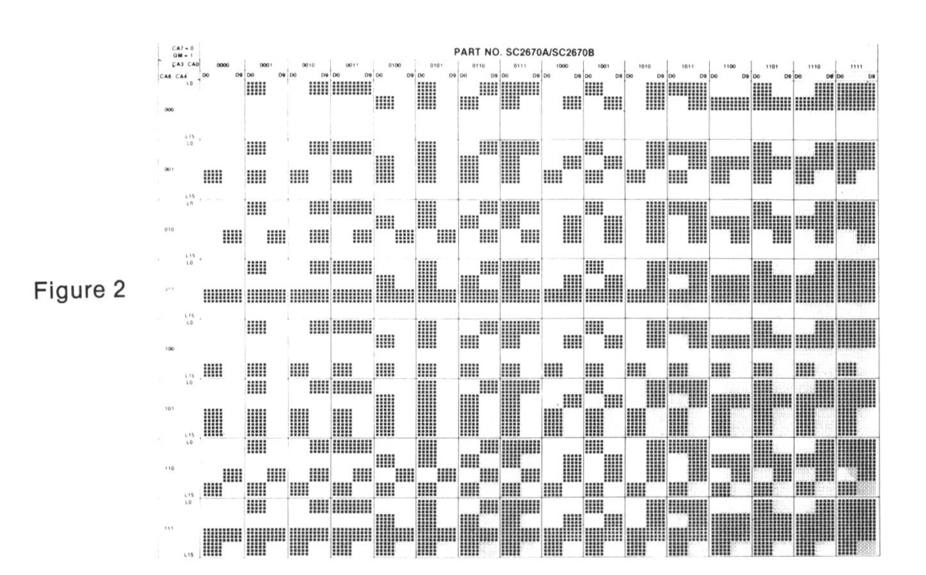

Figure 2

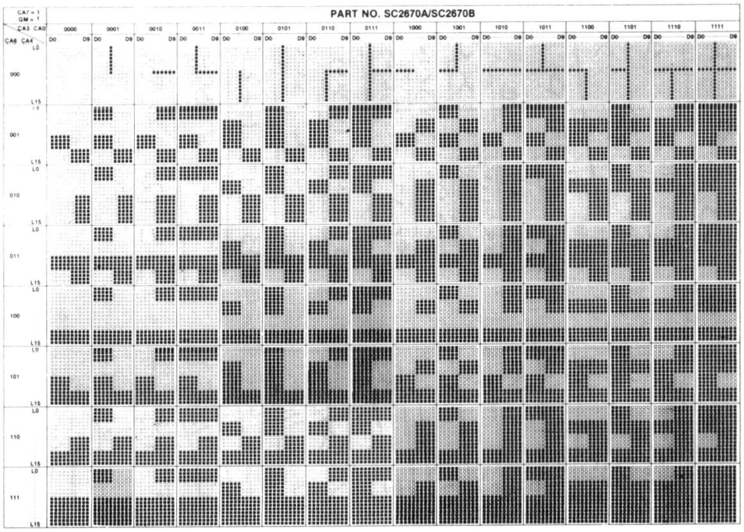

Figure 3

SC2671

Programmable Keyboard & Communications Controller

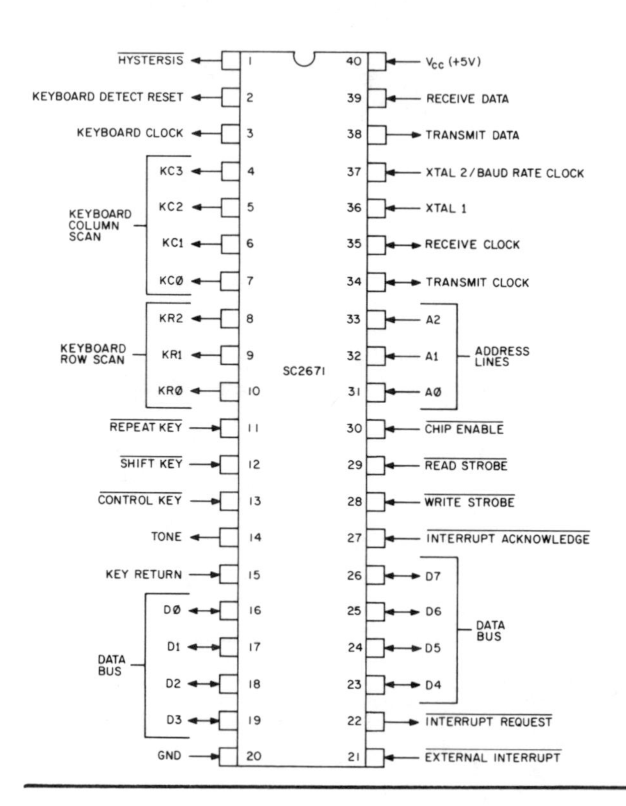

The SC2671 Programmable Keyboard and Communications Controller, by Signetics, is a versatile keyboard encoder plus a full duplex asynchronous communications controller on a single chip. The keyboard interface is capable of encoding up to 128 keys with four keyboard rollover modes. The communications controller section of the chip comprises a full duplex asynchronous receiver/transmitter (UART) with a baud rate generator.

Processor Interface

Operation of the SC2671 controller is determined by 10 internal registers under software control. Communications between the CPU and the chip is established through eight bidirectional DATA BUS lines (pins 16-19, 23-26). The registers are accessed using three ADDRESS LINES (pins 31-11) inputs. The direction of the data flow is controlled by the READ STROBE (pin 29) and the WRITE STROBE (pin 28) inputs, in conjunction with the CHIP ENABLE (pin 30) line.

Several conditions may be programmed into the SC2671 which will effect an INTERRUPT REQUEST (pin 22) to the CPU. An interrupt is subsequently answered with an INTERRUPT ACKNOWLEDGE (pin 27, LOW), placing the address of the current interrupt on the DATA BUS. An EXTERNAL INTERRUPT (pin 21) input queries the SC2671's interrupt priority resolver.

Keyboard Encoder

The keyboard encoder consists of a matrix scanner which detects the presence of a keystroke. The keyboard is continuously scanned by the KEYBOARD COLUMN SCAN (pins 7-4) and the KEYBOARD ROW SCAN (pins 10-8) outputs. A key closed condition causes the current scanning address to be encoded and loaded into the *keyboard holding register,* whereupon the Keyboard Ready Flag is generated.

Four levels of key encoding correspond to the separate SHIFT KEY (pin 12) and CONTROL KEY (pin 13) input combinations. Any key can be made to repeat 15 times per second by enabling the REPEAT KEY (pin 11) input LOW, or selected keys can be repeated automatically using software.

For capacitive keyboards, the high-frequency KEYBOARD CLOCK (pin 3) can be used to gate the KEYBOARD COLUMN SCAN to the keyboard (see fig. 2). The KEY DETECT RESET (pin 2) output resets an external analog detector prior to scanning each key location. The output from the analog multiplexer is sensed and then locked in the analog detector, which is input to the KEY RETURN (pin 15) and the address encoded. The analog detector signal is debounced by the HYSTERSIS OUTPUT (pin 1); if a logic 0 is present, hystersis will be introduced—when the output signal is HIGH, no hystersis exists.

A TONE output (pin 14) of 1-KHz or 2-KHz at bursts of 25-ms and 100-ms is available to the user for an audio keystroke indicator.

Communications Controller

The communications controller section of the SC2671 is a full duplex UART that is totally independent of the keyboard controller. It also may be internally tied into the keyboard under software control. In addition, the receiver and transmitter are separate entities and can operate independently of each other. The UART is extremely versatile, and can be programmed to process words of 5, 6, 7, and 8-bits in length.

The receiver section of the communications controller accepts serial data from the RECEIVE DATA (pin 39) input and converts it into parallel form. Received data is checked for parity (if any), start-stop bits, or break conditions before presenting the assembled characters to the CPU.

The transmitter accepts parallel data from the CPU and converts it to a serial bit stream on the TRANSMIT DATA (pin 38) output. It automatically sends a start bit followed by the data bits, an optional parity bit, and the programmed number of stop bits. Multiple words are transmitted in the sequence entered.

Baud Rate Generator

An internal baud rate generator with 16 divider ratios can be used to derive the receiver and/or transmit clocks. The baud frequencies are programmable between 50-baud and 38400-baud. All timing parameters assume a clock input of 4.9152-MHz—if this frequency is different, the timing rates will vary accordingly.

An external crystal is located between pins 36 and 37, the XTAL1 and XTAL2 inputs, respectively, for frequency control. If desired, the XTAL2 input can be driven by an external TTL clock signal.

The configuration of the baud rate generator, and the baud frequency, is controlled by the *baud rate control register.* The function of the TRANSMIT CLOCK (pin 34) is determined by the 7th bit of this register. When this bit is programmed LOW, the TRANSMIT CLOCK pin becomes an input, and the transmitter's baud rate is set by an external clock. If an internal clock is selected by programming this bit HIGH, the pin becomes an output which reflects a multiple of the actual baud rate as selected by bit 5. The same programming sequence applies to the RECEIVER CLOCK (pin 35) port using bits 6 and 4, respectively.

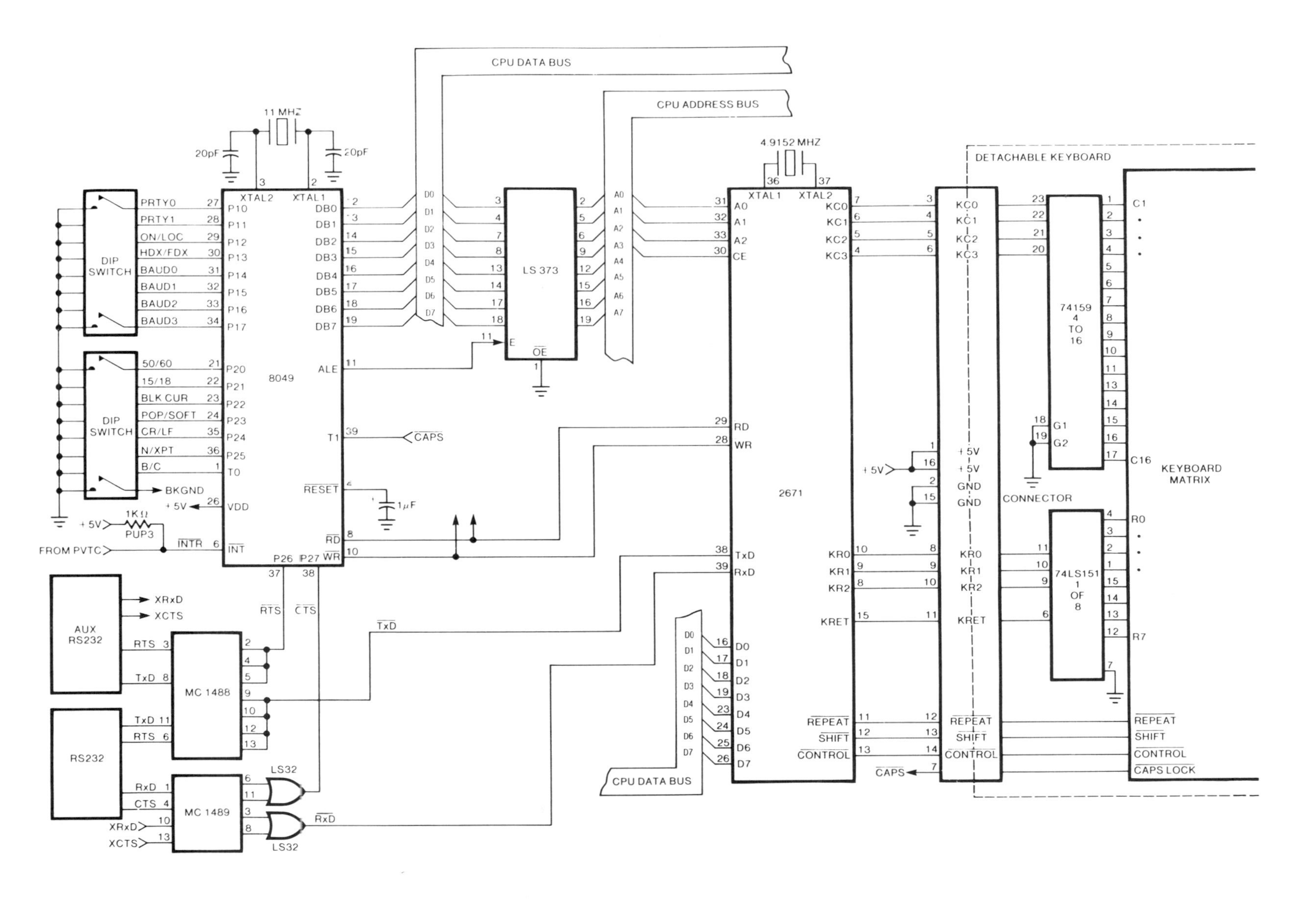

Complete Video Terminal (Sheet 2 of 2). Continues on page 167.

CDP1861

Video Display Controller

The CDP1861, made by RCA, is a Video Display Controller designed for use in CDP1800-series based microcomputer systems. The chip generates both vertical and horizontal timing signals, plus video, and is compatible with most video monitors and RF modulators.

General

The 1861 is wholly compatible with the 1802 microprocessor and will interface to it directly. In fact, the controller utilizes many of the unique features of the 1802 to simplify control and minimize the need for external circuitry. Not the least of which is Direct Memory Access (DMA) coordinated by the 1802 CPU.

Video Display

The video display is a bit-map representation of the video memory. Each bit in the display memory corresponds to one spot on the video screen. The upper left-most pixel that can be displayed on the screen is the most significant bit of the last byte within the memory map. Likewise, the lower right-most dot that can be displayed is the least significant bit of the last byte within the memory map.

The actual video display takes place within a "window" positioned on the CRT screen. The display window begins 62 horizontal scan lines from the top of the screen and is 128 scan lines tall. The remaining 54 lines before the bottom of the screen are simply blanked, as are the top 62. For each of the 128 horizontal scan lines, 8 bytes of video are required. Therefore, the highest resolution possible without external hardware modification is 128 vertical by 64 horizontal segments. This resolution requires 1024 bytes of memory for display.

An INTERRUPT REQUEST (pin 3) is generated by the 1861 once per field, two scan lines before the raster has reached the display window. This request alerts the CPU that the controller will be needing video data from the memory, and to prepare for DMA activity. Beginning on the third CPU machine cycle of each of the display scan lines, and lasting for 8 cycles, the 1861 asserts the DMA REQUEST output (pin 2) to request a sequence of 8-bit bytes that will be used to generate the video output.

DMA access occurs only during the display window portion of the screen as controlled by the DISPLAY ON (pin 10) and DISPLAY OFF (pin 11). Normally driven by the N1 and N0 CPU lines, these inputs inhibit DMA access for all times other than window display. A DISPLAY STATUS output (pin 9) occurs for a period of four horizontal cycles prior to the beginning and end of the 128-line display window, and is commonly used by the software to define the display area boundaries.

CPU Timing

The CONTROL A (pin 22) and CONTROL B (pin 21) signals are the CPU's way of letting the 1861 know its intentions. When CONTROL A is LOW AND CONTROL B is HIGH, the CPU is ready to perform DMA operations. The starting location of the display buffer is specified by the CPU in the Interrupt routine, and may be anywhere in addressable memory, be it ROM, RAM, or whatever. Each assertion of the LOAD (pin 5) input is a DMA REQUEST acknowledgement, causing the 1861 to read a memory byte from the DATA BUS (pins 13-20) and immediately shift it out for video display. A logical HIGH bit generates a lighted spot on the screen.

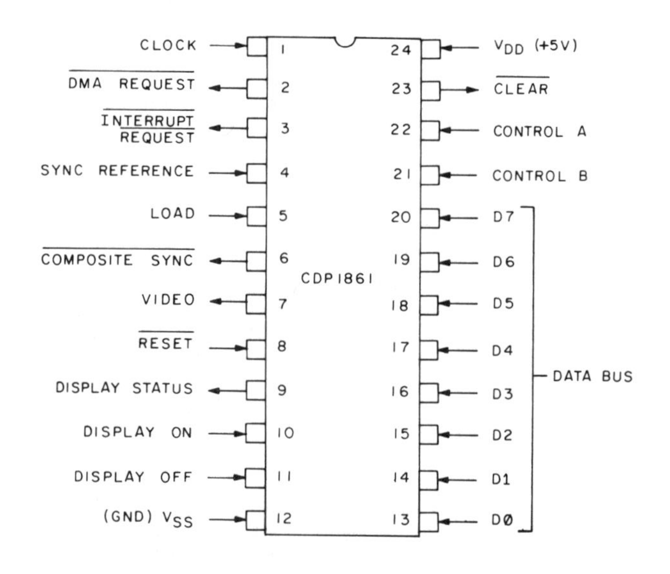

Exactly six machine cycles must be executed beyond the eight DMA cycles during each line, and an even number of cycles must be executed from the start of one display window to the start of the next. These requirements insure that the DMA bursts will not be delayed while waiting for an instruction to finish—a delay that could cause jitter on the screen.

Timing is simplified by operating the microprocessor and video controller at a CLOCK (pin 1) frequency of 1.76064-MHz. A CLOCK source of 3.58-MHz, the NTSC color frequency, may also be used for some applications when divided by two; deviations from the standard NTSC frequencies are shown in fig. 2. When selecting a CLOCK frequency, the user should be aware that, in general, video CRTs are more sensitive to line frequency accuracy than they are to field frequency.

Video Signals

The 1861 generates composite vertical and horizontal sync, COMPOSITE SYNC (pin 6), that can be combined with the VIDEO (pin 7) output to create an NTSC-compatible composite video signal. The SYNC REFERENCE (pin 4) input is used as the clock for the horizontal line counter; this signal is normally derived from the CPU's TPA output, which generates one pulse for every eight clock pulses. Vertical sync is derived from the horizontal sync by dividing it by 262.

The COMPOSITE SYNC creates a 262-line-per-field, 60-field-per-second non-interlaced video picture. The non-interlaced frame generated for display consists of two even fields of 262 horizontal lines apiece. This format differs slightly from the NTSC standard, which has 525 interlaced lines of one odd field and one even field.

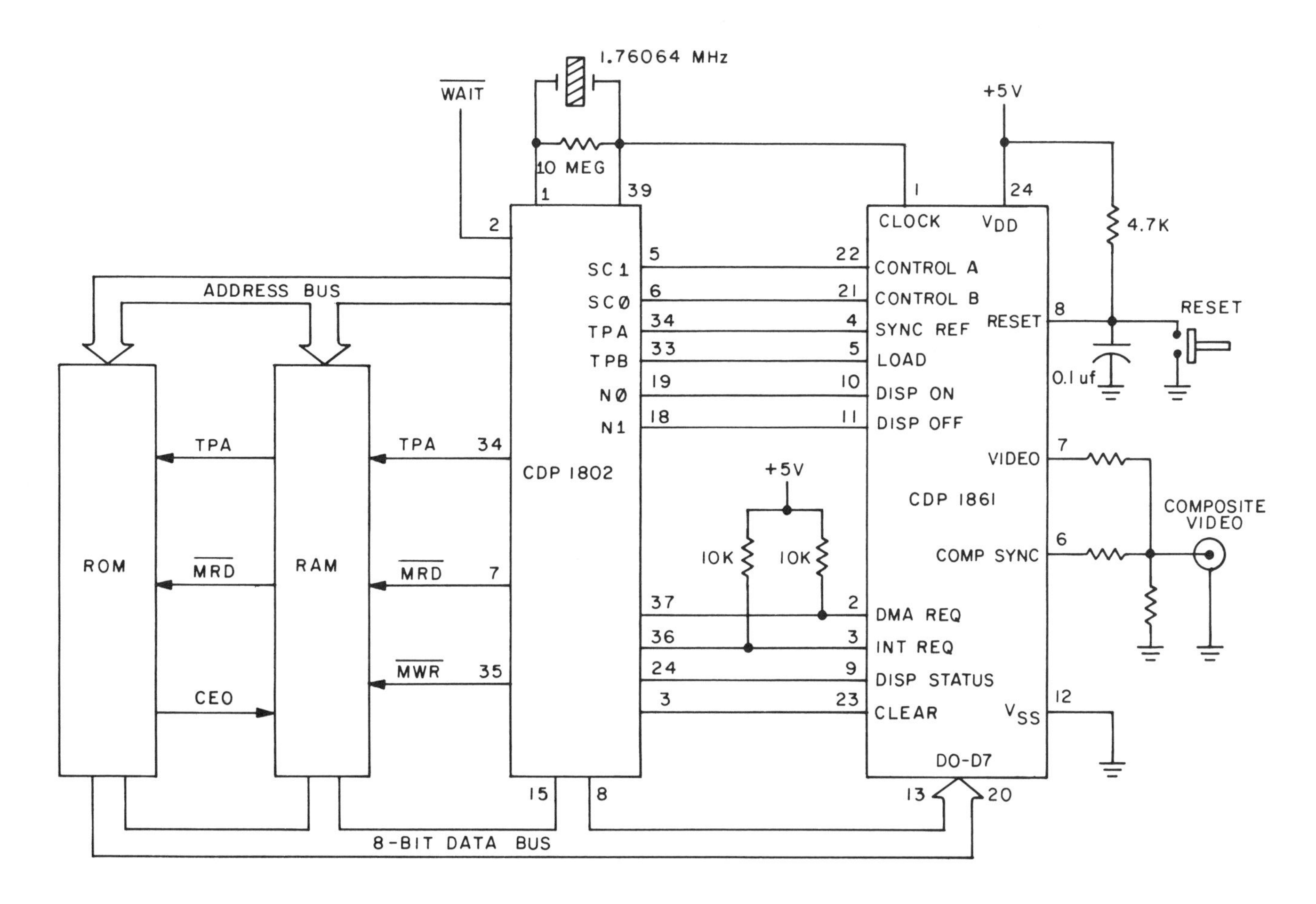

Other display formats, using less memory and having lower resolution, can be created by reproducing a scan line two or more times. For instance, a 64 by 64 field is produced when reading the same scan line memory space twice.

Figure 1

NTSC STANDARD		CLOCK FREQUENCY (MHz)		
		1.76064	1.76400	3.7579545/2
LINE FREQ	15750	15720	15750	15980
FIELD FREQ	60	60.00	60.11	60.99

Figure 2

CDP1862

Color Generator Controller

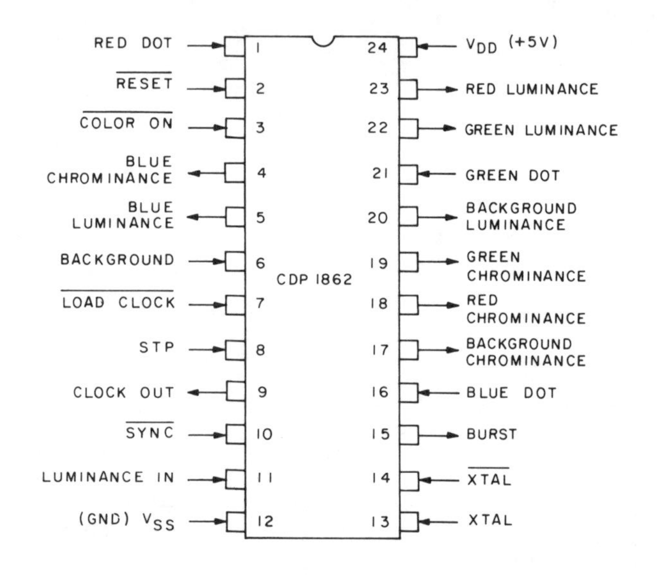

The CDP1862, by RCA, is a Color Generator Controller designed for use in CDP1800-series microprocessor systems. It is primarily intended for use with the RCA CDP1861 video display controller, and directly interfaces with it. Utilizing many of the features of the CDP1802 and CDP1861 to simplify control and minimize external circuitry, the CDP1962 chip generates NTSC color-compatible signals.

Operation

Essentially, the 1862 color generator adds color to the video information generated by the 1861 video controller chip. Each 8-bit byte of horizontal video information can be represented by a color using a color mapping scheme.

Following the same bit-map concept of the 1861 chip, each byte in the video memory is assigned a color, the code of which is stored in a separate external memory. This color map appears as a write-only memory to the CPU, occupying a unique, unused block of memory space. A total of 3072 memory bits, organized 1K × 3 bits for 1024 color blocks, are required when using the system's maximum resolution of 128 × 64.

As the video memory is read for display, the color map, is accessed at the same time. This is accomplised by decoding the video memory address to correspond to the address of the color map location containing the color scheme for that byte, thus pairing the two with only one DMA command.

Background Color

The 1862 begins the display by providing a choice of four background colors: blue, black, green, and red. These colors represent the hue of the unmodulated screen, and are contained within the chip in sequential order. The first color is blue.

Should the user decide to select a new background color, the BACKGROUND input (pin 6) is raised HIGH and the STP input (pin 8) strobed. From the original blue, this operation changes the background to black. A subsequent STP strobe switches it to green, and another puts it red. If the STP is pulsed one more time, the screen reverts back to blue and begins the cycle anew.

Dot Color

The video information generated by the 1861 chip is input to the color generator through the LUMINANCE IN (pin 11) input, where it is encoded with color according to the color map. Altogether, there are eight colors to choose from. These colors are created by combining three signals from the color map RAM—RED DOT (pin 1), BLUE DOT (pin 16), and GREEN DOT (pin 21)—inside the color controller.

By varying the configuration of these inputs, one of the eight colors listed in fig. 2 will be displayed on the screen whenever the LUMINANCE IN input is HIGH. Otherwise, the background color takes precedence. Latching the color scheme into the 1862 chip occurs when the LOAD CLOCK input (pin 7) is strobed while the STP pin is held HIGH.

Clock

A 7.15909-MHz crystal-controlled oscillator is used to generate multiple phases of the 3.579545-MHz color burst clock frequency necessary for NTSC compatible color. The oscillator is controlled by an external crystal in parallel with a resistance, typically 10 Megohms, connected across the XTAL (pins 13 and 14) inputs. An external 7.15909-MHz clock may also be used to drive the oscillator.

Two inputs, STP and SYNC (pin 10), are used to maintain system synchronization. The crystal frequency is divided by two and supplied to BURST output (pin 15), where it is externally mixed with the SYNC pulse for color reference. The color BURST is further reduced by a factor of two to provide a 1.789773-MHz CLOCK OUT (pin 9) signal for the 1861 system timing.

Video Outputs

Using the three primary colors (red, blue, and green) to create the eight controller colors, the 18762 outputs three color signals to the CRT monitor. Color luminance is supplied by the RED LUMINANCE (pin 23), BLUE LUMINANCE (pin 5), and GREEN LUMINANCE (pin 22) outputs; while the color chrominance values are specified by the RED CHROMINANCE (pin 18), BLUE CHROMINANCE (pin 14), and GREEN CHROMINANCE (pin 19) outputs. Background control is handled by the BACKGROUND LUMINANCE (pin 20) and BACKGROUND CHROMINANCE (pin 17) outputs.

These color signals are processed by the 1862's internal clock according to NTSC standards, and combined by an external RC network with the BURST and SYNC signals to generate composite video output.

Reset

A RESET (pin 2) pulse resets the 1862 controller, displaying white dots on a blue background. The chip will remain in this state until the COLOR ON input (pin 3) is pulsed LOW. The COLOR ON input, which is normally connected to the gated MWR line from the CPU, provides a means of inhibiting erroneous color data while the color map is configured.

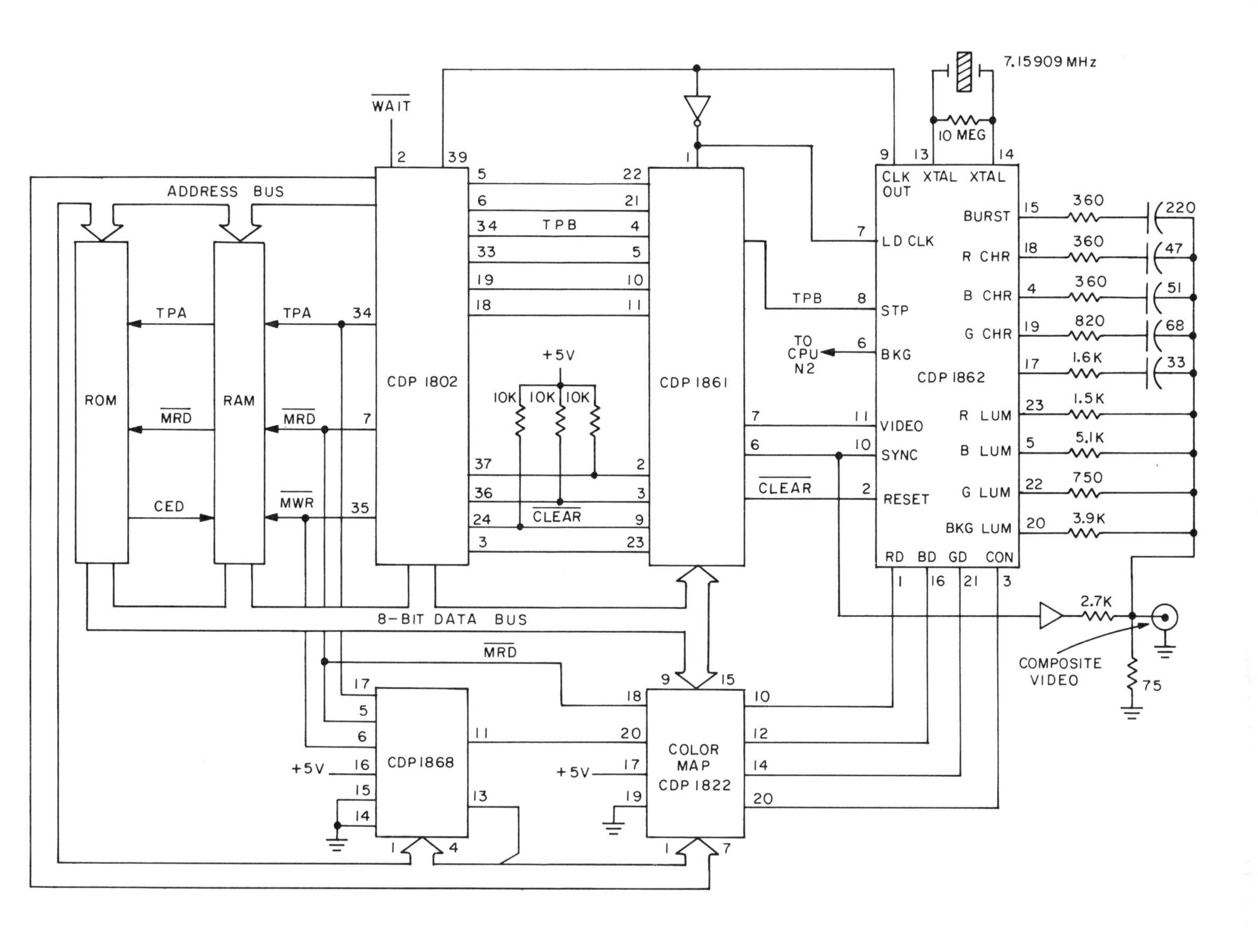

Figure 1

COLOR TABLE

RD	BD	GD	COLOR
0	0	0	BLACK
0	0	1	GREEN
0	1	0	BLUE
0	1	1	CYAN
1	0	0	RED
1	0	1	YELLOW
1	1	0	PURPLE
1	1	1	WHITE

Figure 2

CDP1864

PAL Compatible Color TV Interface

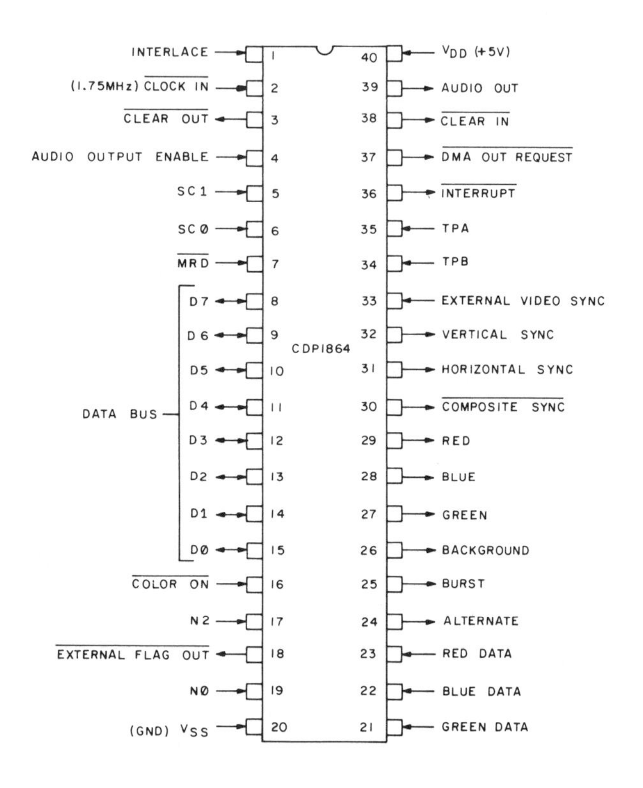

The CDP1864, by RCA, is a PAL-compatible video controller designed for use in CDP1800 microprocessor systems. The chip interfaces directly with the CDP1802 and CDP1804 to generate color or black and white video displays.

Video Display

The video generated by the 1864 is a bit-mapped display with a maximum resolution of 182 lines vertically and 64-dots horizontally. This resolution, however, is seldom used because of the poor aspect ratio of the resultant picture element. An approximately square picture is obtained by repeating each horizontal scan line six times, using software instructions, for a 32-row by 64-dot display. This lower resolution requires 256-bytes of refresh RAM.

Video refresh is accomplished via the DMA OUT REQUEST (pin 37) channel of the CDP1800 microprocessor. The EXTERNAL FLAG OUT (pin 18) signal goes LOW four horizontal lines prior to the start of the display, and again four lines before the end of the display window. This signal alone can be used by the CPU to initialize R(0) (an external CPU routine) for DMA refresh.

Alternatively, the INTERRUPT output (pin 36), which goes LOW two lines prior to the start of the display, may be used to enter an interrupt routine that initializes R(0), and the EXTERNAL FLAG OUT can be used to indicate the end of the display. This combination allows for an interrupt routine to oversee DMA refresh and repeat horizontal lines for configurations less than the maximum 192-line resolution.

CPU Interface

The controller connects to the CPU and memory through an 8-bit DATA BUS (pins 15-8). The 1864 has been designed around the 1800 signals to simplify control and minimize external components. The MRD (pin 7), N0 (pin 19), N2 (pin 17), TPA (pin 35), and TPB (pin 34) lines are a part of these signals, and they connect directly to their CPU counterparts.

SC1 (pin 5) and SC0 (pin 6) are used to provide CPU-to-CDP1864 synchronization for jitter-free display. During every horizontal sync pulse, the controller samples these two lines. Detection of a fetch cycle (both inputs LOW) causes the 1864 to skip cycles to attain synchronization. Once a lock is secured, certain software precautions must be observed to maintain sync.

Video Signals

Latched into the controller on the falling edge of COLOR ON (pin 16), the RED DATA (pin 23), BLUE DATA (pin 22), and GREEN DATA (pin 21) inputs carry color information from the color memory RAM. The bit-mapped video information, input from the video memory through the DATA BUS, is also latched into the chip by the COLOR ON strobe. The chip then combines the raw video material according to PAL standards and outputs BURST (pin 25), RED (pin 29), BLUE (pin 28), GREEN (pin 27), and BACKGROUND (pin 26) video signals.

The background color is program-selected to be either blue, black, green, or red. The BACKGROUND output may be used to lower the luminance background color. This would, for instance, enable a blue spot to be displayed on a blue background, yet still be visible.

The 1864 generates both COMPOSITE SYNC (pin 30) and separate HORIZONTAL SYNC (pin 31) and VERTICAL SYNC (pin 32) pulses. The phase alteration is performed by the ALTERNATE (pin 24) output. The video and sync outputs may be used directly inside a TV monitor, or they may be mixed to produce a composite video signal.

A HIGH level at the INTERFACE input (pin 1) results in the generation of a 625 line-per-frame interlaced display, while a LOW input produces a 312 line-per-frame non-interlaced display.

Tone Generation

The chip also contains a programmable tone generator designed to produce 256 frequencies that range from 107-Hz to 13,672-Hz. The base frequency for this tone generator is derived from the TPB input. An 8-bit programmable up-counter selects the frequency output for the tone.

The audio tone is halved by an internal flip-flop to generate a square wave, and output through the AUDIO OUT (pin 39) port. An AUDIO OUTPUT ENABLE input (pin 4) controls the audio signal. When it is HIGH, the audio is allowed to freely exit the chip; a LOW input to this line clamps the AUDIO OUT line LOW.

Further Reading

The architecture of the CDP1864 is similar to that of the CDP 1861 and CDP1862 combined. The reader would do well to review those sections for a better understanding of the total operation of the chip.

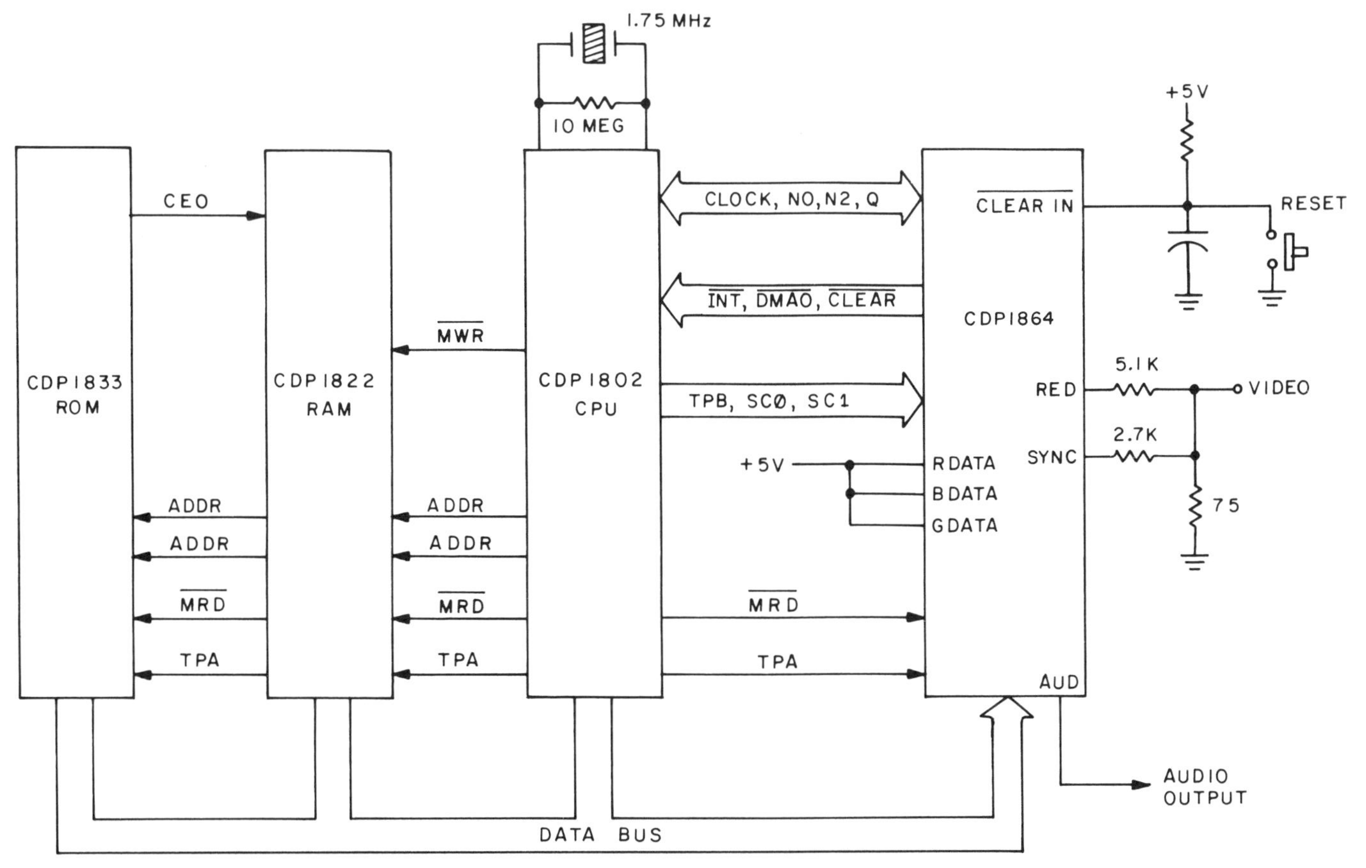

Figure 1

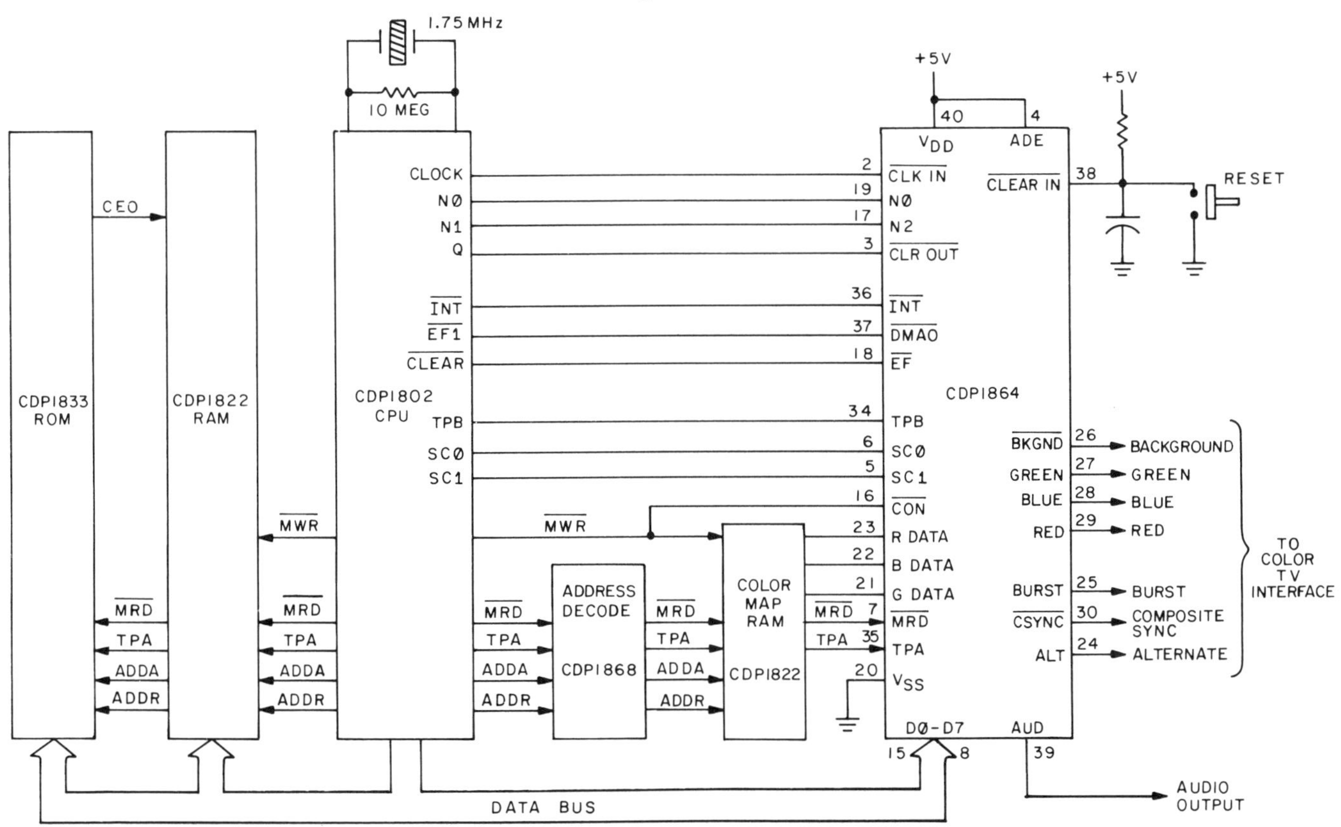

Figure 2

TMS 9918

Video Display Processor

The TMS 9918 Video Display Processor, by Texas Instruments, is an LSI device designed to display video data on a raster-scanned home color television set or color monitor. The chip generates all the necessary video, control, and timing signal requirements, while tending to the chores of memory management, information retrieval, and screen refresh.

General Information

The 9918 has four video display modes which can be best described by envisioning the visual display as having 35 overlapping picture planes. Each display plane starts off transparent, and the user then adds to each leaf the appropriate information. When the entire display is assembled, layer upon layer, the superimposed graphics create the desired effect.

The four display modes include Graphics I, Graphics II, Multicolor, and Text modes. Since the description of the display is extensive—and because there are three chips in the 9918 family (9918, 9928, & 9929)—the discussion will be broken into three sections, with each chip revealing a portion of the whole.

CPU Interface

The 9918 interfaces to the CPU using a standard 8-bit bidirectional DATA BUS (pins 17-24). Through the bus, the CPU can read or write to video memory, program the registers, and check the chip's status. The performance of the DATA BUS is controlled by three pins, two of which determine the direction of the data flow. When a write operation is to be performed, the WRITE STROBE input (pin 14) is pulsed LOW. A read operation, on the other hand, requires that the READ STROBE input (pin 15) be strobed instead. The READ STROBE and WRITE STROBE should never be activated simultaneously—if both are LOW, invalid data will enter the registers.

The MODE control (pin 13) determines the source of the read or write transfer, and is used extensively for data manipulation. This input is normally tied to a low-order CPU address line. These three pins together provide access to all internal registers. It should be further noted that the CPU can communicate simultaneously and asynchronously with the 9918 during TV screen refresh operations, even in the middle of a raster line.

CRT Monitor Interface

All the necessary horizontal and vertical timing signals, as well as NTSC-compatible luminance and chrominance information, is output through a single COMPOSITE VIDEO (pin 36) port. While the video signals aren't exact equivalents to the standard NTSC colors, the differences are slight and can be easily adjusted with the color and tint controls of the monitor.

The output buffer is a source-followed MOS transistor that requires an external pull-down resistor. A value of 470-ohms is typically recommended for 1.9-volt output. To drive a standard television receiver that is not equipped with a composite video input, the signal can be processed by a suitable RF modulator and fed into the antenna terminals.

Graphics I

The Graphics I Mode breaks the picture plane into 768 individual pieces, which are arranged in a grid of 32 columns by 24 rows.

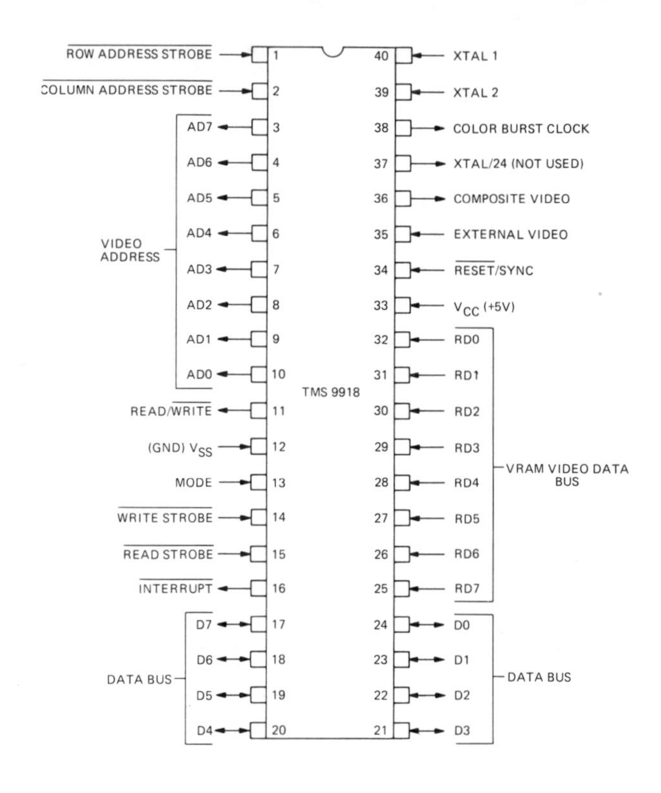

Graphics II

Like Graphics I Mode, the Graphics II *pattern name table* contains 768 entries which correspond to the 768 pattern positions on the display screen. However, the *pattern generator table* has been expanded from 256 to 768 patterns, permitting each and every pattern block on the screen to have its own unique pattern. In addition, the color scheme allows a different pair of color bits for *each* of the eight pattern block rows.

In order to accomplish this, though, the controller must divide the screen into three segments to accommodate the increased capacity. When entering pattern information, the user must also specify which section of the screen will display the pattern.

Each piece is a square 8 pixels wide and 8 pixels high, that we shall call a pattern block.

Configuring the pattern block is done pixel by pixel, and the 9918 contains a programmable *pattern generator table* which can store up to 256 unique block patterns. The pattern encoding is laid out one row at a time; each pixel is assigned a 0 or 1 logic to identify the color of that particular pixel, creating a pattern. As you can see, only two colors are allowed in this mode, and they are defined by software in the *pattern color table.* Altogether, there are 15 colors available to the user, plus eight shades of gray for monochrome displays—but the user is allowed to choose only two per pattern block. However, the colors may be varied from one block to another.

After each 8-bit row of the pattern block has been defined (eight rows in all, for 8 bytes total), the pattern is given an 8-bit name. This name is entered into the *pattern name table*, and every time it is called up, the *pattern name table* will access the *pattern generator table* for the complete 8-byte code.

The Pattern Name entry also contains the location of the requested pattern on the screen, as well as the pattern's colors located in the *pattern color table*, Once the *pattern name table* has been loaded with the pattern set, images can be formed using the pattern plane. Rough motions of objects is achieved by merely updating entries in the *pattern name table.*

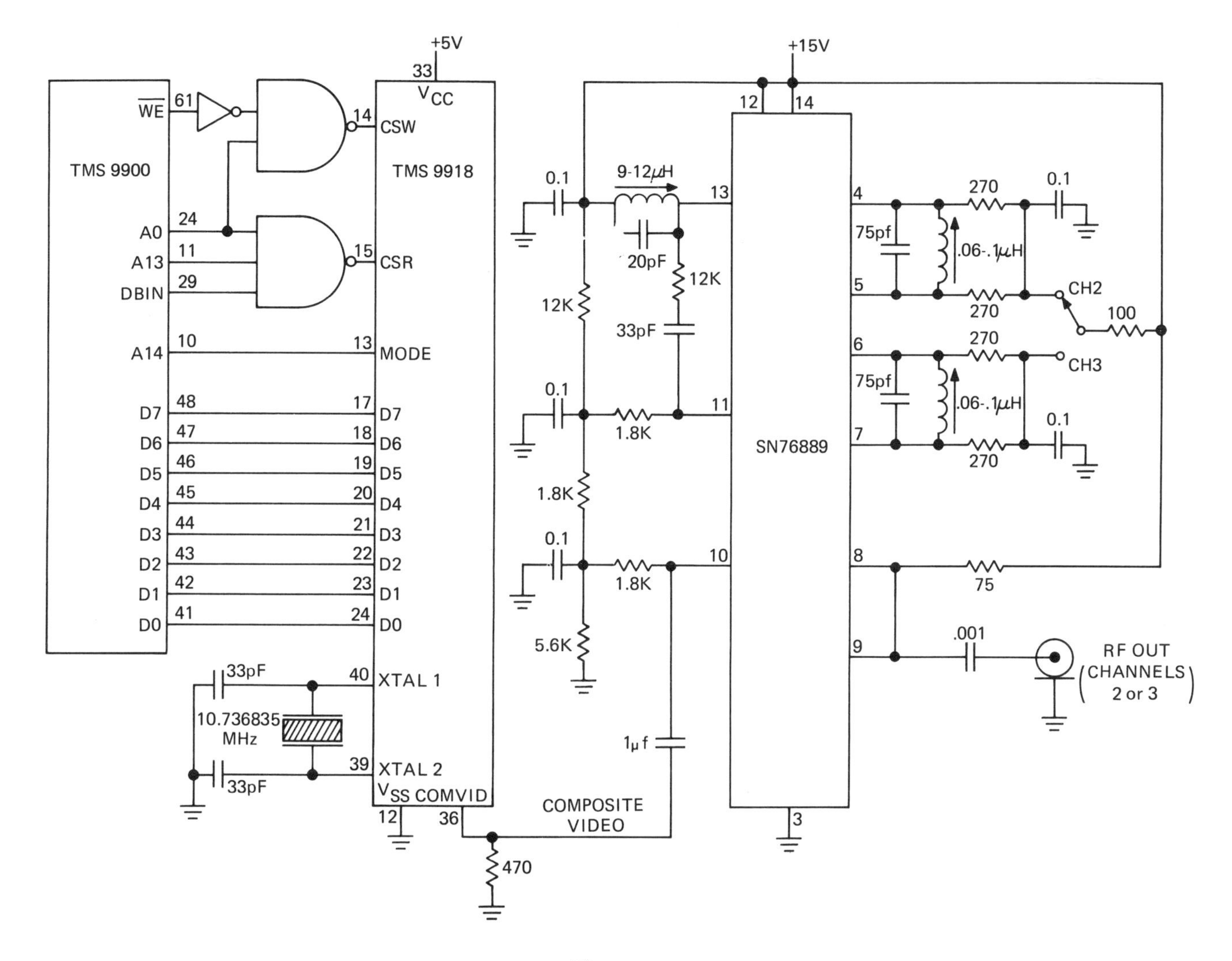

Figure 1

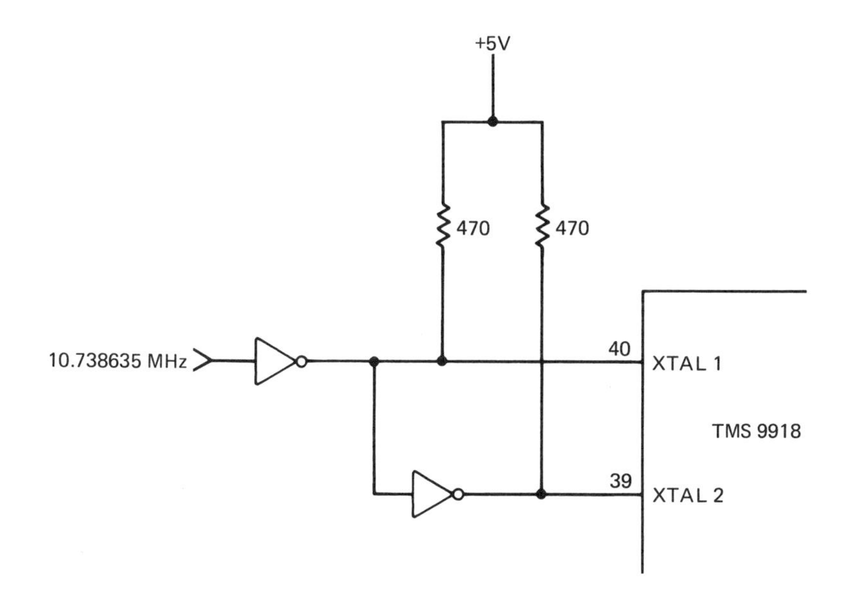

Figure 2

Alternate clock circuit utilizes external 10-MHz generator. The XTAL inputs are not TTL compatible and require pull-up resistors.

TMS 9928

Video Display Processor

The TMS 9928, by Texas Instruments, is a Video Display Processor functionally identical to the TMS 9918 except that the NTSC color encoding circuit has been removed and replaced with luminance and color difference signals.

CRT Monitor Interface

The video output signals are available at the Y LUMINANCE (pin 36), B-Y (pin 35), and R-Y (pin 38) outputs. This approach allows the user to directly drive the color guns of the monitor using a suitable buffer, or encode the signals to NTSC or PAL standards. The Y LUMINANCE signal contains composite sync in addition to the luminance levels, and can be used to drive a black & white monitor.

Register Programming

The 9928 has eight write-only registers and one read-only *status register*. The write-only registers control all video procssor operations and establish video memory modes. The *status register* contains Interrupt, Sprite coincidences, and Fifth Sprite status flags.

Programming the registers is done using two word bytes and two write operations. The first 8-byte contains the program data for the register; the second byte specifies the register to be written. The register is loaded by holding the MODE (pin 13) input HIGH for both bytes while strobing the WRITE STROBE (pin 14) input.

If you wish to rewrite the program data after the first byte is entered, but before the address byte is asserted, the CPU must perform a *status register* read operation by strobing READ STROBE (pin 15) LOW. This procedure resets the internal logic so the new word will be accepted as data, and not as a register address, and should be performed whenever the status of the write parameters is in question.

Clock & Timing Signals

The 9928 is designed to operate with a 10.738638-MHz crystal across the XTAL1 (pin 40) and XTAL2 (pin 39) inputs. The master clock is divide by two to generate the pixel clock (5.3-MHz) and by three to provide the 3.58-MHz COLOR BURST CLOCK (pin 38, 9918 only; not available in 9928/9929).

The INTERRUPT (pin 16) pin generates an output at the end of each active display scan, about once every 1/60 second, or when the *status register* is requesting service. Interrupts are cleared when the *status register* is read. The RESET (pin 34) input initializes the chip when held LOW for 3-microseconds or longer.

Sprites

The greatest attribute of the 9928 processor is the inclusion of sprite patterns in all modes except Text. To better understand what a sprite is, envision the screen as a set of transparent video planes sandwiched together. The first 32 planes are dedicated to sprites, behind that lies the Graphics pattern, backdrop, and external planes.

A sprite is a special animation pattern, which exists on each of the first 32 planes (one sprite to a plane) to provide smooth motion and multilevel pattern overlaying. In other words, you can create backgrounds, foregrounds, and smoothly flowing objects by simply superimposing the sprite pattern over the video image,

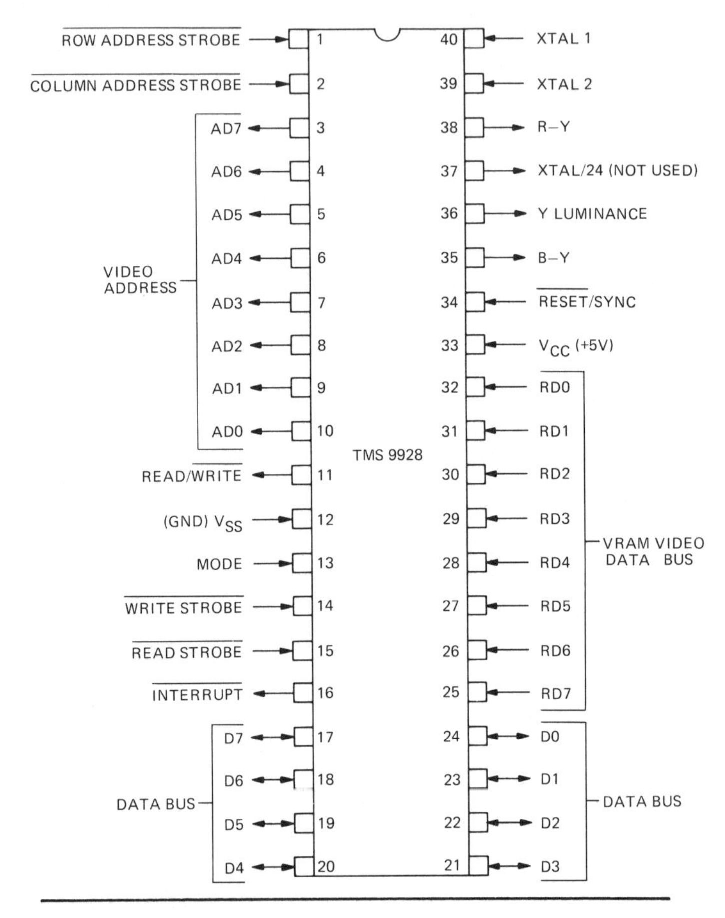

layer by layer. When viewed as a whole, the viewer sees a composite image.

Objects on the planes closest to the viewer have priority over the planes behind it. In cases where two entities on different planes occupy the same spot on the screen, the entry on the plane closest to the viewer will take precedence and mask the spot behind it, setting a Sprite Coincidence flag in the *status register* as it does. All unused sprites are transparent, as is the plane surrounding it.

The sprite is an 8 × 8 pixel block, very similar to the Graphics pattern block. In fact, it is programmed in very much the same way. Each row of the sprite block is programmed pixel by pixel to represent a color or transparency. The shape of the sprite is entered into the *sprite generator table*, and an 8-bit name is assigned to it in the *sprite attributes table*. As with its *pattern name table* counterpart in Graphics, the *sprite attributes table* retrieves the sprite pattern and locates it on the screen.

The sprite can be easily and accurately moved one pixel at a time by redefining the sprite origin. This provides a simple but powerful method of quickly and smoothly moving special patterns.

However, the maximum number of sprites that can be displayed on one horizontal line is limited to four. If a fifth sprite is requested on that line, only the four sprites with the highest priority will be displayed. Sprites are also limited to one color apiece; multiple colors require separate planes.

Sprites come in two sizes: 8 × 8 and 16 × 16 pixels. In the 16 × 16 mode, the rows are longer and they are defined using 31 bytes. Sprites can also be magnified, as specified by *register 1*. In the Magnification Mode, the sprite doubles in size; i.e., an 8 × 8 becomes 16 × 16. However, the resolution does not increase—it remains the same. So now we have sprites of 16 × 16 and 32 × 32 with 2 × 2 pixel resolution. The sprite mapping, however, doesn't change; it can still be moved pixel-by-pixel.

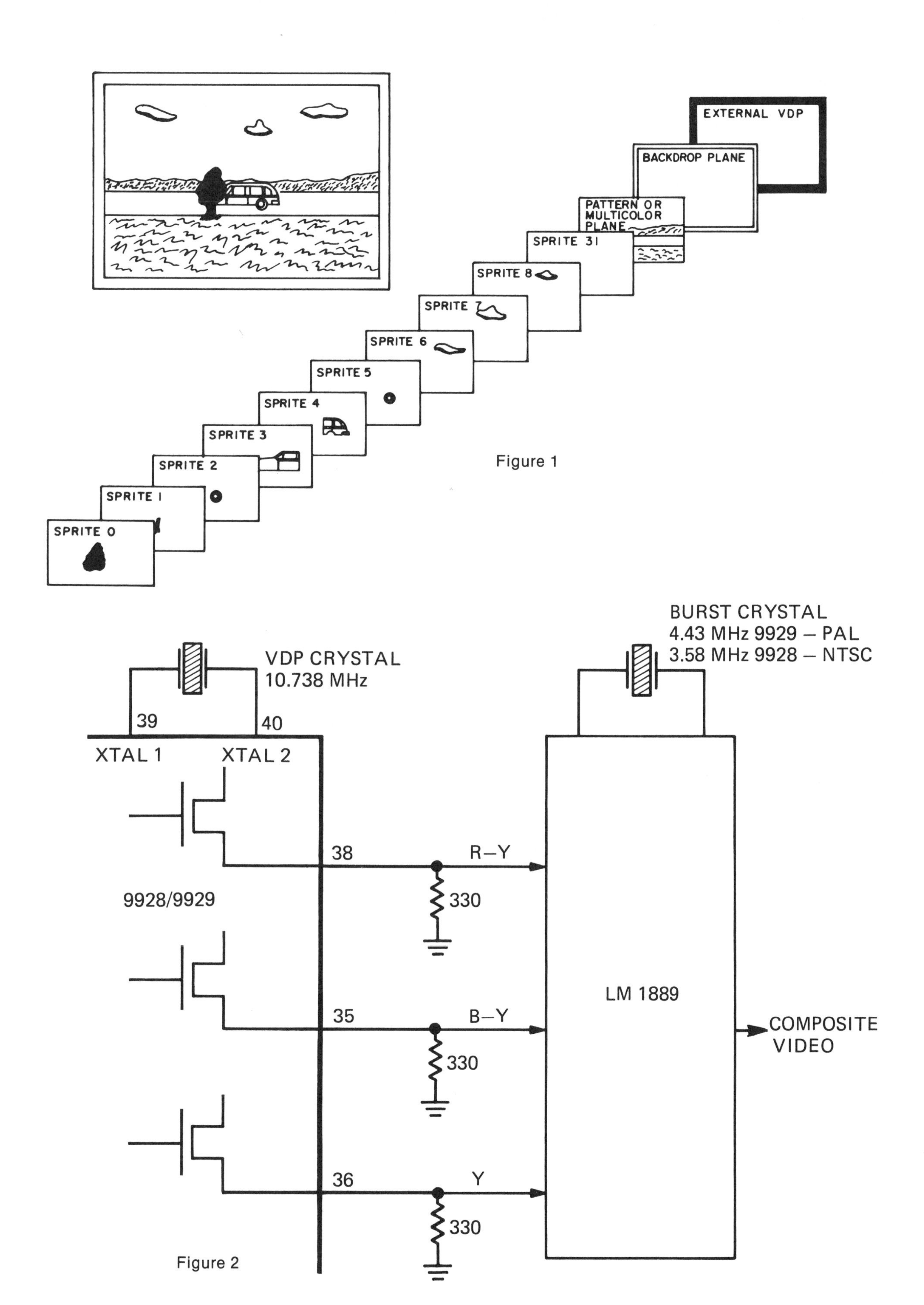
EXTERNAL VDP
BACKDROP PLANE
PATTERN OR MULTICOLOR PLANE
SPRITE 31
SPRITE 8
SPRITE 7
SPRITE 6
SPRITE 5
SPRITE 4
SPRITE 3
SPRITE 2
SPRITE 1
SPRITE 0
VDP CRYSTAL
10.738 MHz
39
40
XTAL 1
XTAL 2
9928/9929
38
R–Y
330
35
B–Y
330
36
Y
330
BURST CRYSTAL
4.43 MHz 9929 – PAL
3.58 MHz 9928 – NTSC
LM 1889
COMPOSITE VIDEO

Figure 1

Figure 2

TMS 9929

Video Display Processor

The TMS 9929 is the third in the 9918 family of Video Display Processor chips by Texas Instruments. This device is pin compatible with the TMS 9928; however, the 9919/9928 devices have been formatted with 525 lines for U.S. television standards while the 9929 has a 625 line field for use with the European PAL system.

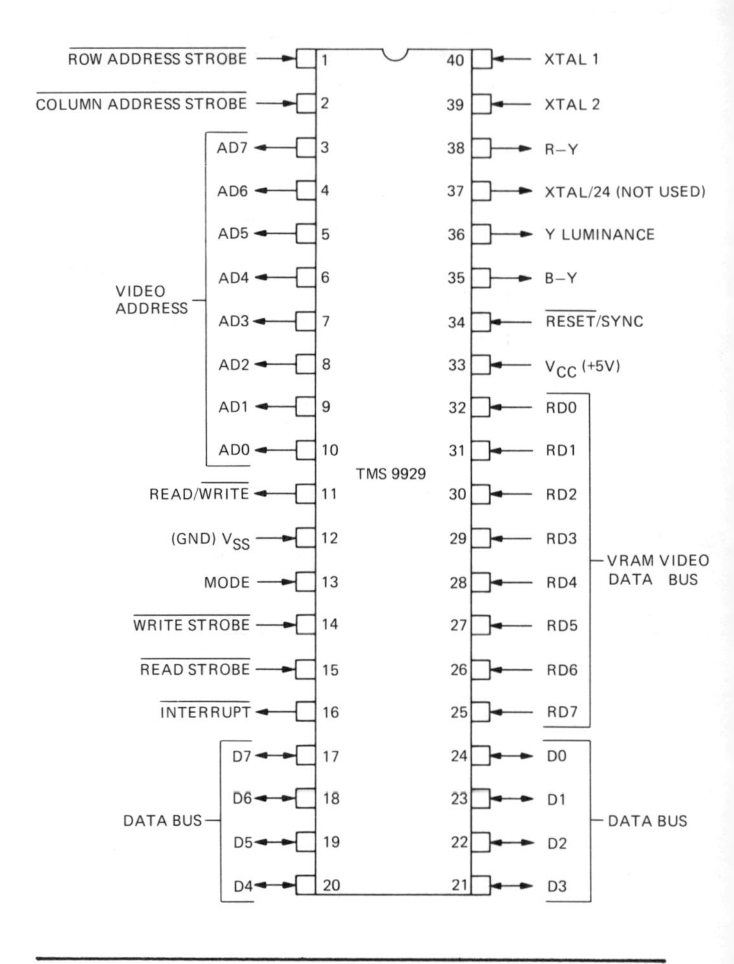

Video Memory

The 9929 can access up to 16K of video memory using two unidirectional 8-bit data buses and three memory control lines. The memory is addressed using multiplexed column/row techniques by the VIDEO ADDRESS outputs (pins 3-10). When the ROW ADDRESS STROBE (pin 1) is actively LOW, the VIDEO ADDRESS outputs address rows A6 through A13. Alternately, the column addresses, A0 through A5, are generated when the COLUMN ADDRESS STROBE (pin 2) is forced LOW.

Data is written from the CPU into the video memory using the same eight VIDEO ADDRESS output ports and an internally executed memory contention scheme. The memory location is first addressed using the column/row addressing procedure described above. This requires two bytes of CPU input. The third byte is memory data which enters the video memory after the MODE (pin 13) control line is changed LOW, and takes place on the LOW strobe of the READ/WRITE (pin 11) output. After the transfer cycle, the *address register* is automatically incremented to the next memory location. Consequently, only one byte is required to enter data into the next sequential memory location. However, a random address write operation must begin with the three step process anew.

The 9929 fetches video memory data on the VRAM VIDEO DATA BUS (pins 25-32) lines. Normally, memory read operations are transparent, and the display processor chip systematically addresses the memory and reads it for screen refresh. When the CPU wishes to read video memory, however, a three-step sequence similar to the write operation is required. The read cycle begins by entering the address of the memory location using two bytes, as before. This time, though, the two MSBs of the second byte are set to 0 to indicate a read operation. The MODE input is then pulled LOW and the READ STROBE exchanged for the WRITE STROBE. Data will be available to the CPU on the next data transfer cycle.

Video Memory Types

With transparent memory refresh available, there are several RAM chips that will fill the bill. The 4027-type 4K, 4108-type 8K, or 4116-type 16K are recommended. However, there is a minor difference in the way the 4027 and 4108/4116 chips are wired to the 9929. In the 4027, all CHIP ENABLE pins are returned to ground. With the 4108/4116, on the other hand, the A6 address lines (same pin as CHIP ENABLE on 4027) are all tied to the AD 1 address on the 9929. A simple jumper connection can be used to select between the memory types.

Multicolor Mode

The Multicolor Mode provides a field of unrestricted color blocks arranged in a 64 × 48 pattern. Each color square consists of a 4 × 4 block of pixels. The color of each square can be any one of the 15 video colors available to the user, plus transparent, and all 15 colors may be used simultaneously.

The color squares are ordered up in blocks of four, with an 8 pixel by 8 pixel border. Only two bytes are required to specify the four colors for each pattern. The color designs are contained in the *pattern generator table*, the same table used for the Graphics modes, and the *multi-color name table* places the blocks on the screen.

However, the screen has been divided into four row locations, spaced four blocks apart, with the color pattern of the block keyed to its position on the screen. Therefore, color patterns located in the *pattern generator table* must be related to the screen location of the displayed pattern, not unlike the Graphics II Mode memory contention scheme.

Text Mode

In the Text Mode, the screen is divided into a text page of 40 columns by 24 rows. Each character block contains a 6 × 8 pixel field. The Text Mode is virtually identical to the Graphics I and Graphics II modes, except that the pattern blocks are reduced in size. Defining the pattern is exactly the same as before, with the *pattern generator table* capable of listing and displaying up to 256 characters.

As the name implies, the Text Mode is intended mainly for textual applications. With care, however, the same text patterns can be displayed in the Graphics I Mode. The advantage of the Text Mode is that eight more characters can be fitted onto each line, and the character blocks are proportionally better suited for letter symbols. The disadvantages are that sprites cannot be used, and only two colors are available for the entire screen.

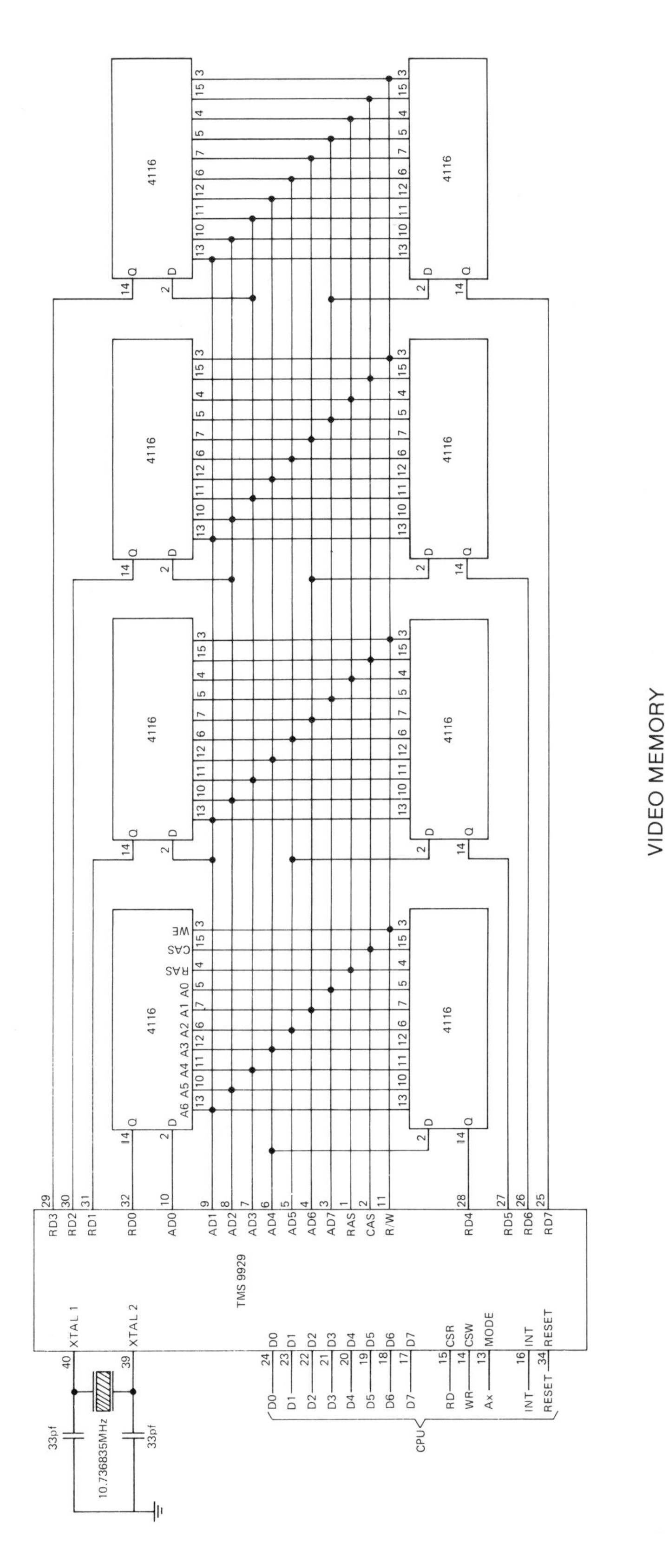
VIDEO MEMORY
4116
TMS 9929
XTAL 1
XTAL 2
10.736835MHz
33pf
33pf
CPU
RD0
RD1
RD2
RD3
RD4
RD5
RD6
RD7
AD0
AD1
AD2
AD3
AD4
AD5
AD6
AD7
RAS
CAS
R/W
D0
D1
D2
D3
D4
D5
D6
D7
CSR
CSW
MODE
INT
RESET
RD
WR
Ax
A6 A5 A4 A3 A2 A1 A0
WE

Figure 1

CRT 9007

Video Processor and Controller

The CRT 9007, by Standard Microsystems Corporation, is a Video Processor/Controller which affords the user a wide range of video attributes while supplying all the timing and control signals needed for the presentation of video data on a CRT monitor. The format of the chip is totally software programmable and capable of defining over 200 characters per data row and up to 256 data rows per frame—and includes scrolling.

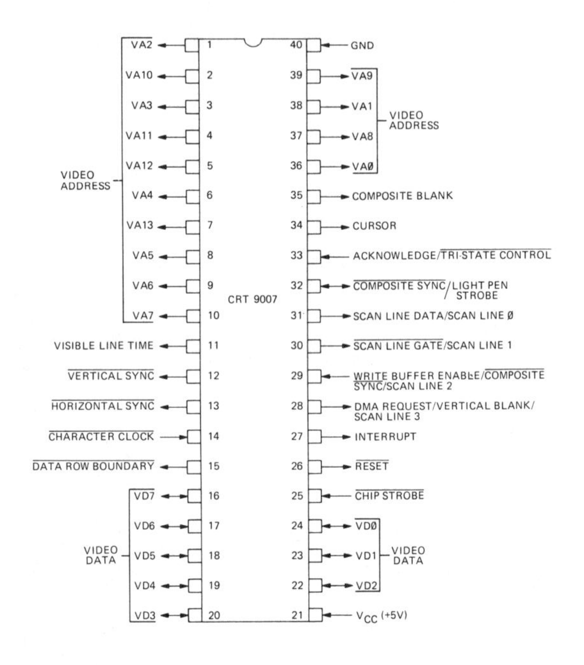

Operation

The chip contains 30 programmable registers that define display operations. These registers are selected by VIDEO ADDRESS lines VA0 through VA5 (pins 36, 38, 1, 3, 6, and 8). VIDEO ADDRESS VA5 (pin 8) performs the read/write control function. When VA5 is HIGH, only the read registers are addressed; when it it LOW, the write only registers are selected. Data is transferred through eight bidirectional VIDEO DATA bus lines (pins 24-22, 20-16) upon receipt of a CHIP STROBE (pin 25) pulse.

Altogether there are 14 lines of VIDEO ADDRESS (pins 1-10, 36-39) that can directly address up to 16-K of video memory. This is equivalent to eight pages of 80 × 24 CRT display. The memory chips are paralleled with the VIDEO ADDRESS lines and the CPU to permit DMA (direct memory access) by both. The controller works with a variety of memory contention schemes. They include no buffer, single and double row buffer, repetitive addressing, and attribute assembly.

Memory Modes

When configured in the Single Row Buffer Mode, the buffer is loaded with an entire row during the top scan line of each data row. The DATA ROW BOUNDARY output (pin 15) is LOW for one full scan of the top scan line of each new data row, and indicates to the buffer the particular retrace time that the controller outputs VIDEO ADDRESSes. During the remainder of the data row scanning, the CPU has full access to the memory.

The Double Row Buffer Mode relaxes the transfer timing requirements by stealing memory cycles from the CPU to fill the write buffer. The controller first sends a DMA REQUEST (pin 28) signal to the CPU, then awaits an ACKNOWLEDGE (pin 33, HIGH) response before addressing the memory. Data is transferred when the WRITE BUFFER ENABLE (pin 29) input is activated.

The user has the ability to program the number of DMA cycles performed during each REQUEST sequence, as well as the delay time between each sequence. The delay evenly distributes the DMA operation, permitting the use of other DMA devices, and allows the CPU to respond to real-time events.

In the Repetitive Memory Addressing Mode, the CRT 9007 will repeat the sequence of video addresses for every scan line of every data row. The VIDEO ADDRESS bus enters its high-impedence state during the horizontal retrace interval, permitting the CPU to access the video memory during retrace periods or to alternately access memory at predetermined time slots.

CRT Interface

The interface to the CRT monitor consists of several pinouts to provide the flexibility needed for differing monitor requirements. There is both a HORIZONTAL SYNC (pin 13) and a VERTICAL SYNC (pin 12) output. These are active LOW outputs with an open drain driver. A COMPOSITE SYNC is available at pins 29 or 32. This signal provides a true RS-170 composite sync waveform, and if coupled through a suitable RF modulator, it will drive a standard television receiver.

There are two blanking signals available: VERTICAL BLANK (pin 28) and COMPOSITE BLANK (pin 35). Notice that pins 28 through 33 serve more than one purpose. The configuration of these pins is determined by the software programming of *register 6,* as shown on the schematic. The ultimate choice of pinouts will depend upon your requirements and the operating mode.

The CURSOR output (pin 34) signal marks the cursor position on the screen. However, external circuitry can strobe the LIGHT PEN STROBE input (pin 32) to load the *light pen register* with a character's address for identification.

Character Generator

The CRT 9007 controls an external character generator for the video text. Proper sequencing of events is controlled by the CHARACTER CLOCK (pin 14) input signal. When in the Single Row Buffer or Repetitive Memory Modes, the character counter is scanned by SCAN LINE 0 through SCAN LINE 3 (pins 31-28). Pin 30 can be programmed as a SCAN LINE GATE output that is used in conjunction with pin 31, SCAN LINE DATA, to serially load an external shift register.

The VISIBLE LINE TIME output (pin 11) is active HIGH during all visible scan lines and during the horizontal trace times at vertical retrace. This signal is used to gate the CHARACTER CLOCK when supplying data to a character generator from a single or double row buffer.

The INTERRUPT output is pin 27; a RESET (pin 26) command redefines the registers when pin 26 is LOWered.

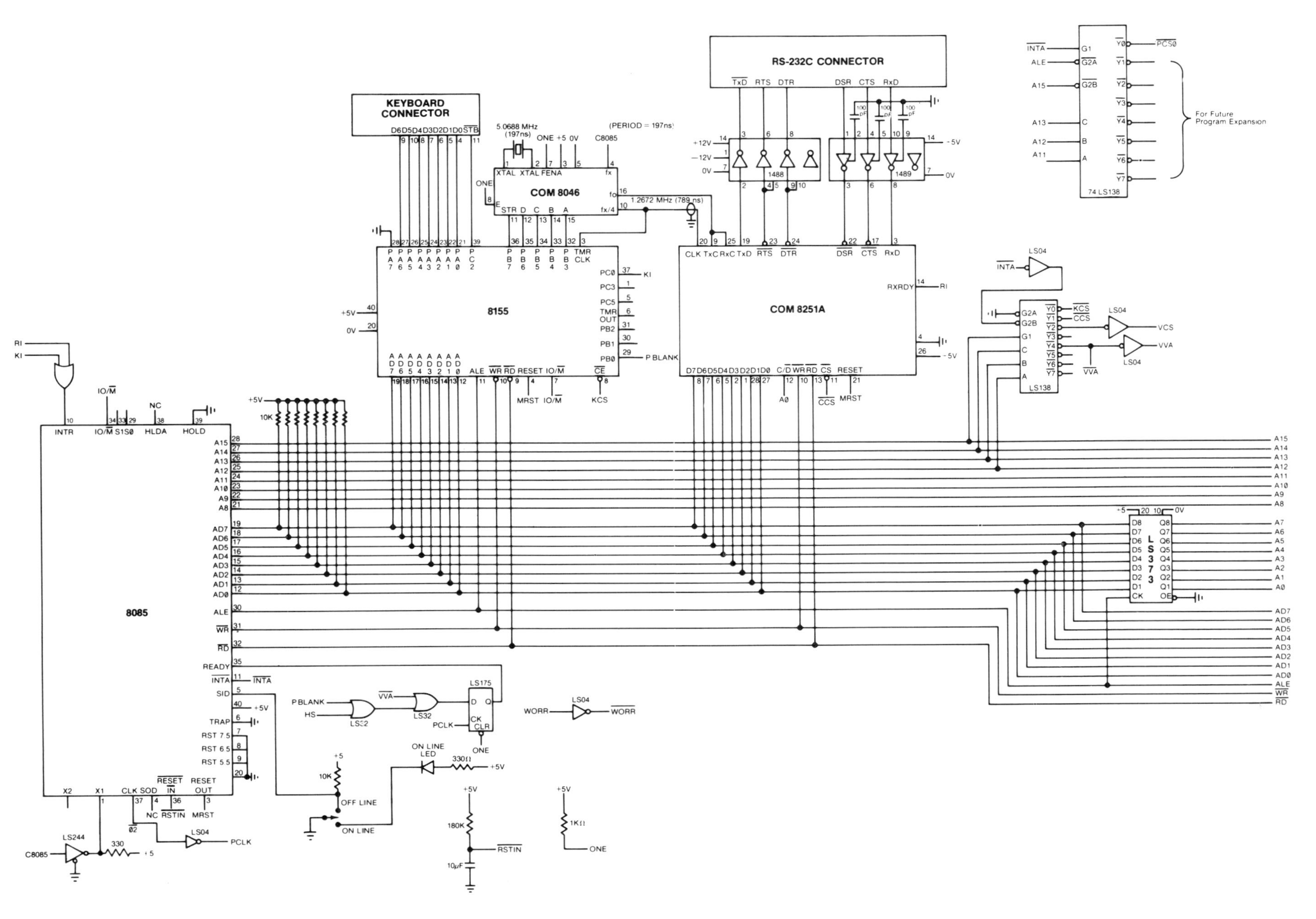

An entire CRT 9007 display terminal is outlined in pages 185, 186, and 187.

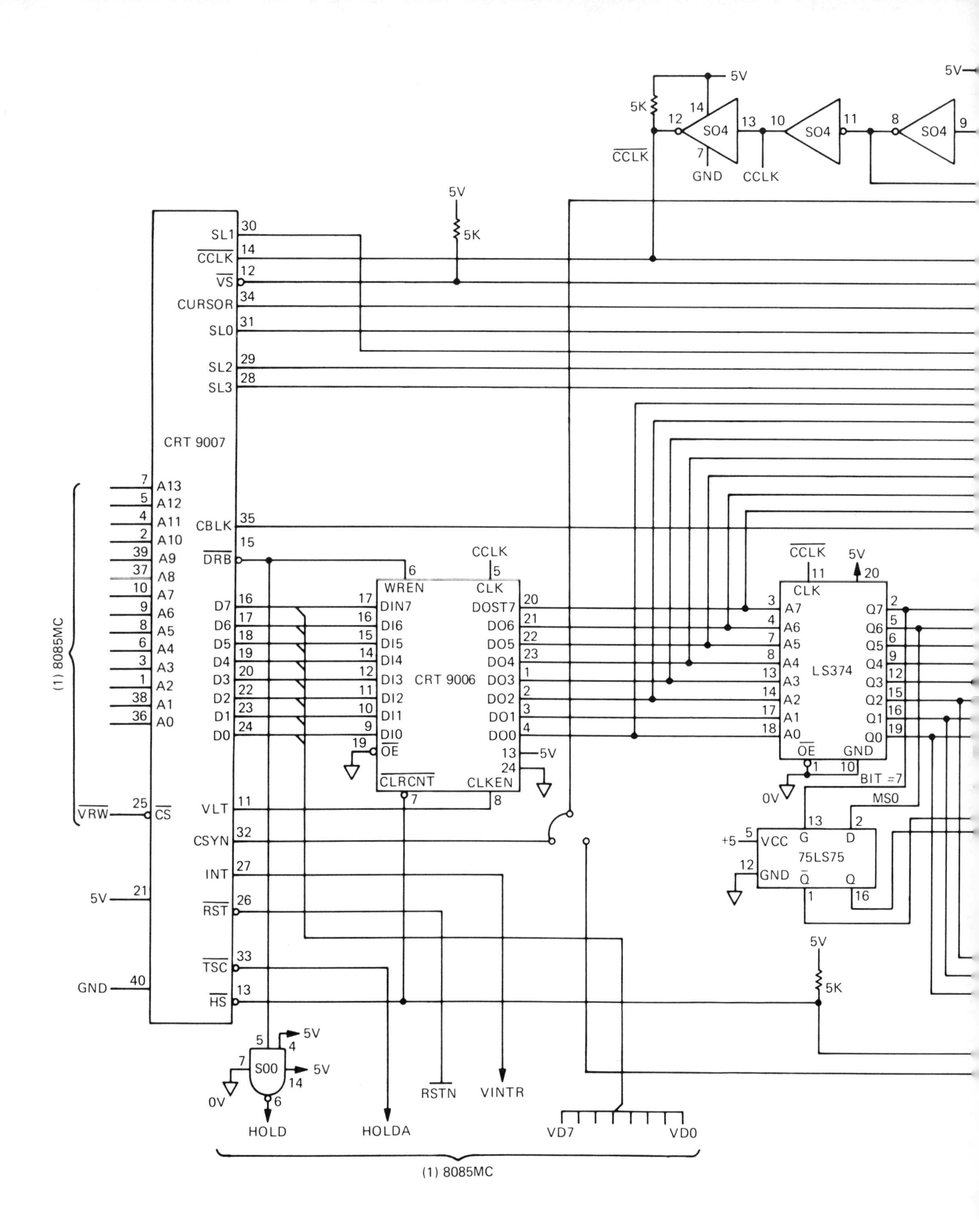

CRT 9007
CRT 9006
LS374
75LS75
SO4
S00
SL1
CCLK
VS
CURSOR
SL0
SL2
SL3
CBLK
DRB
VLT
CSYN
INT
RST
TSC
HS
VRW
CS
(1) 8085MC
HOLD
HOLDA
RSTN
VINTR
VD7
VD0
BIT =7
MS0
5V
5K
GND
0V

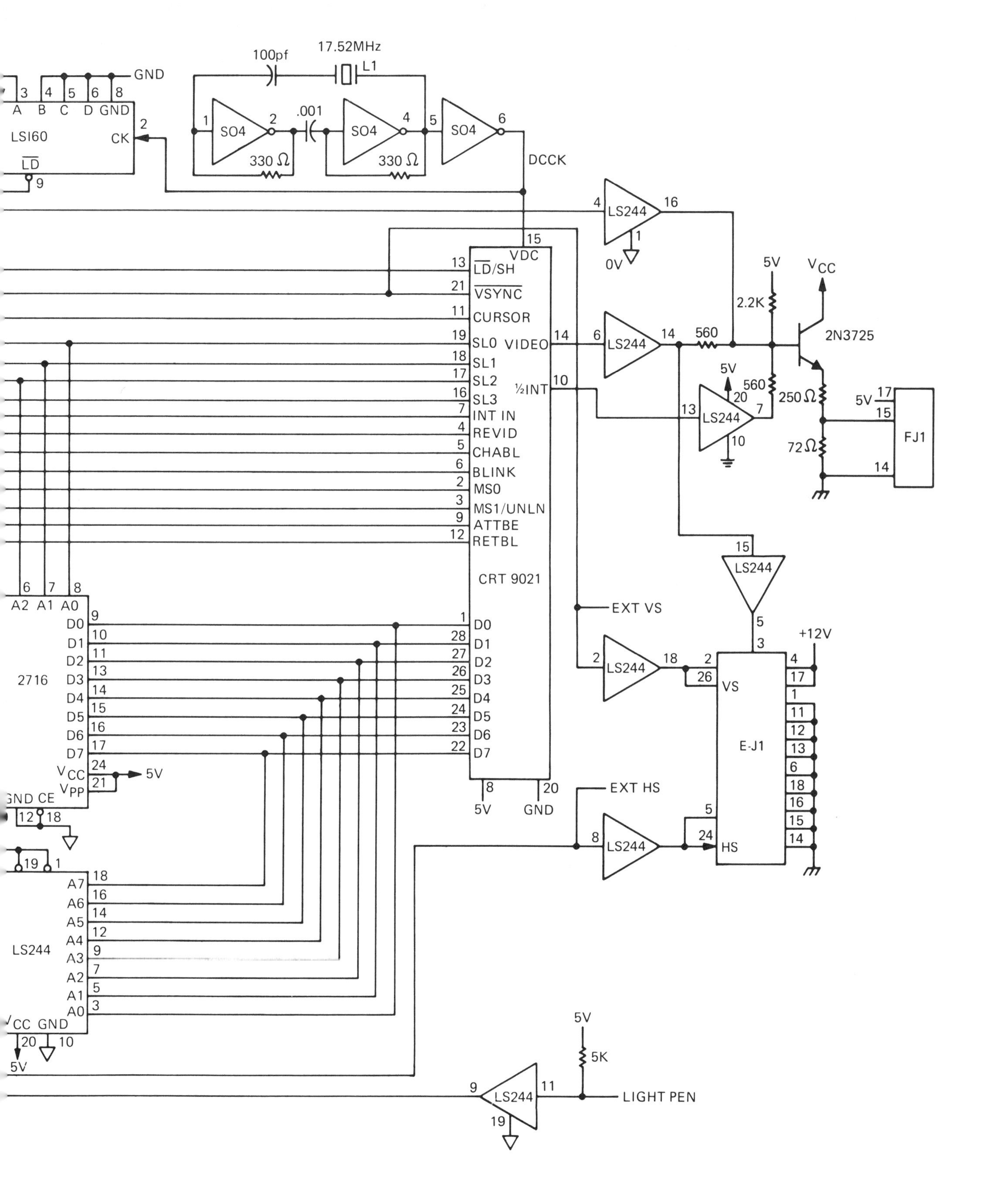

100pf
17.52MHz
L1
GND
LSI60
CK
LD
SO4
330 Ω
.001
DCCK
LS244
0V
5V
VCC
2.2K
2N3725
560
250 Ω
72Ω
FJ1
VDC
LD/SH
VSYNC
CURSOR
SL0
VIDEO
SL1
SL2
SL3
½INT
INT IN
REVID
CHABL
BLINK
MS0
MS1/UNLN
ATTBE
RETBL
CRT 9021
EXT VS
+12V
VS
E-J1
HS
EXT HS
2716
GND
LS244
5K
LIGHT PEN

8275

Programmable CRT Controller

The 8275 Programmable CRT Controller, by Intel, is a single chip device interfacing CRT raster scan displays to microprocessor systems. Its primary function is to refresh the display by buffering the video information from the main memory, while keeping track of the display position on the screen. The flexibility designed into the chip allows a simple interface to almost any CRT display with a minimum of external hardware and software overhead.

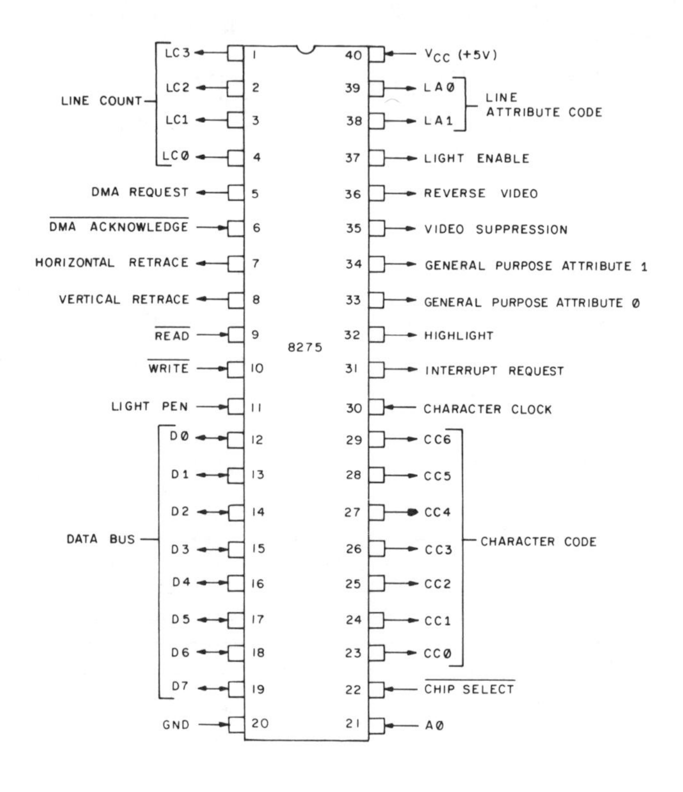

Operation

First, the 8275 must generate all the timing signals needed to properly sequence the raster on the CRT screen. Totally software programmable, the HORIZONTAL RETRACE (pin 7) and VERTICAL RETRACE (pin 8) signals control raster timing. A VIDEO SUPPRESSION output (pin 35) doubles as a blanking signal, inhibiting beam current during the retrace intervals.

Second, the controller supplies the video information for the display. The video images are stored, by the CPU, in an external memory. The 8275 simply retrieves them from the memory when they are demanded on the screen. To this end, the 8275 contains a unique memory contention scheme. Inside the chip are two 80-byte *character row buffers.*

CRT characters are "painted" onto the screen one row at a time, and each *character row buffer* is capable of storing an entire row of characters. While the contents of one buffer is being presented to the CRT video input, the other buffer is loaded with the next row of characters. After the present row is fully displayed, the controller begins reading from the buffer just loaded, and proceeds to load the alternate buffer with the next character row. This switching back and forth between *character row buffers* permits the CPU maximum access to the video memory.

The buffers are loaded using a simple DMA operation. Data is requested by a DMA REQUEST output (pin 5) and transferred on a DMA ACKNOWLEDGE (pin 6) input. DMA operations can also be programmed to transfer 2, 4, or 8 characters at a time, thus tailoring the DMA overhead to the system's needs. Although the DMA cycle is designed to operate with a DMA controller, such as the 8257, these lines may be used to synchronize the controller to a dedicated CPU, as shown by the schematic on the opposite page.

The actual generation of the display character is performed by an external character generator. The 8275 sequences the operation of the character generator through the CHARACTER CODE (pins 23-29) and LINE COUNT (pins 4-1) outputs at the rate established by the CHARACTER CLOCK (pin 30).

Special Functions

The 8275 provides an abundance of video signals which are best defined as visual attributes. Visual attributes are special codes which affect the visual characteristics of a character. Two types of visual attributes exist: character attributes and field attributes.

Character attributes are used to generate graphic symbols without the use of a character generator. This is accomplished by selectively activating the LINE ATTRIBUTE CODE (pins 39, 38), VIDEO SUPPRESSION, and LIGHT ENABLE (pin 37) outputs. Altogether, there are eleven special graphic symbols.

The field attributes affect the visual characteristics of an entire field of characters, starting with the character immediately following the field attribute code on up to, and including, the character preceeding the next field attribute command—or until the end of the frame. There are six field attributes.

The Blink attribute flashes the characters at a rate equal to the screen refresh frequency divided by 32, using the VIDEO SUPPRESSION output to alternately blank the character. All characters following the HIGHLIGHT code (pin 32) are effectively intensified on the screen. An Underline attribute makes use of the LIGHT ENABLE output to create a character underline. Of course, any of the above attributes may represent a cursor.

There are two additional GENERAL PURPOSE ATTRIBUTE outputs (pin 33, 34) which are independently programmable. These attributes may be used to select colors or perform other desired screen control functions. All attributes are reversed when the REVERSE VIDEO (pin 36) output is asserted.

The visual attributes, character or field, are actually commands which are stored in the video memory and accessed by the controller in its natural course of memory scanning. As you can see, many of the attributes require extensive external logic in order to perform the functions for which they were intended.

CPU Interface

The controller interface to the CPU is represented by three primary register types: status, command, and parametric. The *status* and *command registers* are accessed when the A0 (pin 21) ADDRESS is HIGH; the *parameter registers* are accessed when A0 is LOW. All data flows through the DATA BUS (pins 12-19) according to the READ (pin 19) and WRITE (pin 10) controls when the CHIP SELECT input (pin 22) is enabled.

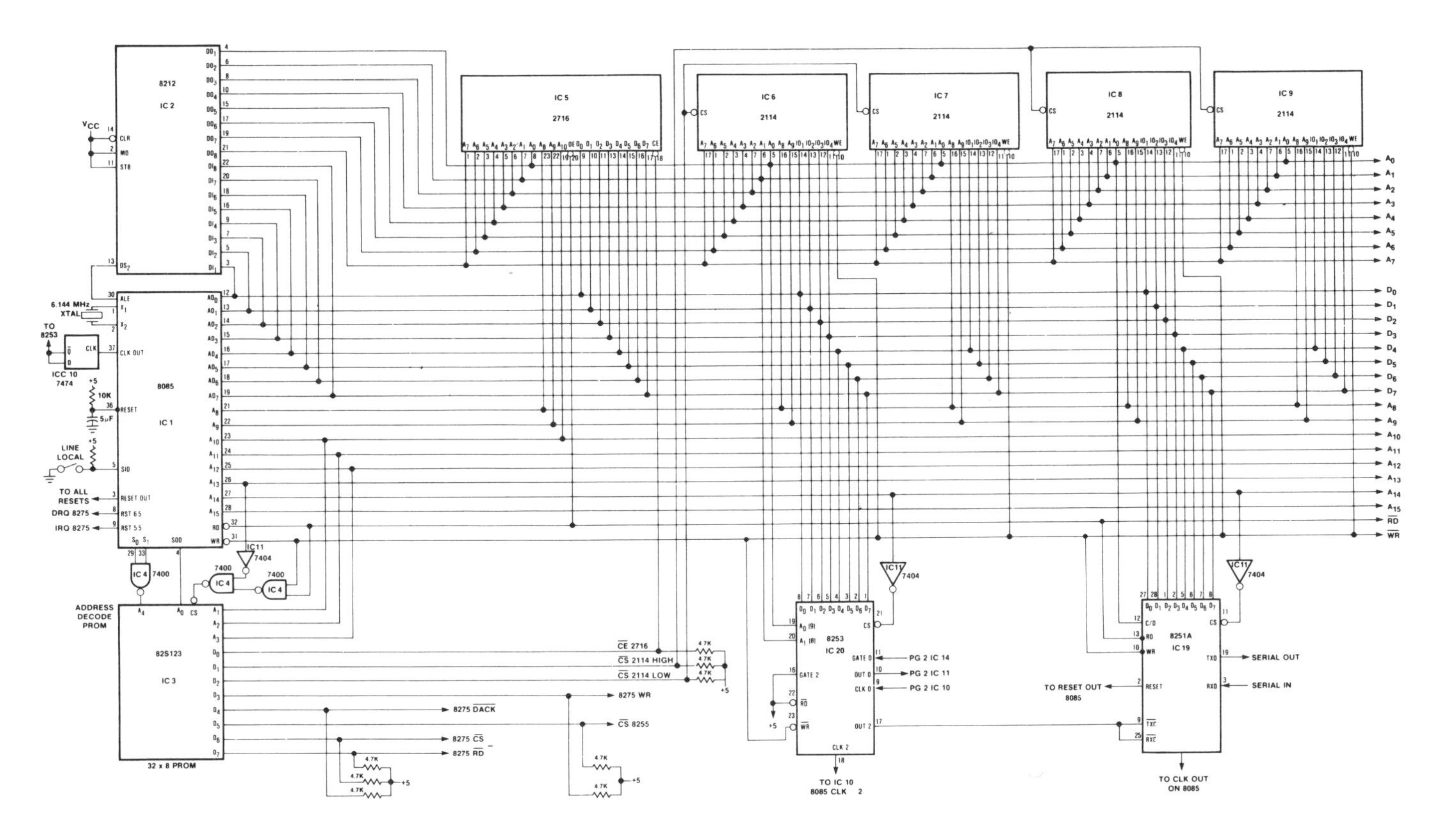

Figure 1

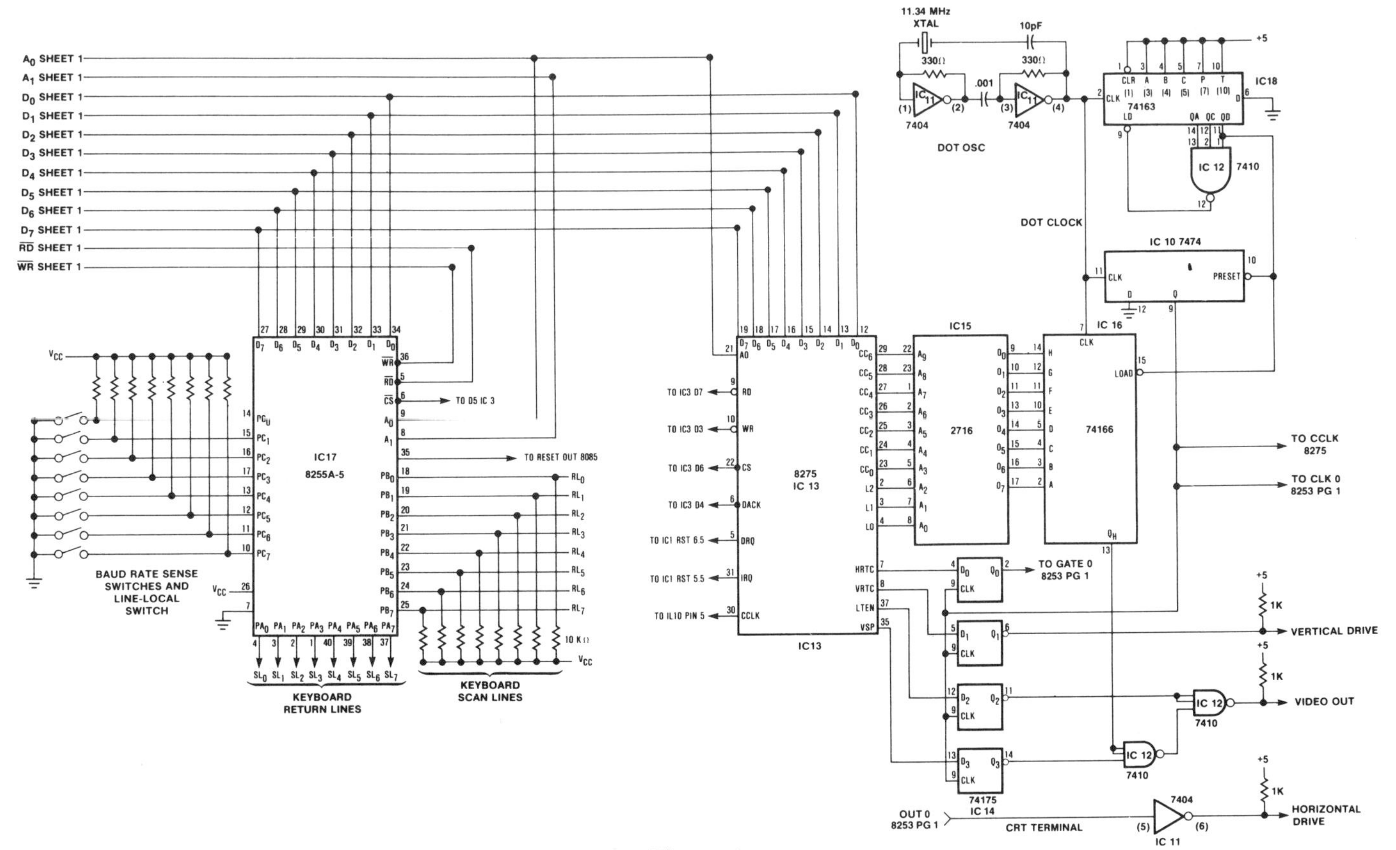

Figure 2

CRT 96364

CRT Controller

The CRT 96364, made by Standard Microsystems Corporation, is a CRT Controller which controls all the functions associated with a 64 × 16 line video display. Along with CRT refresh, character entry, and cursor management, the chip includes functions such as erase page, erase line, and erase to end of line. The controller easily interfaces to any computer or microprocessor, and is capable of stand-alone applications.

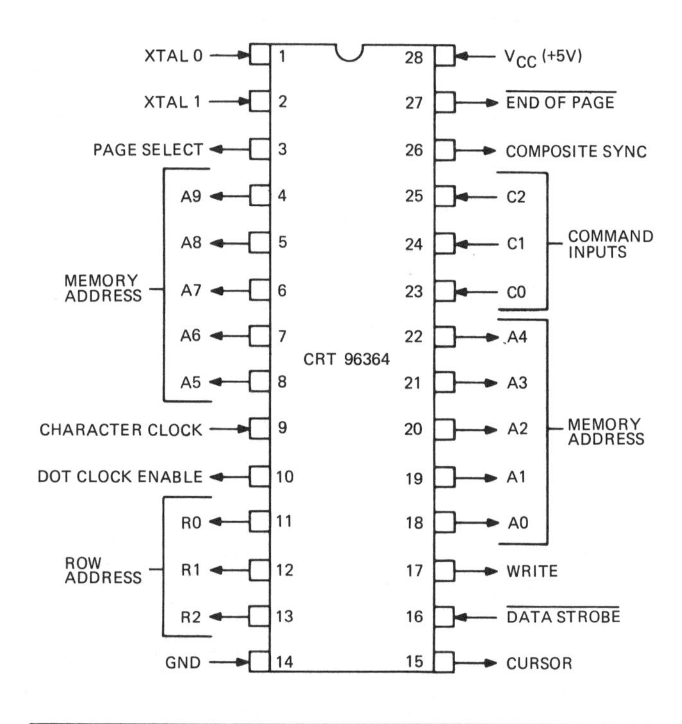

Operation

The controller is programmed to respond to eight commands: Page Erase, Cursor Home; Erase To End of Line, Cursor Return; Line Feed; No Operation (blanking); Cursor Left; Erasure of Cursor-Line; Cursor Up; and Normal Character. These instructions are entered through three COMMAND INPUTs (pin 23-25) upon receipt of a DATA STROBE (pin 16) pulse.

It is possible to extract the COMMAND INPUT control word from the 8-bit data bus by decoding the command in a PROM which has been programmed to recognize non-conflicting ASCII character words. Moreover, using the fourth bit of the decoder PROM as a write enable signal gives the user the additional commands of Home Cursor, Return Cursor, and Roll Screen.

Video Memory

Memory access is controlled by the ADDRESS MEMORY lines (pins 18-22, 8-4). MEMORY ADDRESS lines A6 through A9 (pins 7-4, respectively) determine which lines of text are being refreshed or written, while A0 through A5 specify character position. Data is written into memory using the Normal Character Command and a DATA STROBE pulse.

Video text enters the memory when the WRITE (pin 17) output is HIGH. The Write command is generated internally during horizontal retrace, giving the CPU uncontested access to the memory. Immediately following the Write sequence, the video memory addresses revert to screen refresh addressing and the cursor is shifted one character.

Cursor

The cursor position is displayed by the CURSOR (pin 15) video output. Operation of the cursor has been programmed so that this signal alternates between HIGH and LOW at a 2-Hz rate. This results in a screen cursor that blinks from the character position to an underline at the same rate. Writing data into the video memory always takes place at the cursor address.

Multiple Page Display

When displaying more than one page of memory, pages may be linked together in a "moving window" text display (scrolling). Scrolling is automatically performed when the cursor is forced into a carriage return while in the bottom line of the screen, resulting in the entire top line being erased and all remaining lines shifted up one. If a scroll has occurred, the PAGE SELECT (pin 3) output will show the transition from line 15 of page 1 to line 0 of page 2. A LOW output on this pin indicates page 1 is displayed; a logic HIGH indicates page 2.

Alternatively, scrolling can occur using either a Line Feed or Roll Screen command. Roll Screen is accomplished by inhibiting the WRITE signal to the page memory and inputting the Line Feed command.

The END OF PAGE (pin 27) output is used to increment an external page counter. To properly maintain the memory address when displaying more than two pages, the END OF PAGE output pulses LOW when page 1 is completely off the screen. This output will remain LOW for the entire frame since page 2 is now the only displayed page. This signal, in conjunction with the PAGE SELECT output, can be used to display up to 4 pages of text, as shown by the schematic in fig. 2.

Character Clock

In addition to supplying the CRT monitor with dot clock pulses, the character clock is used to sequence all the functions inside the controller chip. The clock signal is input to the CRT 96364 through pin 9, the CHARACTER CLOCK input. A DOT CLOCK ENABLE (pin 10) output is a logic LOW signal used to inhibit oscillation of the dot clock for retrace blanking.

The character clock must also increment the character generator. The timing sequence for this duty is generated by the con-troller chip and is available at the ROW ADDRESS outputs (pins 11-13), R0 through R2. During character entry, R2 (pin 13) controls the erase function.

Crystal Clock and Video Output

The CRT controller also needs a clock independent of the character clock to generate timing signals for proper CRT synchronization. This is accomplished by placing a crystal across pins 1 and 2, X0 and X1, or inputting a TTL compatible signal to pin 1. For standard 60-Hz line operation, a 1.018-MHz crystal must be used in conjunction with the "B" version of the chip; 50-Hz operation, standard in Europe, requires a 1.008-MHz crystal and the CRT 96364A.

CRT synchronization signals are taken from the COMPOSITE SYNC (pin 26) output. This is a combination of both horizontal and vertical sync pulses; however, a vertical sync can be extracted by logically ANDing the COMPOSITE SYNC and the DOT CLOCK ENABLE outputs. The video is, of course, obtained from the character generator output.

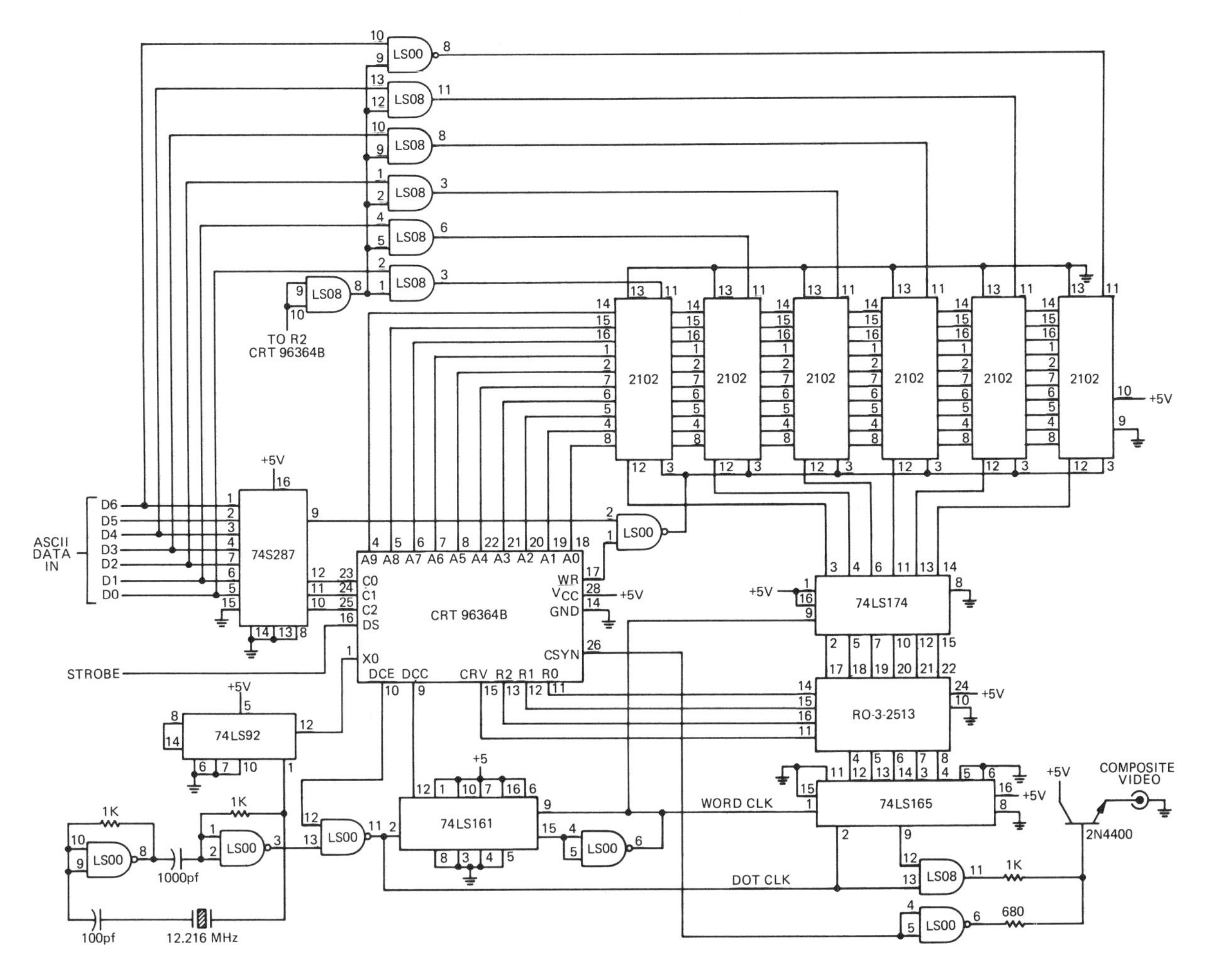

Figure 1

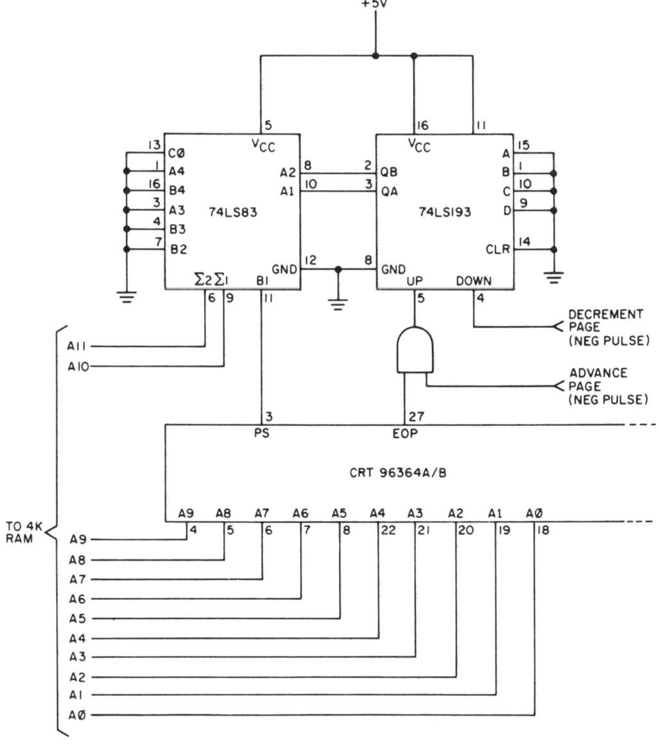

Figure 2

CHAPTER SIX
A/D AND D/A CONVERTERS

AD558

Complete uP- Compatible 8-Bit DAC

The AD558, made by Analog Devices, is a complete voltage-output 8-bit digital-to-analog converter, including full microprocessor interface, data latch, A/D converter, precision reference, and output amplifier on a single chip. No external components or trims are required to interface an 8-bit data bus to an analog system.

Circuit Description

The AD558 has been configured for ease of application, with all reference, output amplifier, and logic connections made internally. The chip consists of four major functional blocks: the main D to A converter section, an internal 1.2-volt reference source, a high-speed buffer amplifier, and the control logic which allows the data latches to be operated from a decoded microprocessor address and write signal.

Pinout of the chip has been arranged so that the digital inputs are on one side of the package and the analog outputs are on the opposite side, thus facilitating circuit board layout.

D/A Converter

The main digital-to-analog section uses eight equally weighted current sources switched into an R/2R resistor ladder to give a direct 0 to 400-mV range to the digital input. The reference currents are supplied by a highly-stable 1.2-volt precision reference source which, unlike the 6.3-volt temperature-compensated zeners commonly employed in competitive designs, allows the chip to operate from a single, low-voltage logic power supply.

Output Amplifier

This digitally generated signal is directed to an internal operational amplifier. The amplifier output stage is an NPN transistor with passive pull-down for zero-based output capability with a single power supply.

The gain of the amplifier is determined by a network of internal feedback resistors. In fact, the only parametric decision that the user can make concerning the operation of the AD558 is the selection of the output voltage range from the buffer amplifier.

Determining the desired output voltage scale is a simple matter of connecting a single jumper wire between two feedback pins. A network of feedback resistors is available to the user at the V_{OUT} SENSE (pin 15) and V_{OUT} SELECT (pin 14), and when used in conjunction with the analog output, V_{OUT} (pin 16), the output voltage will range between 0-V and 2.56-V, or 0-V and 10-V. For 2.56-volt operation, pins 14, 15 and 16 are shunted together; while a link from pins 15 to 16, with pin 14 grounded, designates 10-volt full scale output.

Output Modifications

The AD558 is intended to operate without user trims for gain and offset; therefore, no provisions have been made for such features. However, it is possible to introduce a small scale increase in the output by altering the effective gain of the output buffer. A resistor inserted in series with V_{OUT} SENSE will increase the overall feedback resistance, hence, the output voltage.

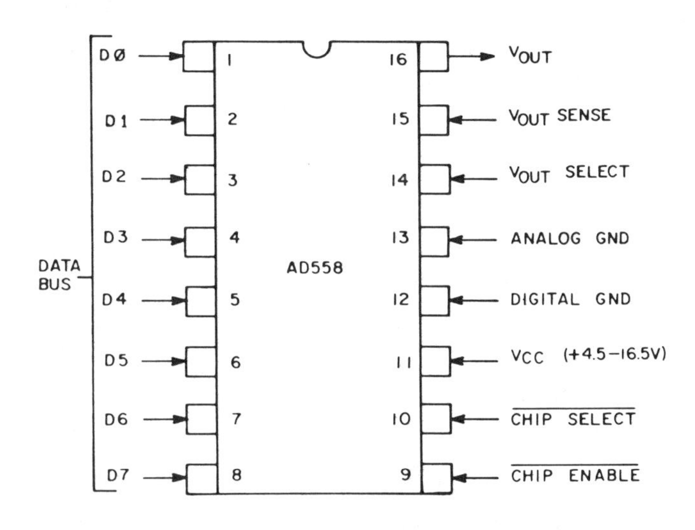

For example, if a 0-V to 10.24-V output range is needed, a nominal resistance of 850-ohms is required. It must be remembered, though, that all internal resistance values are relative, and not absolute. Only constant ratios must be maintained to ensure guaranteed performance, not specific values. Therefore, it will be necessary to individually trim the resistor for each unit when external ranging is used.

This procedure is the only one recommended for modifying the output. Decreasing the scale by inserting a resistor in series with the ground leg will not work properly due to the code-dependent current within the ground return. Nor is adjusting the offset by injecting DC into the ground path recommended for the same reason. The proper way to offset the output, such as required for bipolar output, is to use an amplifier external to the DAC with offsetting capabilities.

Separate access to the feedback resistor allows additional application versatility. The V_{OUT} SENSE line permits the chip to compensate for IR loss by remote sensing, or to increase output current by driving an external current booster (see fig. 1).

Digital Interface

The AD558 interfaces to the CPU through eight DATA BUS (pins 1-8) input ports. While microprocessor control signals vary widely from one architecture to the next, only two conditions are required for D to A conversion. First, the CPU must select the AD558 for operation. An address decoder is used to provide a unique signal for this function and, depending on system complexity, may range from a direct connection to a complete decoding of all memory locations. This signal is input to the CHIP SELECT (pin 10) control pin.

The second control signal is normally derived from the CPU's Write output, and is fed into the CHIP ENABLE (pin 9) input. When both control inputs are active LOW, the DAC is transparent, with the output voltage tracking the DATA BUS. Latching occurs when either or both controls go HIGH.

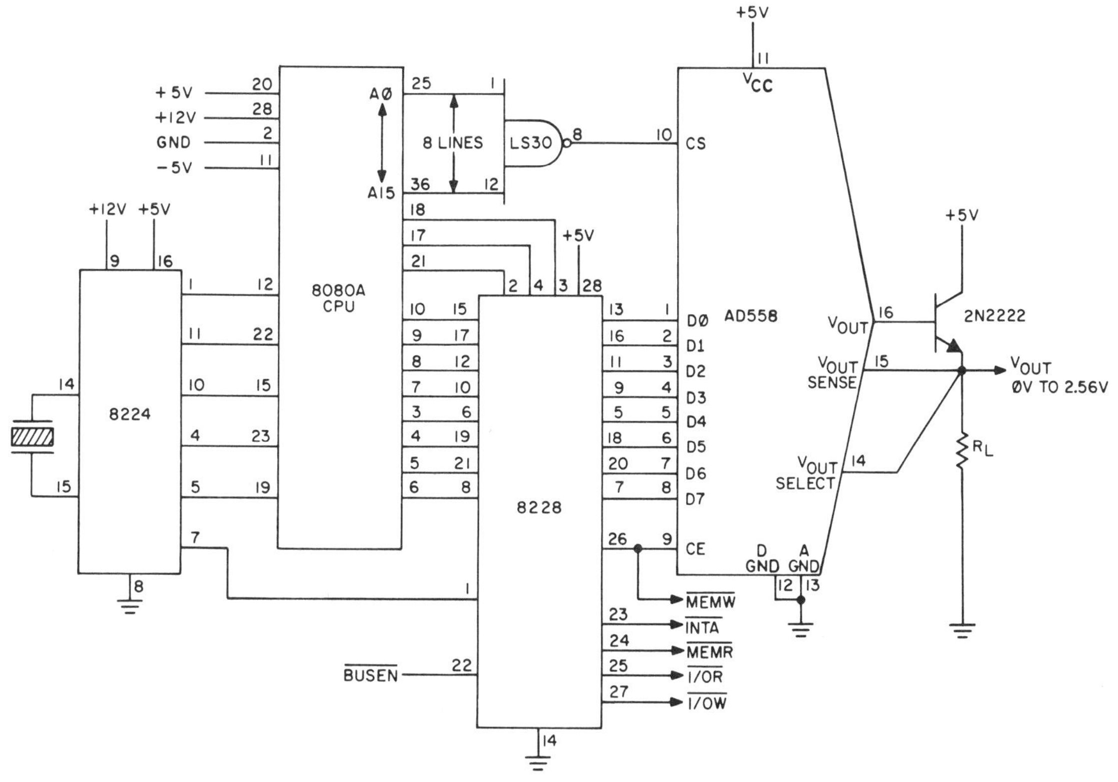

The AD558 output current can be boosted by utilizing the sense line, as shown. Through remote sensing, output is multiplied and line loss becomes negligible.

Figure 1

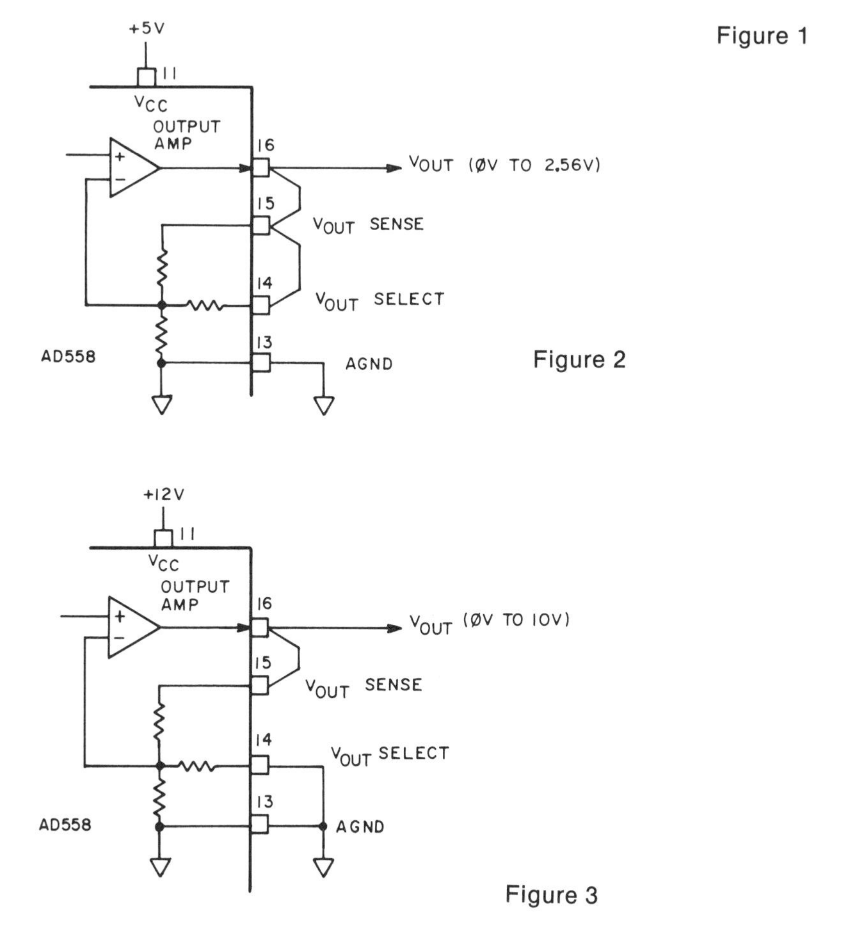

Figure 2

Figure 3

MC6890

Eight-Bit MPU-Bus-Compatible D/A Converter

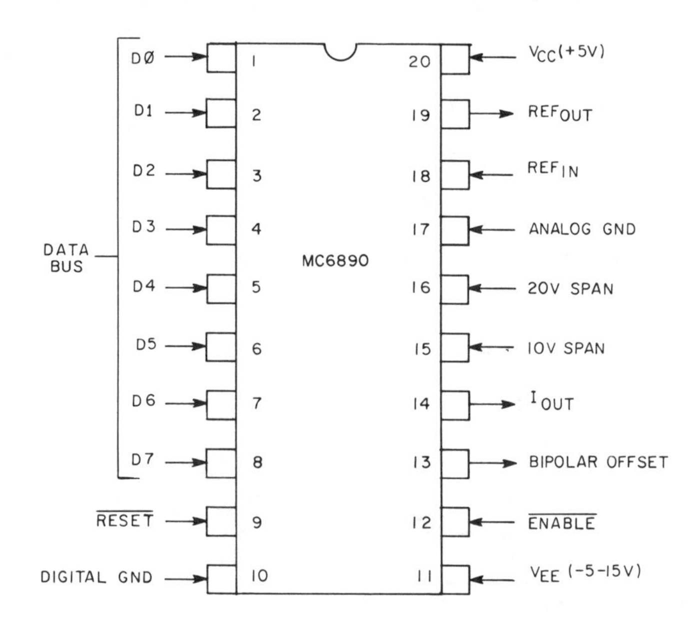

The MC6890, made by Motorola, is an 8-bit double buffered digital-to-analog converter designed for use with microprocessors. Double buffering prevents the output from tracking the data bus during data acquisition. The chip also contains its own internal voltage reference for current sourcing.

Digital Input

The MC6890 interfaces with the processor through eight DATA BUS (pins 1 through 8) inputs. Data is latched into the buffers when the ENABLE input (pin 12) is strobed LOW for longer than 20-ns. During the writing process, the latches are transparent and the binary word must be valid at least 40-ns before the rising edge of ENABLE. Once latched in the buffer, the data will remain indefinitely; however, a hardware RESET (pin 9) can override the stored software, resetting the latches and forcing the output to zero. This input is level sensitive, and the ENABLE pin should be held HIGH when RESET is LOW.

Analog Output

Because it's usually faster (and easier) to switch currents rather than voltage, the output of the MC6890 is a current source that is directly proportional to the binary digital input multiplied by a reference current. The reference current is established through the REF_{IN} (pin 18) input. This input feeds a current-operated operational amplifier through an internal resistor. The total resistance path seen by an external reference voltage is just 50-ohms shy of 5000-ohms.

The chip contains a 2.5-volt internal reference source REF_{OUT} (pin 19), intended for use as the reference supply current. With an external 50-ohm resistor linking the two pins together, 0.5-mA of reference current will flow through the current amplifier input. The REF_{OUT} output can also supply an additional 15-mA of current for external use. However, it's not necessary to use this pin for the reference current; any external voltage source can be used, provided it doesn't exceed ±7.5-volts.

The product of the binary number and the reference current is output to pin 14, the I_{OUT} output. When used with REF_{OUT}, this current (in mA) is equal to the binary input decimal equivalent divided by 128, or 1.0-mA with a 1000 0000 binary input. The settling time is approximately 200-ns. This output current may be converted to a voltage using an external operational amplifier, as is the general case.

Since the relative current gain of the current amplifier is fixed by REF_{OUT}, the full scale range of the output voltage is determined by the gain of the external voltage amplifier. This, in turn, is controlled by a single feedback resistor.

In order to ensure thermal stability throughout the system, the chip contains two preprogrammed internal feedback resistors which can be used to set the full scale range of the output amplifier. The 10-V SPAN (pin 15) pin sets the output voltage between 0 and 10-volts across the entire binary spectrum, while the 20-V SPAN (pin 16) feedback path doubles that range to 20-volts. If desired, intermediate ranges can be obtained by incorporating an external feedback resistor. As a rule of thumb, expect 2-volts output for every 1000-ohms of feedback resistance; i.e., 2.5-k equals 5-volts.

The output voltage will swing bipolar by connecting the BIPOLAR OFFSET (pin 13) output to the REF_{IN} input through a 25-ohm resistor. What this does, in essence, is force 1.0-mA of output current back into the reference input, offsetting the reference source and resulting in bipolar operation. Using the 10-V SPAN feedback configuration, this will produce an output that swings plus and minus 5-volts, with the binary input 1000 0000 representing zero volts.

Power Supply

The circuit requires two power supplies for operation. The first is a +5-volt source to the V_{CC} (pin 20) pin, and it can be derived from a standard 5-volt TTL power source. The second supply is a negative voltage applied to V_{EE} (pin 11). This voltage can assume any value between −5- and −15-volts, depending upon your requirements. As you see, it isn't necessary to have the two supplies of equal value—or even tracking, for that matter.

The chip contains two completely separate ground pins for the converter's analog and digital sections: ANALOG GND (pin 17) and DIGITAL GND (pin 10). It is important that you keep these lines separate. As a general rule, the ANALOG AND DIGITAL grounds are routed individually and tied together at the power source only. Proper bypassing techniques should always be observed.

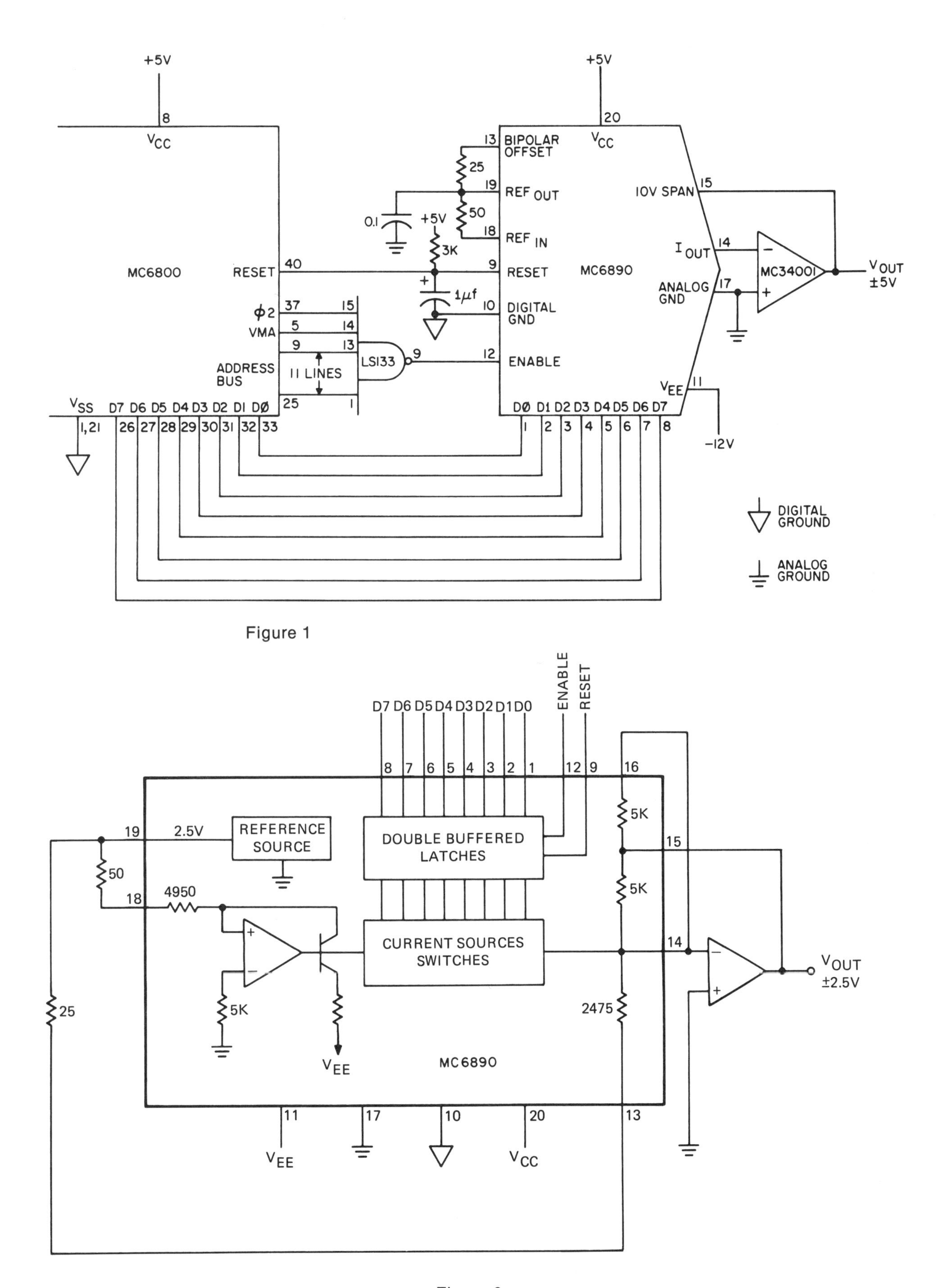

Figure 1

Figure 2

DAC-888

8-Bit High-Speed "Microprocessor Compatible" Multiplying D/A Converter

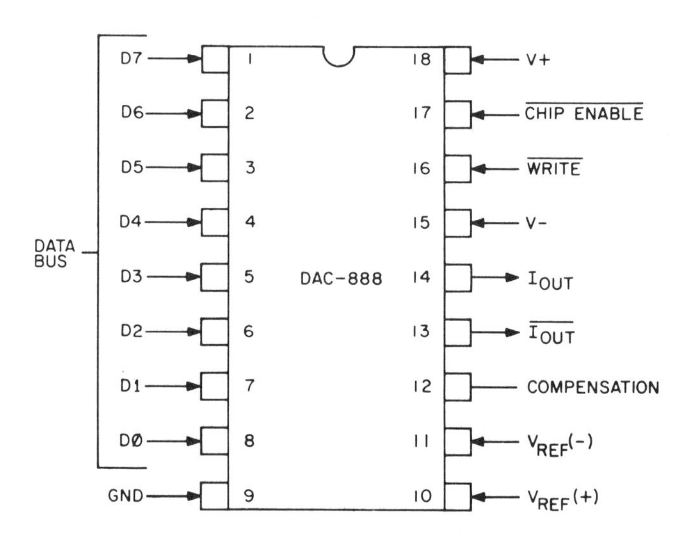

The DAC-888, by Precision Monolithics Inc., is a buffered 8-bit digital-to-analog converter designed specifically for microprocessor applications. The processor oriented control pins provide an easy interface to virtually all available microprocessors.

Digital Considerations

With a data hold time requirement of zero nanoseconds, the DAC-888 essentially interfaces the microprocessor to the outside analog world. The converter accepts 8-bit binary words at the DATA BUS (pins 8-1) inputs. Data is transferred to the *data input latches* when both WRITE (pin 16) and CHIP ENABLE (pin 17) inputs are LOW. When data is fetched, the latches are transparent; therefore, data should be valid at least 100-ns prior to the enable command and until the write cycle is complete. Once either operational input is returned HIGH, the data in the latches will hold the selected output indefinately.

Reference Amplifier

The DAC-888 is a multiplying D/A converter in which the output current is the product of a digital number and the input reference current. The reference current may be fixed or it may be a dynamic signal that varies from nearly 0 to 4.0-mA.

The reference current flows through pins 10 and 11, the $V_{REF}(+)$ and $V_{REF}(-)$, respectively. Due to the high gain of the internal reference amplifier, the voltage at pin 10 is equal to—and tracks the voltage at—pin 11. Therefore, the reference current must be limited by an external resistance.

To minimize the offset bias error, current limiting is normally done with a pair of equal-value resistors to each input, the total sum of which limits the current to the desired value (see fig. 2). However, R_{11} may be eliminated altogether with only a minor increase in error.

The current reference amplifier must be externally compensated with a capacitor across the COMPENSATION (pin 12) pin and $-V$ (pin 15). When the reference source is an AC signal, the value of this capacitor depends on the impedence presented to pin 10 (see fig. 3). For fastest response to a pulse, low values of R_{10} should be used, enabling the use of smaller compensating capacitors. For fixed reference operation, a 0.01-uf capacitor is recommended.

Analog Output Currents

The full scale output of the converter is a linear function of the reference current. Quadrant multiplication is determined by the relative polarity of the reference signal.

In positive unipolar applications, an external positive reference current flows through R_{10} into pin 10, while pin 11 is returned to ground through R_{11}. Alternatively, a negative reference may be applied to pin 11, with pin 10 grounded, in which case the output current assumes a negative value. The negative reference connection has the advantage of a very high input impedence present at pin 11. True bipolar operation is accomplished by offsetting the voltage at either input with an appropriate bias supply. A 5-volt TTL logic supply should not be used as a reference. Notice that pin 10 is always positive in relationship to pin 11.

Both true and complemented output sink currents are available, I_{OUT} (pin 14) and $\overline{I_{OUT}}$ (pin 13), where $I_0+\overline{I_0}=I_{FR}$. Current output of pin 14 appears as true output to each logic input. As the binary count increases, the current proportionally increases. The complementary output current, on the other hand, decreases as the binary count increases, giving the user an inverted logic output. If one of these outputs is not required, it must still be connected to ground or to a point capable of sourcing full scale current. Do not leave an unused output pin open.

Power Supply

The circuit operates over a wide range of power supply voltages. However, at least 8-volts is required to ensure turn-on of the internal bias network. The acceptable voltage range is between 9-and 17-volts; symmetrical supplies are not required, as the device is quite insensitive to variations in supply voltage. When operating with supplies of $\pm$5-volts or less, the reference current is limited to 1.0-mA.

A/D Conversion

The D/A converter can also be used to convert analog data into digital using the successive approximation method. The successive approximation method is a software intensive scheme that converts the analog input into digital form by comparing the magnitude of each bit to the analog value.

First, the MSB is compared to the analog input. If the analog value is larger than the digital equivalent, the bit is retained; if not, it is discarded. The next MSB is then added to the first results and the sum compared to the analog value. Again a decision is made. This branching continues until each successively smaller bit has been tried, with the end result a nearly perfect digital approximation of the analog input.

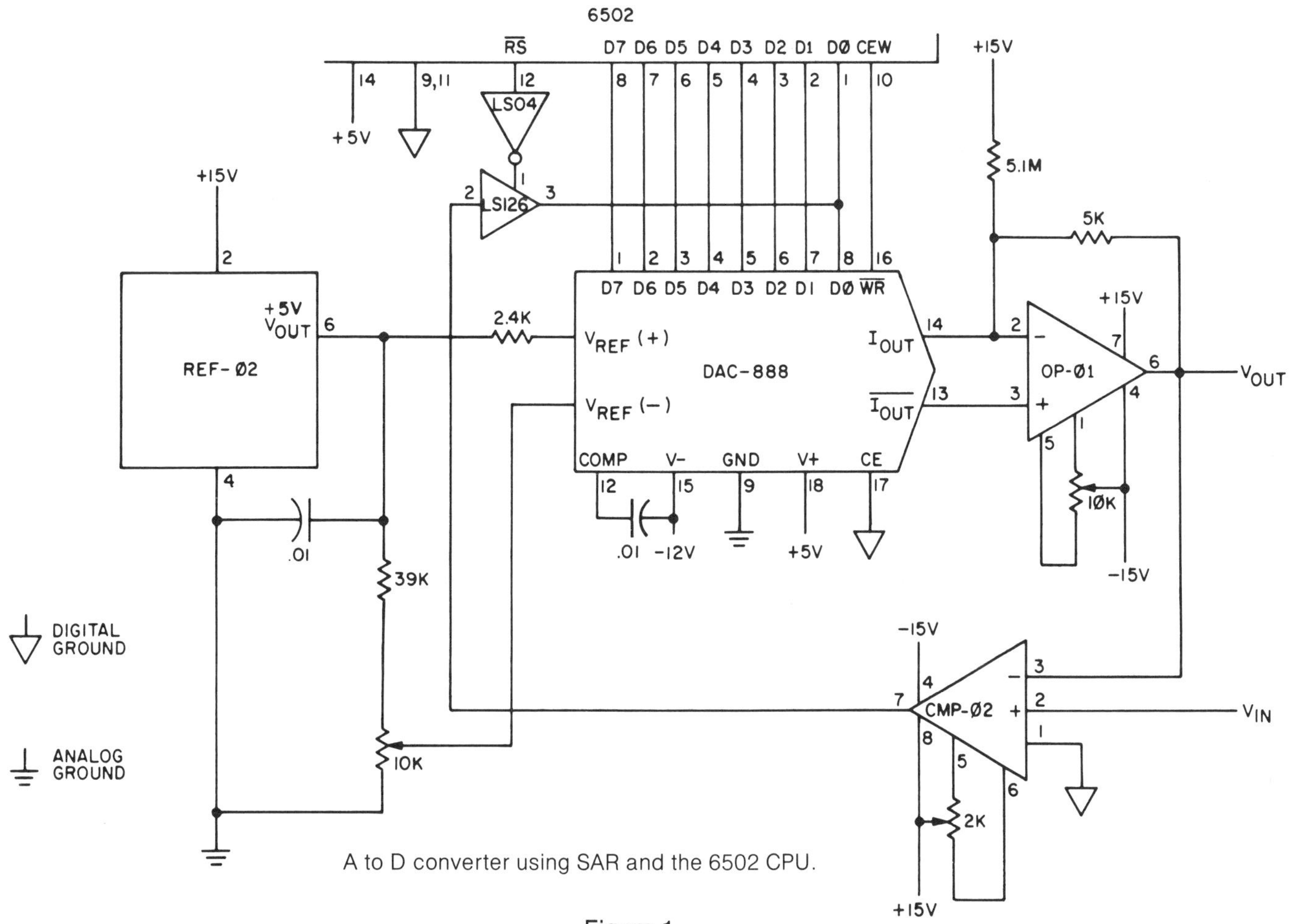

A to D converter using SAR and the 6502 CPU.

Figure 1

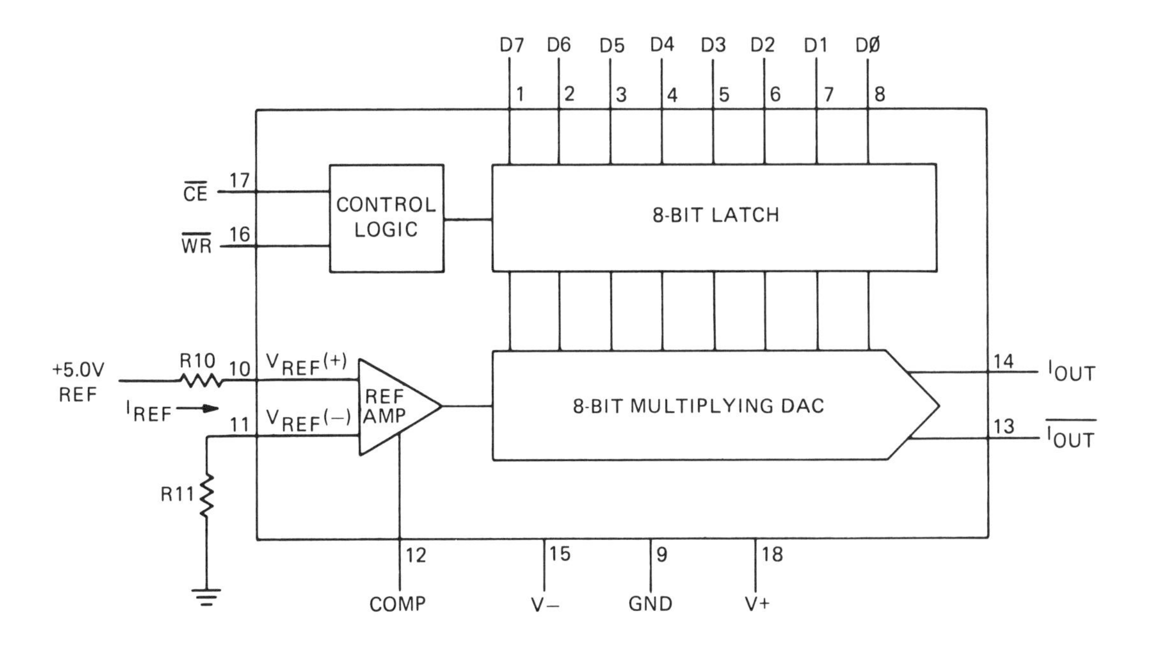

Figure 2

AD7524

8-Bit Buffered Multiplying DAC

The AD7524, made by Analog Devices, is a low-cost 8-bit monolithic digital-to-analog converter designed for direct interface to most microprocessors via their data bus or output ports. The converter offers 1/8-LSB accuracy with a typical power dissipation of less than 10 milliwatts.

Circuit Operation

The circuit consists of a highly stable thin-film resistor ladder, eight SPDT N-channel current steering switches, and an 8-bit latching driver. The resistor ladder is arranged in an inverted R-2R ladder structure. In other words, the binary weighted currents are switched between two output bus lines, as shown in fig. 2. With this configuration, a constant current is maintained in each ladder leg, regardless of the stage of the switch.

Each switch is driven by a respective logic bit, which is latched into the switch driver *data latches.* When a voltage is applied to the top of the ladder, a current will flow through the output bus lines proportional to the position of the switches and the significance of the logic bit.

Digital Input

Data can be written into the chip by enabling the CHIP SELECT (pin 12) and WRITE (in 13) inputs. The transfer takes place at the DATA BUS (pins 11-4) inputs when both controls are LOW. In this mode, the AD7524 behaves like a transparent D/A converter, and the output will follow the digital input.

When either CHIP SELECT or WRITE is returned HIGH, the data present on the DATA BUS at that time is latched into the *data latches,* and the analog output assumes a corresponding value. The digital input should be valid at least 150-ns before the write cycle; analog settling time is less than 200-ns.

Analog Output

The analog output is a current which is equal to the product of the digital data input (decimal equivalent) and the reference current. The reference current is input at the V_{REF} (pin 15) pin, and is derived from a stable voltage reference source. This voltage can assume any value between + 10 and − 10-volts, and can even be a dynamic signal for multiplying applications. The chip is extremely versatile and has excellent multiplying characteristics in both two and four quadrants.

For most applications, though, the output current is converted to a voltage using an external operational amplifier, and the digital conversion is expressed as a ratio of reference voltage to output voltage. Two analog bus line outputs supply the analog signal, the OUT1 (pin 1) and OUT2 (pin 2) outputs, which feed the inverting and non-inverting inputs of the external output amplifier, respectively. The gain of the amplifier is controlled by an internal feedback network available at pin 16, the FEEDBACK pin. By incorporating the bulk of the feedback resistance on the same chip as the D/A converter, it is possible to guarantee good temperature characteristics.

It is important to realize that the output impedence at the OUT1 and OUT2 bus lines is a function of the resistor ladder and the internal switch positions. When only one ladder leg is shunted

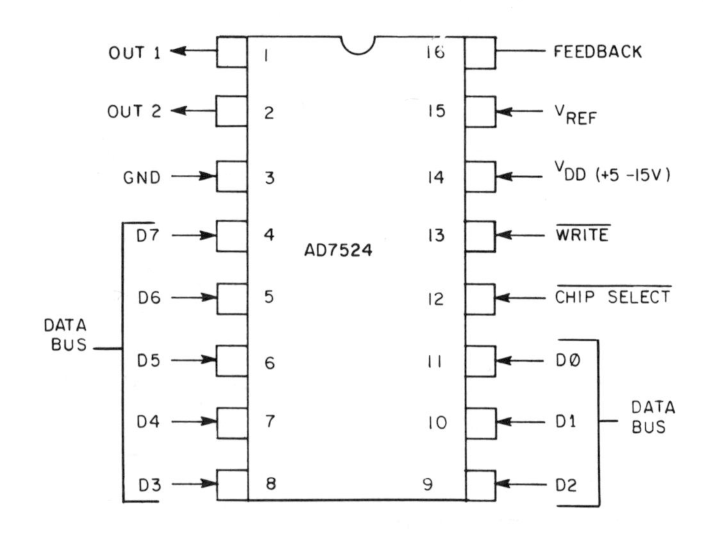

to an output bus, the impedence is greater than 30-k, while four bits of logic on a single output line drops the output impedence to approximately 10-k. Since the impedence seen by the operational amplifier is constantly changing, offset bias errors will be introduced. Therefore, it is imperative to select an amplifier which has an offset compensation feature and very low offset currents to avoid tracking errors.

Offset errors can be introduced by including the usual bias current compensating resistor in the amplifiers non-inverting input terminal. This resistor should not be included in the design. Instead, the amplifier should have a bias current which is low over the temperature range of interest, and most certainly must not exceed 75-nA in any case.

When using a high-frequency high-gain amplifier, phase lag within the system may cause ringing or oscillation. Shunting the amplifier's input to output with a 5 to 20-pf feedback capacitor will alleviate the problem and ensure stability.

The output characteristics of the AD7524 are virtually identical to the AD7528, and additional insight into operational amplifier performance can be gained by reviewing that section.

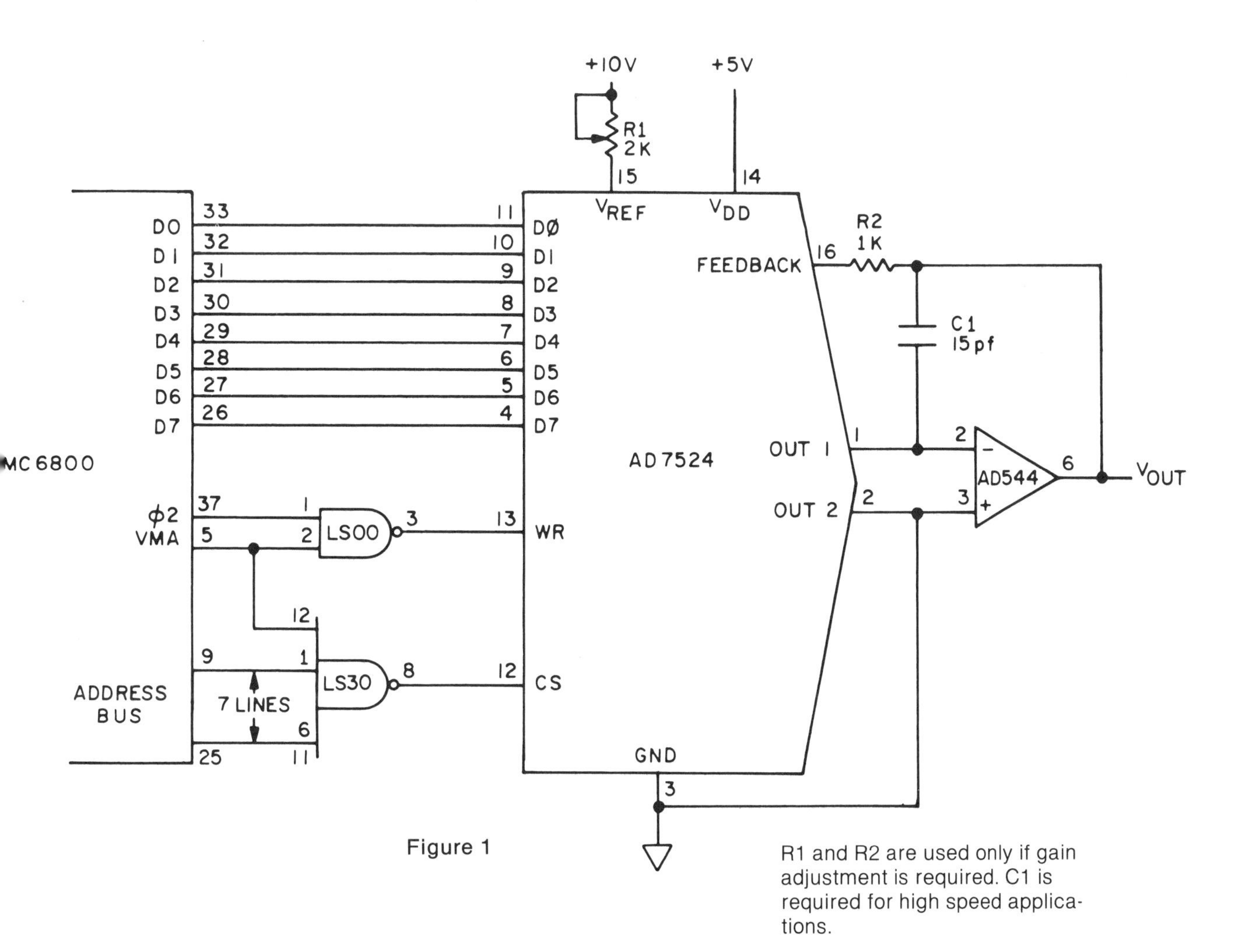

Figure 1

R1 and R2 are used only if gain adjustment is required. C1 is required for high speed applications.

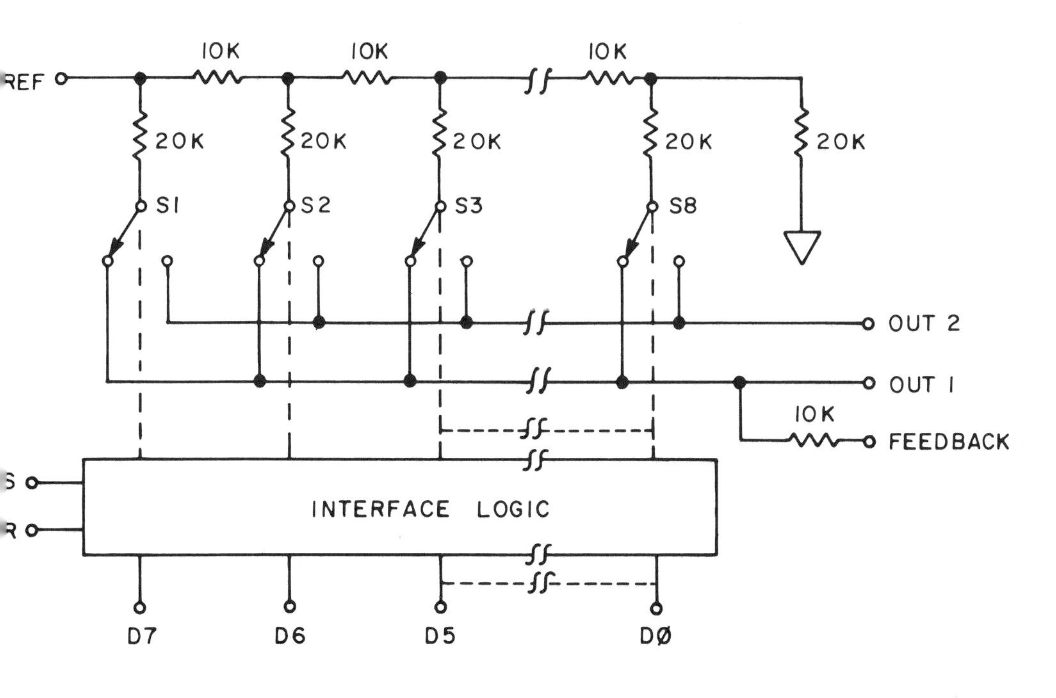

Figure 2

AD7528

Dual 8-Bit Buffered Multiplying DAC

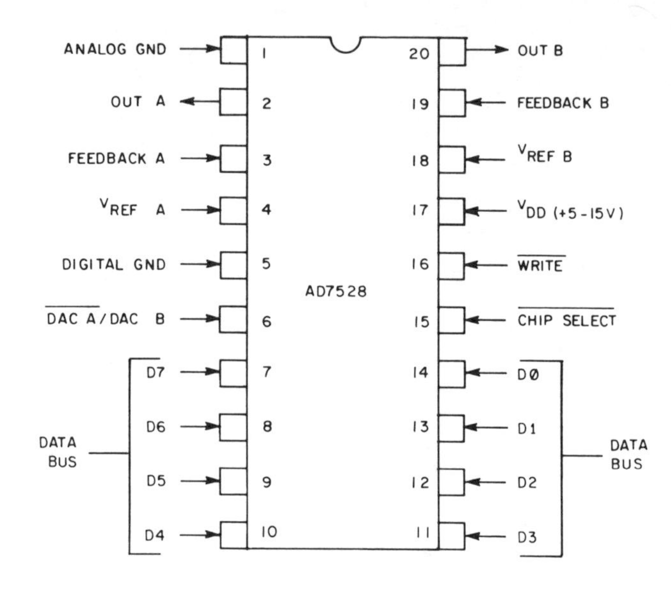

The AD7528, made by Analog Devices, is a dual 8-bit digital/analog converter featuring excellent DAC-to-DAC matching. The converter's load cycle is similar to the write cycle of a random access memory and the device is bus compatible with most microprocessors, including 6800, 8080, 8085, and Z80.

Since both of the DACs are fabricated at the same time on the same chip, precise matching and tracking between the two is inherent. Precision matching opens up a whole new world of possible applications, particularly in the audio, graphics, and process control areas.

Digital Interface

The AD7528 contains two identical 8-bit multiplying D/A converters: DAC A and DAC B. Data is transferred into either of the two DAC latches via a common 8-bit DATA BUS (pins 14 through 7) input port. The control input DAC A/DAC B (pin 6) selects which DAC can accept data from the DATA BUS input. When pin 6 is LOW, DAC A is selected; when it is HIGH, DAC B is actively engaged.

The transfer takes place during the write cycle, when both CHIP SELECT (pin 15) and WRITE (pin 16) inputs are forced LOW. The input latches of the selected converter are transparent during the write operation, and its analog output responds to activity on the DATA BUS. Information is latched into the selected DAC when either control pin, CHIP SELECT or WRITE, assumes a HIGH state. Both analog outputs remain at the value corresponding to the data in their respective latches.

Analog Section

Each D/A converter consists of a highly stable thin-film R-2R resistor ladder and eight N-channel current steering switches. The resistor ladder is configured in an inverted R-2R structure so that a constant current is maintained in each leg of the divider independent of the switch state. This arrangement requires two output lines per converter, one of which is internally tied to the analog ground. The other output is available at pin 2 for converter A and pin 20 for DAC B.

The output signal is a current which is proportional to the product of the digital input and the reference current. The reference current is established by applying a voltage to the top of the resistor ladder, which is pinned out at V_{REF} A (pin 4) and V_{REF} B (pin 18) for DAC A and DAC B, respectively.

Since most applications require the current to be converted to a voltage using an external operational amplifier, the digital conversion is usually expressed as a ratio between the reference voltage and the output voltage. Due to the manner in which the resistor ladder is switched, however, the output impedence of the DAC is a function of the digital input code; the equivalent output resistance of the device varies from 0.8R to 2R, and is typically 11-k.

Therefore, the external operational amplifier experiences a code-dependent input impedence source, which in turn causes amplifier dynamic noise gain. The effect is code-dependent differential nonlinearity at the amplifier output of a maximum equal to 0.67-V_{OS}, where V_{OS} is the amplifier input offset voltage. This differential nonlinearity adds to the R/2R nonlinearity. To minimize the effect, it is recommended that V_{OS} be no greater than 10% of 1 LSB over the desired temperature range.

The gain of the amplifier(s) is set by an internal feedback resistor, accessible at FEEDBACK A (pin 3) and FEEDBACK B (pin 19). Placing the feedback resistors on the same chip as the DACs guarantees precise tracking over a wide range of temperatures. When using high-frequency amplifiers, however, the output capacitance of the DAC works in conjunction with the feedback resistance to add a pole to the open loop response, leading to ringing or oscillation. Stability can be restored by adding a phase compensating capacitor (5-20-pf) in parallel with the feedback resistor.

Power Supply

The AD7528 may be operated with any supply voltage in the range of +5 to +15-volts on V_{DD} (pin 17). When V_{DD} is equal to 15-volts, however, the input logic levels are no longer TTL compatible and must conform to CMOS levels; i.e., 1.5-volts LOW and 13.5-volts HIGH.

The chip has separated the ANALOG GND (pin 1) and DIGITAL GND (pin 5) returns, with the reference ladders connected to the ANALOG GND. This arrangement is particularly attractive for single supply operation because ANALOG GND may be biased at any voltage between DIGITAL GND and V_{DD}. Figure 2 shows a circuit which provides a +5- to +8-volt analog output by biasing ANALOG GND +5-volts above DIGITAL GND. The two DAC reference inputs are tied together and driven by a single voltage source without a buffer amplifier, thus taking advantage of the constant and matched impedences exhibited by the two reference inputs.

Refer to AD7524.

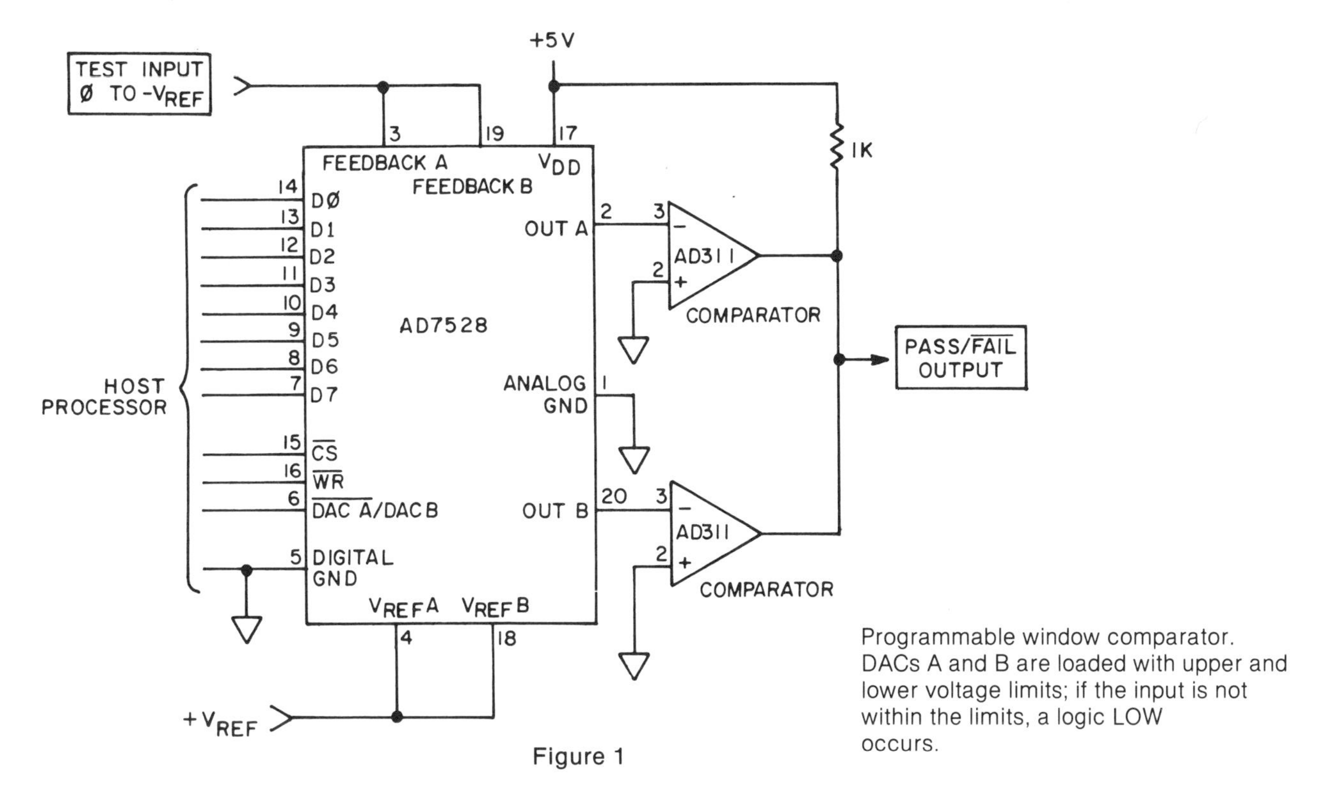

Figure 1

Programmable window comparator. DACs A and B are loaded with upper and lower voltage limits; if the input is not within the limits, a logic LOW occurs.

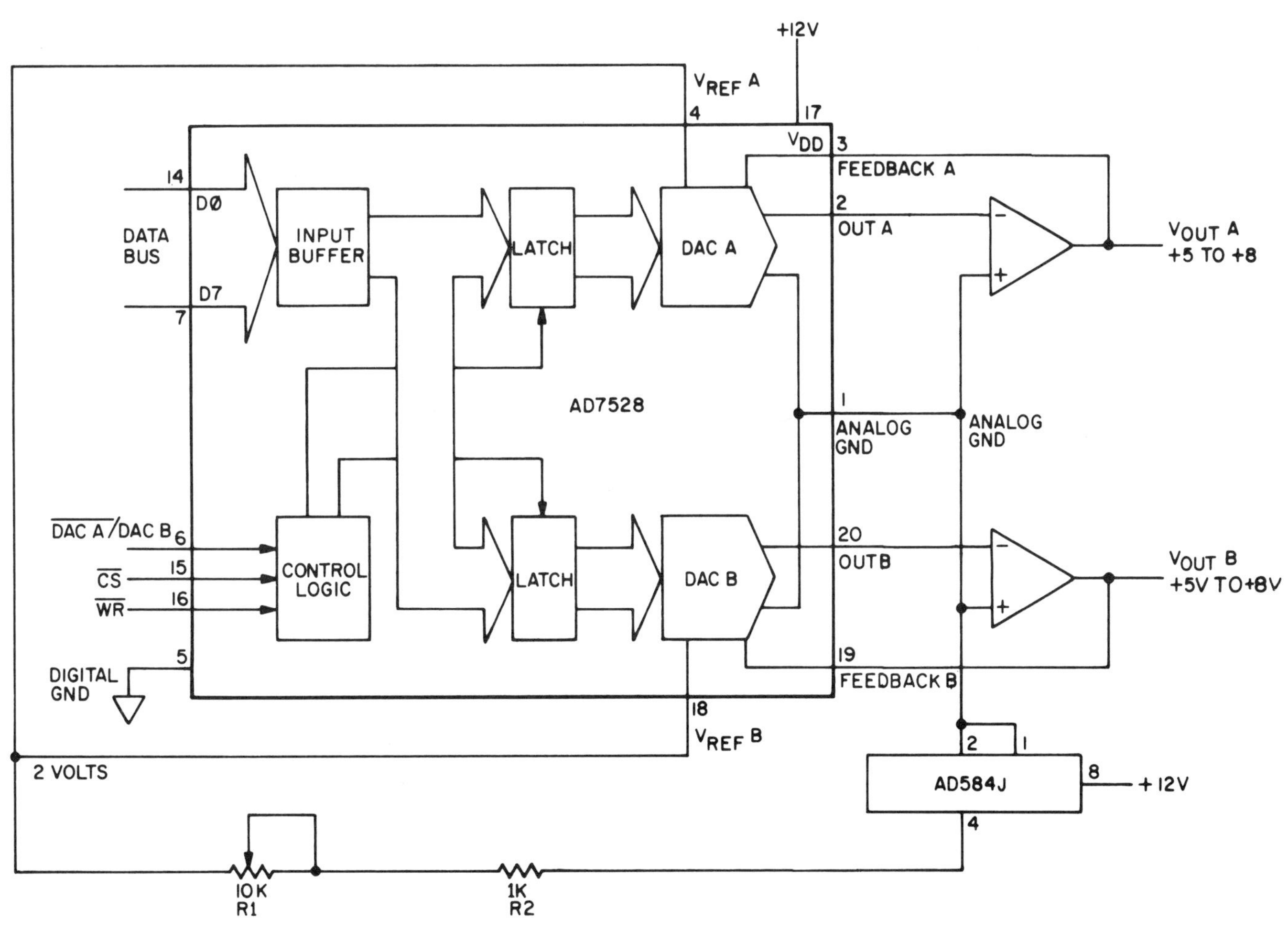

Figure 2

Am6108

Microprocessor Compatible 8-Bit A/D Converter

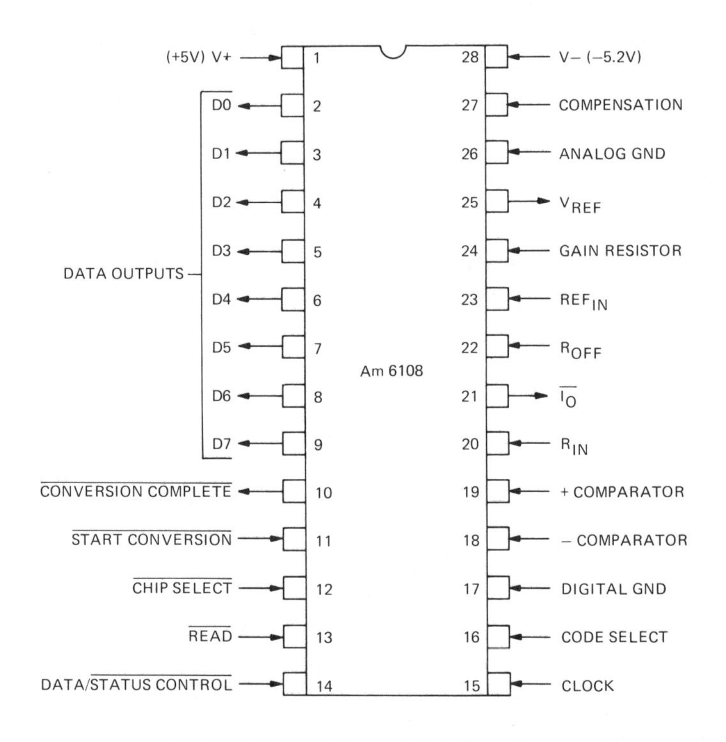

The Am6108, introduced by Advanced Micro Devices, is a Microprocessor Compatible 8-Bit Analog-to-Digital Converter designed to interface with most existing microprocessors. The chip includes a precision reference, DAC, comparator, Successive Approximation Register (SAR), scale resistors, 3-state output buffers, and control logic.

Operation

The Am6108 converts analog voltages into digital signals using the successive approximation method. This procedure (which is explained in detail within the DAC-888 text) uses a DAC in conjunction with a *successive approximation register*.

The DAC in the Am6108 is made up of eight current sources that can be switched between the I_O and $\overline{I_O}$ outputs by eight digital signals from the internal *SAR*. The sum of the two output currents, I_O (accessible through pin 19) and $\overline{I_O}$ (pin 21), is equal to the DAC full-scale current.

Full-scale output current is determined by the reference current supplied to the DAC's positive input. The GAIN RESISTOR input (pin 24) is a 2.5k current limiting resistor in series with the DAC input that will convert the 2.5 volt reference voltage, V_{REF} (pin 25), into a 1-mA reference source. Full-scale output current is four times the reference current, or 4.0-mA when pin 24 is tied to V_{REF}. The REF_{IN} (pin 23), on the other hand, connects directly to the DAC input and allows the user to utilize external scaling resistors for establishing the reference current.

The digital output of the *successive approximation register* is fed into the DAC inputs, where it is converted into a current. The output current from the DAC is then compared to the current generated by the analog input voltage. Based on this comparison, the *SAR* either keeps the bit under test or discards it before proceeding to the next approximation.

Analog Input

The Am6108 can be operated with either a unipolar or bipolar input signal. Two inputs are provided for unipolar operation. The R_{IN} input (pin 20) has a 2.5k resistor connected between it and the comparator summing node. When the analog signal is input to this pin, the full-scale analog voltage range is 0V to +10V. The R_{OFF} input (pin 22) is identical to R_{IN}, except the value of the input resistor is 1.25k, and provides an analog input range of 0V to +5V.

Bipolar operation of the A/D converter requires a half-scale offset current to be supplied to the comparator summing node. This can be accomplished by connecting the R_{OFF} input to the V_{REF} output, thus forcing a 2-mA offset current through the comparator, which gives the analog input (R_{IN}) a −5V to +5V range. When desired, the comparator input can also be accessed directly through the +COMPARATOR input (pin 19), allowing the user to provide any external scaling network necessary.

Digital Conversion

A conversion cycle begins by taking the CHIP SELECT (pin 12) LOW while simultaneously strobing the START CONVERSION (pin 11) input. This resets the *SAR*. Conversion begins when the START CONVERSION pin returns HIGH. After completing the conversion, the Am6108 drives the CONVERSION COMPLETE output (pin 10) LOW, allowing the 3-state outputs to be enabled. The data is read at the DATA OUTPUTS (pin 2-9) by enabling the READ (pin 13) and CHIP SELECT inputs. The CODE SELECT input (pin 16) chooses between binary offset and two's complement digital output formats.

A minimum of nine CLOCK (pin 15) cycles are required to complete a conversion. When operating with a 10-MHz clock, the conversion takes place within 1-us, which is short enough to allow most microprocessors to accept data immediately following a conversion request. However, when working with a slower CLOCK, it may be necessary to check on the progress of the conversion. The conversion status may be determined by holding the DATA/STATUS CONTROL (pin 14) LOW during a READ cycle and evaluating the D7 (pin 9) bit.

Power Supply

Two power supplies are required to operate the converter chip. A +5-volt source connects to V+ (pin 1) and a −5.2-volt source supplies the V− (pin 28) input.

The ANALOG GND (pin 26) and DIGITAL GND (pin 17) have also been isolated inside the chip to avoid the possibility of ground loops. Under normal conditions, these two points are returned to the same ground line on the circuit board, but care should be taken to prevent external ground loops from forming. In certain instances, the ground divisions can ease the interfacing requirements of the system, and they need not always be common to each other.

Additional Information

The Am6108 is an expanded version of the Am6148 A/D converter, and additional information pertaining to its performance can be found in that section.

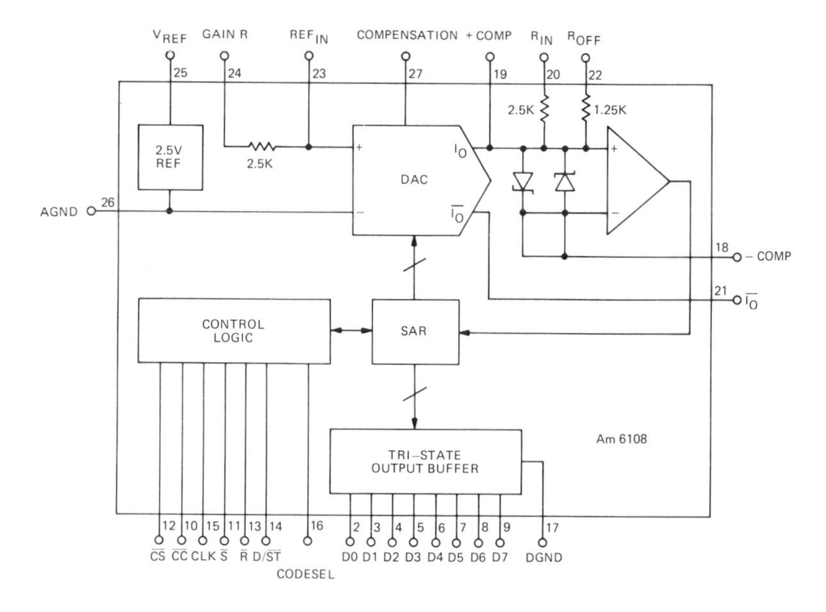

Figure 1

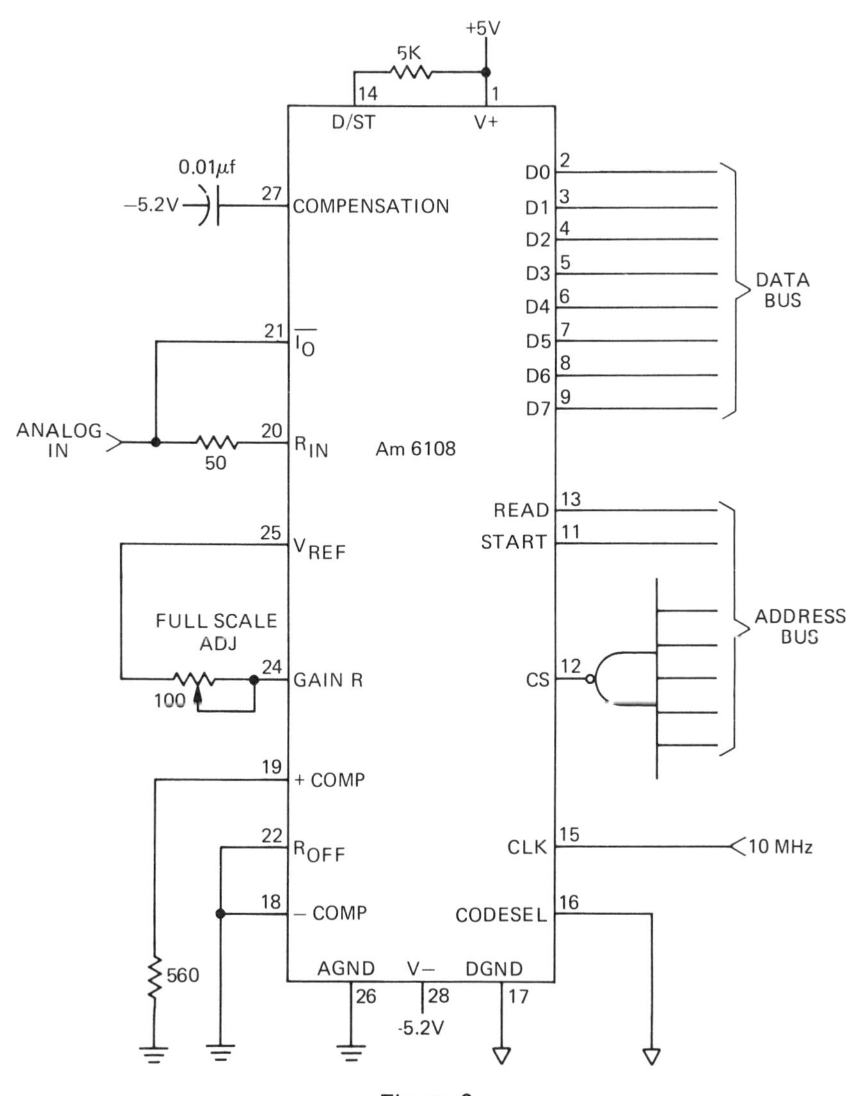

Figure 2

Am6148

Microprocessor Compatible 8-Bit A/D Converter

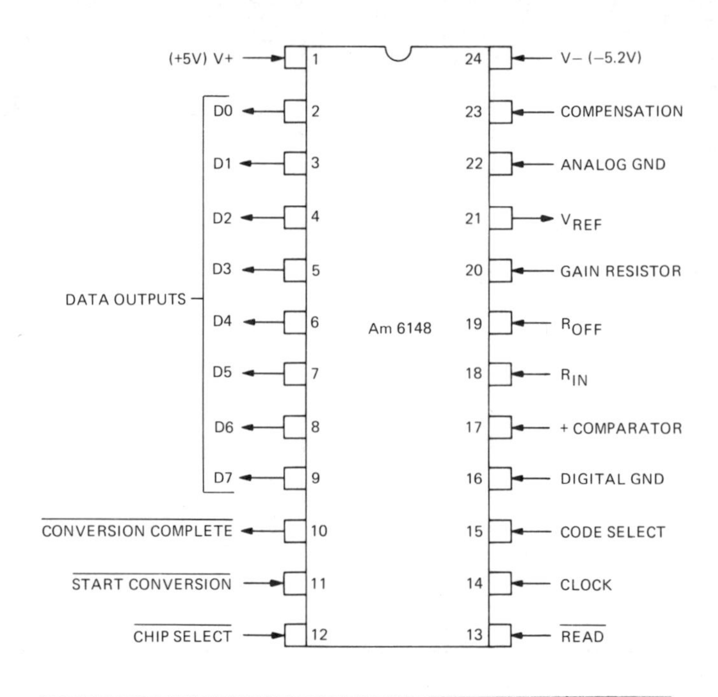

The Am6148, by Advanced Micro Devices, is a Microprocessor Compatible 8-Bit Analog-to-Digital Converter. The Am6148 is capable of completing an 8-bit conversion in under one microsecond, and is useful in microprocessor-based systems or stand alone applications.

Basically, the Am6148 is identical in operation to the Am6108, also by Advanced Micro Devices, and much of the following description is applicable to both devices. However, the Am6148 is housed in a slim 24-pin DIP package, creating significant differences between the two devices.

Probably most noticeable is the elimination of the REF_{IN} input, which was used to directly access the positive input of the DAC in the Am6108 chip. The only access the user has to the DAC input on the Am6148 is through a 2.5k resistor at the GAIN RESISTOR (pin 20) input. The complementary $\overline{I_O}$ output has also been eliminated, and is tied directly to the ANALOG GND (pin 22) internally.

Another notable revision to the Am6148 is the fact that the CPU is no longer able to ascertain the progress of the conversion by querying the D7 DATA OUTPUT bit (pin 9). The control pin which provided this function (DATA/STATUS CONTROL) has been dropped. The only indication that a conversion has been completed is regulated by the CONVERSION COMPLETE (pin 10) output.

Operating Parameters

Internally, the speed of the A/D conversion is dependent upon three types of delay. However, only the DAC settling time is of any consequence, and is the primary limiting factor. With no external components connected to the + COMPARATOR input (pin 17), it takes about 105-ns for the DAC output to settle to ± ½ LSB accuracy after each change of the *successive approximation register* (*SAR*). Since the DAC output must settle during the CLOCK (pin 14) LOW period, the total CLOCK cycle must be at least 155-ns long to satisfy the minimum 50-ns CLOCK pulse timing requirements. This amounts to no less than 1.4-us per conversion.

Nevertheless, the DAC settling time can be significantly improved by reducing the effective load impedence at the + COMPARATOR input. This is handily accomplished by connecting a resistor from the + COMPARATOR pin to the ANALOG GND. As the value of this resistor is reduced, so is the DAC output settling time. Unfortunately, a reduction in the load impedence increases the response time of the comparator, which leads to differential nonlinearity.

Therefore, the value of the comparator resistor is a compromise between fast settling times and differential nonlinearity. The optimum value for effective load impedence is 330-ohms, and an external resistor of 560-ohms should yield ± ½ LSB accuracy with conversion results in less than 1-us, when used with a 10-MHz clock.

The effects of bias currents at the inputs of the comparator will also contribute to conversion errors. To minimize the bias current effect, a resistor can be connected between the − COMPARATOR input (pin 18, Am6108 only) and the ANALOG GND. To be most effective, this resistor should be equal to the value of the virtual impedence at the comparator summing node. To save on board space and costs, though, this pin can be returned to ANALOG GND with little noticeable increase in error. In fact, that is just what is done in the Am6148; the − COMPARATOR input has been tied to the ANALOG GND internally and the input pin eliminated.

Negative Supply Loss Protection

The maximum ratings for the Am6108/Am6148 indicate a positive limit on the negative supply V − (pins 28 & 24, respectively) of + 0.3-volts. Like most ICs, the removal of the negative supply can cause the substrate to float above the ground potential, turning on the substrate diode and creating a catastrophic condition.

The solution to this potential problem is to connect a diode with low forward voltage drop, such as a Shottky Barrier Rectifier 1N5818, between the negative supply and DIGITAL GND (pin 16). Now if the negative supply shorts to ground for some reason, the device will not experience the potentially disastrous failure situation.

An examination of the Am6108 chip will reveal more details concerning the Am6148 operations, and should be reviewed.

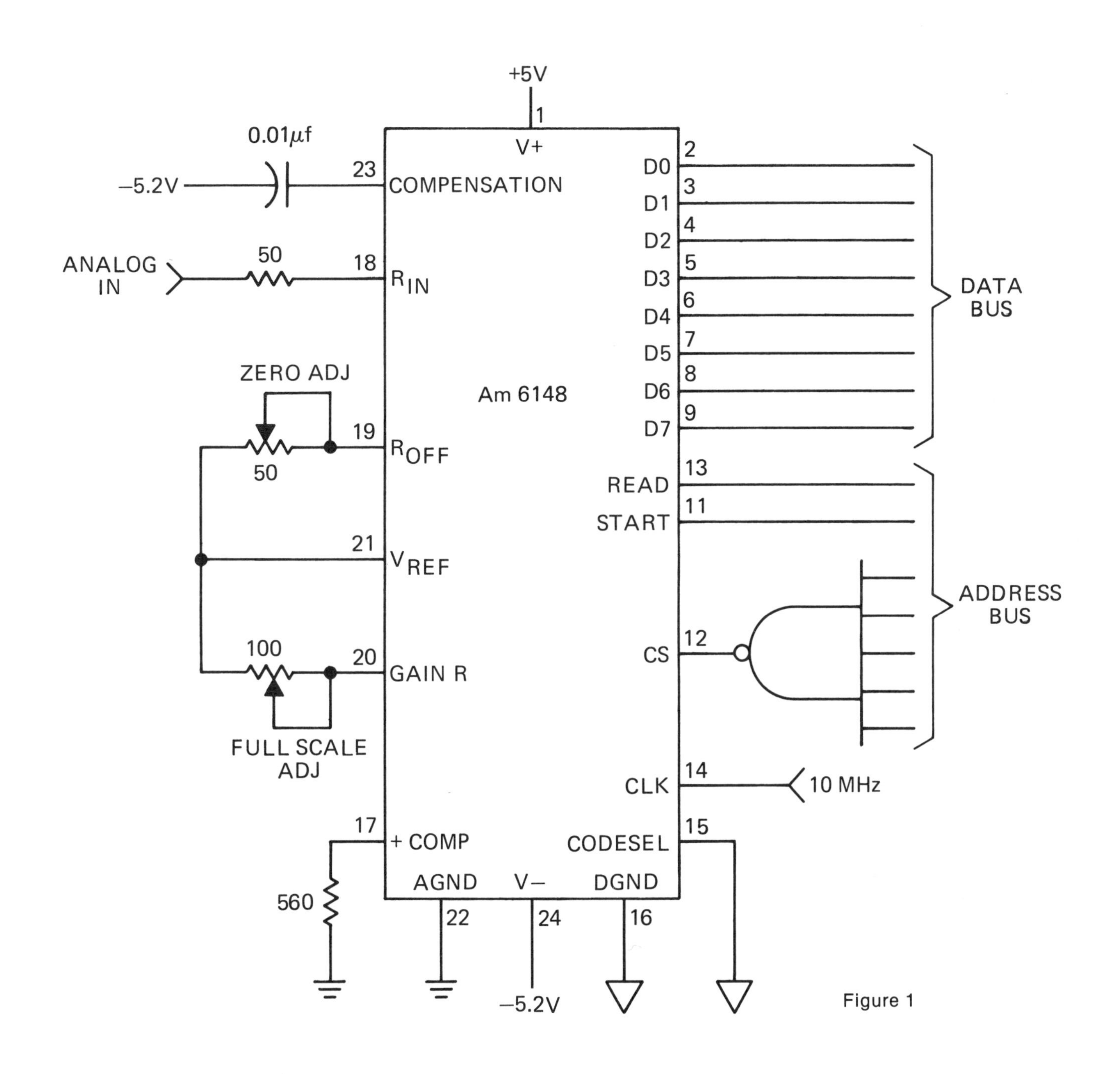

Figure 1

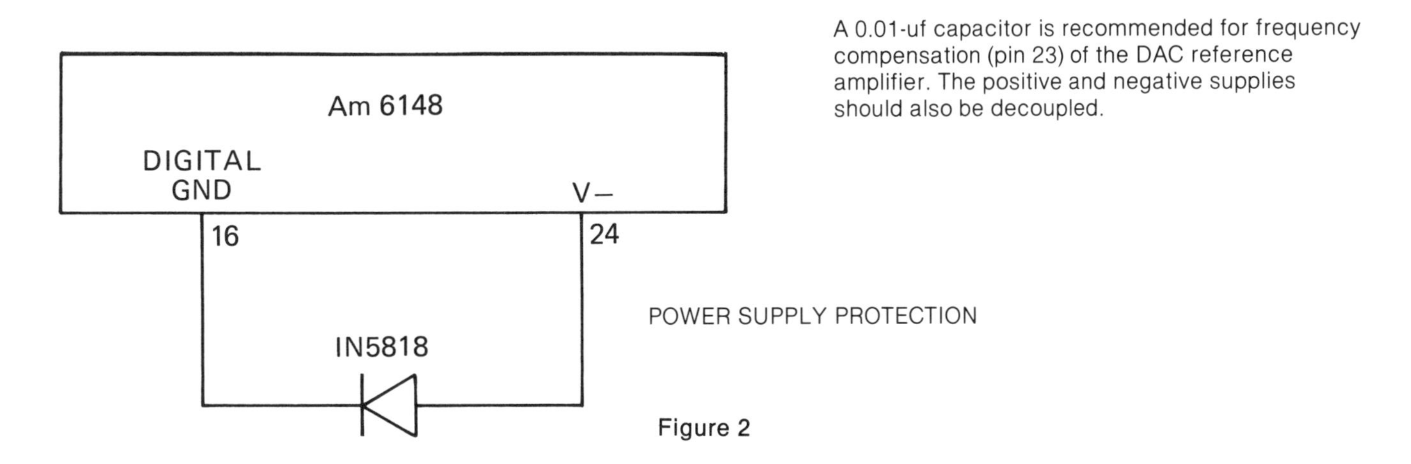

Figure 2

A 0.01-uf capacitor is recommended for frequency compensation (pin 23) of the DAC reference amplifier. The positive and negative supplies should also be decoupled.

DAS-952R

16-Channel, 8-Bit Data Acquisition System

The DAS-952R Data Acquisition System, by Datel-Intersil, provides versatile analog interface between the CPU and the outside world. The chip can randomly access up to 16 channels of analog data and convert it into 8-bit digital form for input to the microprocessor.

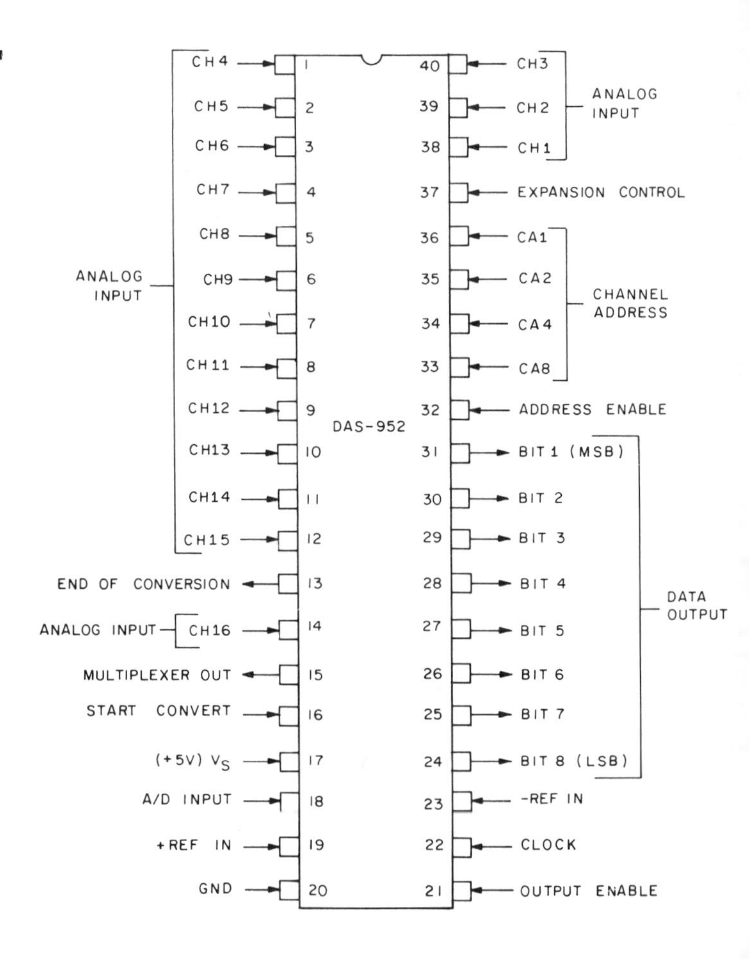

General Information

The DAS-952R is a ratiometric data acquisition system that expresses an analog input voltage as a percentage of the full scale output voltage range. In other words, it will digitally represent an analog voltage as a ratio of the full-scale system voltage to the actual analog input voltage and display it in terms of percentage.

The chip has two distinct functions to it, the first of which is a 16-channel analog multiplexer. It is the duty of a multiplexer to establish a link between the desired analog channel and the digital output.

The digital conversion takes place in the second chip function: the A/D converter. The two functions are completely isolated from each other and connection of the multiplexer output to the converter input must be done externally. This permits conditioning of the analog signal, such as amplification, linearization, etc., before it is input to the A/D converter.

Analog Multiplexer

The analog input multiplexer allows random access to any one of 16 single-ended ANALOG INPUTS (pins 38-40, 1-12, 14). The desired channel is selected by an address decoder, which is available at the four CHANNEL ADDRESS (pins 36-33) inputs. The CHANNEL ADDRESS inputs specify the ANALOG INPUTS in straight binary form, and are latched into the decoder on the low-to-high transition of the ADDRESS ENABLE (pin 32) input. For a valid transfer to take place, the address inputs must be stable for at least 50-ns preceding and following the ADDRESS ENABLE strobe.

Once the ANALOG INPUT has been selected, it is available at the MULTIPLEXER OUT (pin 15) within 2.5-us. The typical ON input resistance per channel is 1.5k, and all channels are matched to within 75-ohms of each other.

A/D Converter

The 8-bit A/D converter section uses a *successive approximation register* in harmony with a chopper-stabilized comparator to encode the digital word. By employing a chopper-stabilized comparator, the converter is extremely tolerant of thermal effects and long term input-offset drift errors.

In ratiometric conversion, the converter expresses the A/D INPUT (pin 18) as a percentage of the +REF IN (pin 19) and −REF IN (pin 23) reference input. The analog input voltage range is equal to the full-scale range of the reference input voltage, and may be varied from +0.512-volts to +5.25-volts.

The A/D converter uses a 256R resistor ladder network that divides the full-scale range into 256 steps. The size of each step is determined by the reference voltage. Halfway through its 256-step range, however, the analog switch tree changes from N-channel switches to P-channel switches. Therefore, the center-point of the reference voltage range, measured from pin 19 to pin 23 and divided by two, must be held to ±0.1-V of the center of the power supply voltage — $V_S/2$ — otherwise erratic switching operation will occur.

This condition is automatically satisfied when +REF IN is tied to V_S (pin 17) and −REF IN is returned to GND (pin 20). For configurations other than the above, −REF IN must be offset from ground by the same amount that +REF IN differs from the V_S input. Pin 19 must always be positive with respect to −REF IN, and in no case can +REF IN exceed V_S or −REF IN be less than GND.

Raising the START CONVERT input (pin 16) to HIGH will clear the *successive approximation register* and begin the conversion process on its trailing pulse edge. It requires 64 CLOCK (pin 22) periods to resolve the analog signal voltage level. The time it takes for a conversion, and the throughput data rate, is dependent upon the external CLOCK frequency, which can range from 10-KHz to 1.2-MHz. For a CLOCK frequency of 640-KHz, a typical conversion takes 100 us.

When the conversion is completed, the END OF CONVERSION output (pin 13) is forced HIGH. For continuous conversions, the END OF CONVERSION output is tied to the START CONVERSION input, with an external pulse starting the process after power up.

The digital conversion results are available on the DATA OUTPUT lines (pins 24-31) when the OUTPUT ENABLE control (pin 21) is HIGH. These outputs are driven by 3-state latches, whose contents remain valid until the next conversion is complete, whether the DATA OUTPUTs are enabled or not.

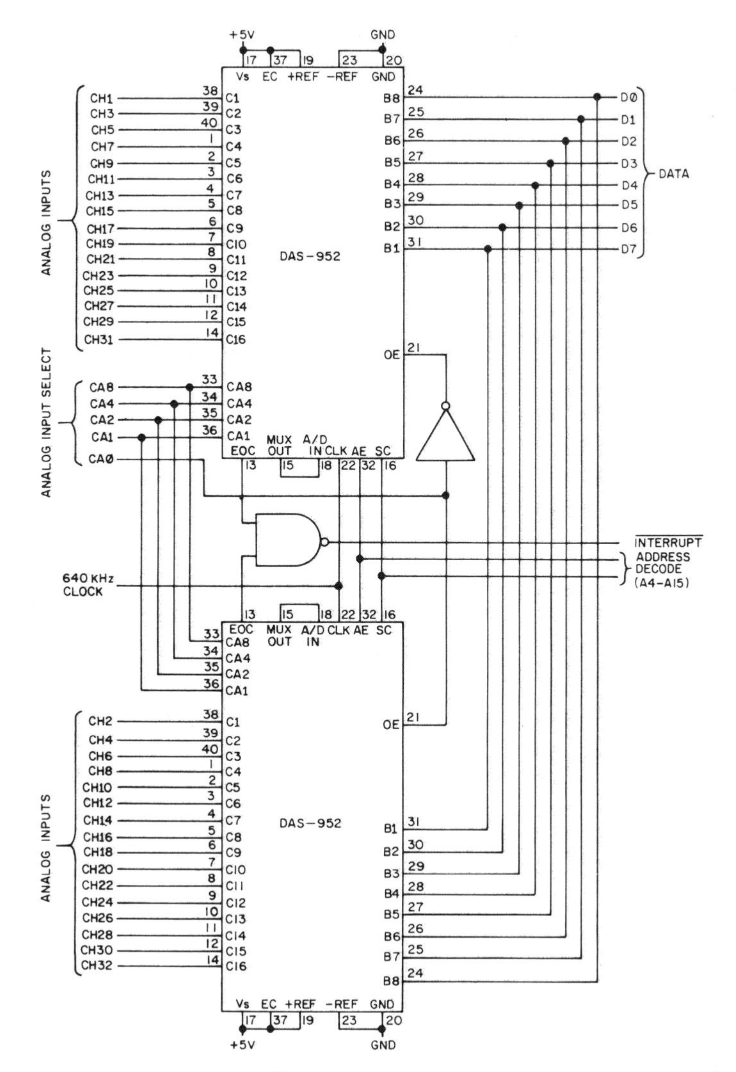

Figure 1

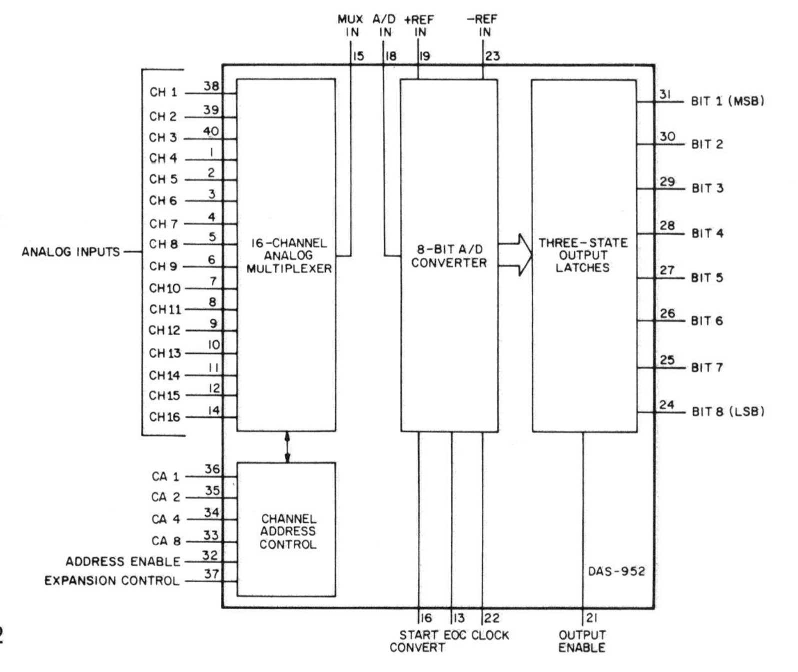

Figure 2

CA3300

Video Speed 6-Bit Flash Analog-to-Digital Converter

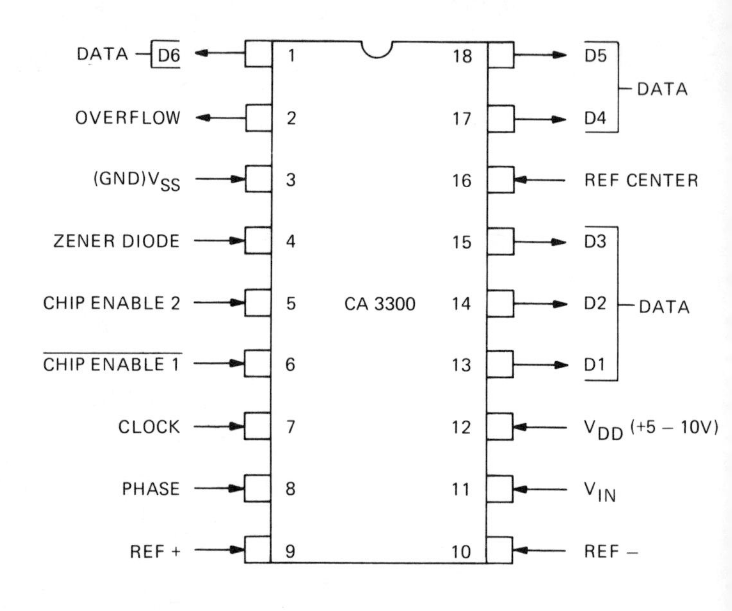

The CA3300, made by RCA, is a CMOS parallel analog-to-digital flash converter designed for applications demanding both low power consumption and high-speed digitization. The circuit operates from a single supply voltage and dissipates less than 150-mW at 15-MHz.

Clock

A sequential parallel technique is used by the CA3300 converter to obtain its high-speed operations. This sequence is keyed to a split-phase conversion CLOCK (pin 7) pulse. Although only one clock pulse is required for conversion, it has been divided into the Auto Balance ø1 phase and Sample Unknown ø2 phase. The ø1 phase occurs during the HIGH period of the CLOCK cycle, and the ø2 phase occurs when the CLOCK cycle is LOW.

To expand the usage of the converter, a PHASE control (pin 8) is provided which effectively inverts the CLOCK input logic, swapping the ø1 phase for the ø2 phase. The following discussion will assume the PHASE input to be in the LOW state.

Operation

Sixty-four paralleled auto-balanced voltage comparators measure the input voltage with respect to a known reference to produce the binary-encoded parallel outputs. The reference consists of a resistor voltage divider with 64 equally spaced taps on it. This resistor ladder is accessible through the REF+ (pin 9) and REF– (pin 10) pins. Since the device is single-supply operated, REF– is normally returned to ground. The chip also contains an on-board 6.4-volt ZENER DIODE (pin 4) for use as a reference voltage for REF+. However, the REF+ input may be any stable voltage from 2.4-volts up to the DC supply voltage.

During the Auto Balance (ø1) phase of operation, an internal switch is used to connect 64 commutating capacitors to each one of the 64 resistor ladder taps. The other side of the capacitor is connected to a single stage amplifier, which will later serve as the comparator. Altogether, there are 64 amplifiers, one for each commutating capacitor. The output of each amp is shorted to its input by another transmission switch, which biases the amplifier at its intrinsic trip point—or approximately ½V_{CC} when V_{SS} is grounded. The capacitors now charge to their associated tap voltage, priming the circuit for the next phase.

In the Sample Unknown (ø2) phase, all resistor tap switches are opened, the comparator amplifier shorts are removed, and the 64 commutating capacitors are connected to V_{IN} (pin 11), the analog quantity to be converted. Since the opposite end of the capacitors are effectively looking into an open circuit with the removal of the amplifier short, any voltage that differs from the previous tap voltage will appear as a voltage shift at the comparator. All comparators with tap voltages greater than V_{IN} will drive the comparator outputs to a LOW state, while all comparators with tap voltages lower than V_{IN} will drive the comparator outputs HIGH.

Digital Outputs

The status of the comparator amplifiers is captured at the end of the ø2 phase by a secondary latching amplifier. Once latched, the 64 information bits are encoded and the results clocked into an output *storage register* on the rising edge of the next ø2 clock.

A 3-state buffer is used to drive the DATA (pins 13-15, 17, 18, 1) outputs, D1 through D6. Two CHIP ENABLE signals control these outputs. CHIP ENABLE 1 (pin 6) will disable outputs D1 to D6 when driven HIGH; CHIP ENABLE 2 (pin 5) will independently disable D1 through D6—plus an OVERFLOW (pin 2) output—when in its LOWstate.

The OVERFLOW bit makes possible the connection of two CA3300s in series to increase the resolution from 6-bits to 7-bits. Meanwhile, the parallel connection of two devices doubles the conversion speed; i.e., it increases the sampling rate from 15-MHz to 30-MHz by staggering the clock PHASE.

Pulse Mode Operation

For sampling high-speed nonrecurrent or transient data, the CA3300 may be operated in a pulse mode rather than the conventional continuous clock mode. There are two ways to accomplish pulse mode conversions.

The fastest method is to hold the device in the ø2 phase while in standby. The converter can now be pulsed through the ø1 in as little as 33-ns. The analog value is captured on the leading edge of ø1 and is transferred into the *storage registers* on the trailing edge of ø1. In this mode, another conversion can be started within 33-ns, but no later than 10-us due to the eventual droop of the commutating capacitors.

The second method uses the ø1 phase as the standby state. Although this mode consumes more standby power, the converter can remain in this state indefinitely waiting for a conversion. A conversion is performed by strobing the CLOCK input with two ø2 pulses. The first pulse converts the analog value, the second transfers it to the outputs. Conversion now takes 67-ns, but the repetition rate may be as slow as desired.

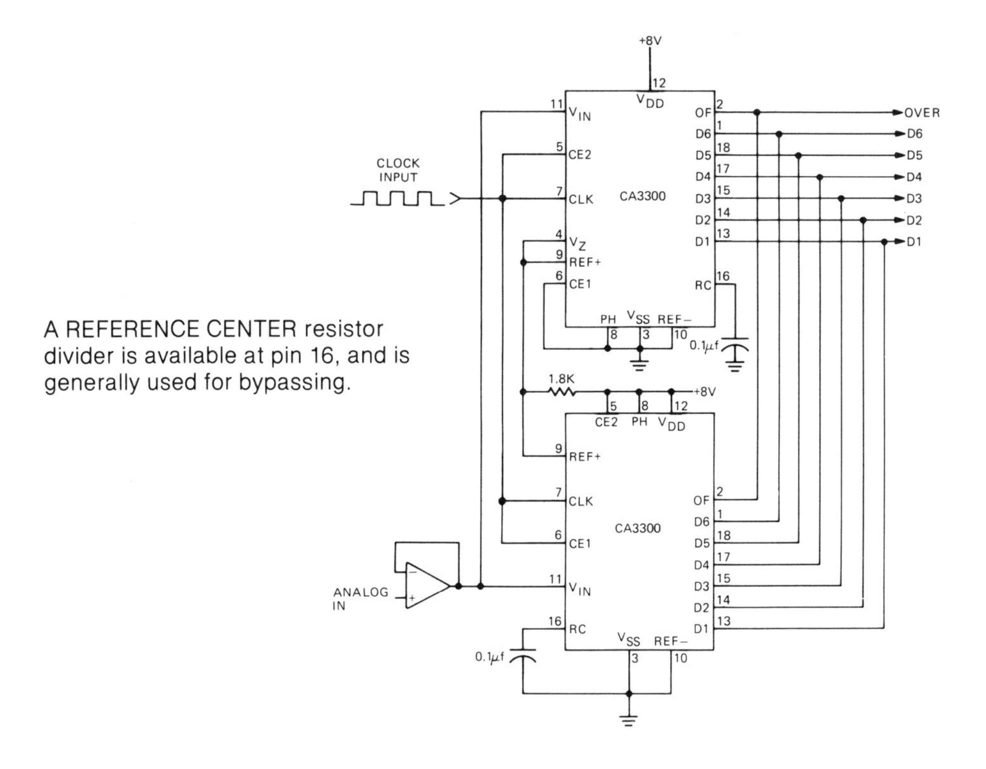

Figure 1

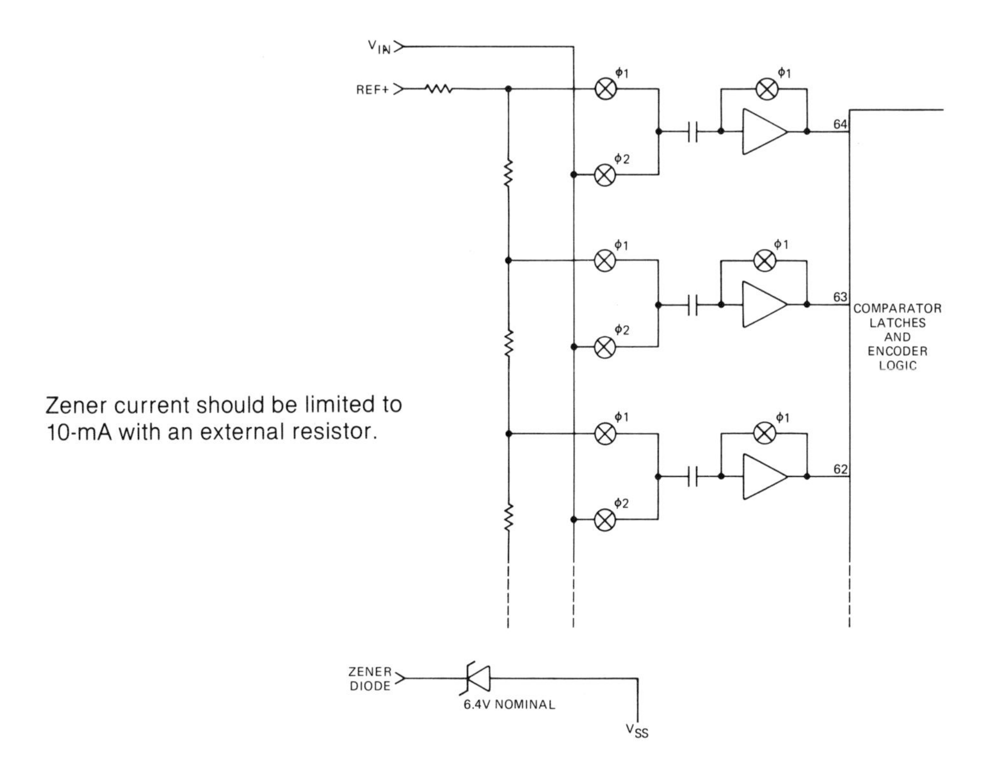

Figure 2

TM1070

Flash Converter

The TM1070, introduced by Telmos Inc., is a 7-bit parallel A/D Flash Converter designed for 15-MHz sampling at low power levels. Conversion is accomplished using only one clock pulse, with data appearing at the outputs following the conversion signal.

Operation

The TM1070 contains 128 comparators whose reference inputs are individually connected to a resistive voltage divider with 128 taps. All 128 sampling inputs, however, are commonly tied, and a signal input to them will successively turn on each comparator as the input voltage exceeds the individual reference voltage, with all 128 comparators actively outputting when the input signal equals the total reference voltage. The comparator outputs are then fed into an 128 to 7 encoder, where they emerge as binary coded 7-bit outputs.

Analog Input

The analog signal to be converted is applied to the comparator inputs through the V_{IN} (pin 6) CMOS input. This signal is compared to the reference voltage established across the $+V_{REF}$ (pin 10) and $-V_{REF}$ (pin 3) resistor divider inputs. Nominal reference inputs are +3.2-V and −3.2-V, respectively, giving a ±3.2-volt input range to the analog signal. However, there are several ways in which the reference voltage can be applied, provided the voltages never exceed their respective power supplies less 1.5-volts (V_{DD} − 1.5-V, for example).

If the $-V_{REF}$ input is grounded, the TM1070 will convert unipolar positive inputs, while grounding $+V_{REF}$ and applying a negative reference voltage to $-V_{REF}$ will encode negative unipolar inputs. The reference can also be offset by any ratio, keeping in mind that pin 10 is always positive with respect to pin 3.

The internal resistor divider also has three proportional taps which have been made accessible to the user at $+V_{REF}/2$ (pin 8), REF CENTER (pin 7), and $-V_{REF}/2$ (pin 5). Their primary purpose is to provide high-frequency bypassing of the resistor chain to improve stability and lessen the chance of oscillation. A 0.01-uf cap to each is recommended for most applications. However, these taps can be used to tailor the performance of the referencee voltage, such as correcting small errors in linearity. It is even possible to create pseudo-logrithmic scales by applying consecutively higher-ordered voltages across adjacent resistor legs (see fig. 2).

Digital Output

The analog-to-digital conversion takes place using a single CLOCK (pin 15) pulse. When the CLOCK input is HIGH, the comparators are in the Reset Mode and their outputs are disabled while the inputs track the analog signal. Data is sampled on the falling edge of the CLOCK pulse, and propagates to the outputs when the CLOCK line is LOW.

The digital output appears on the DATA BUS (pins 23-17) output lines, which are driven by open-drain CMOS transistors. Because of the open drain configuration, pull-up resistors (1-K, typically) are required. The CHIP ENABLE (pin 13) control line enables the digital outputs when HIGH; if this control input is LOW, the DATA BUS outputs are floated by disabling the driver transistors. During the Disable and/or Reset Modes, all data outputs are forced HIGH by the pull-up resistors.

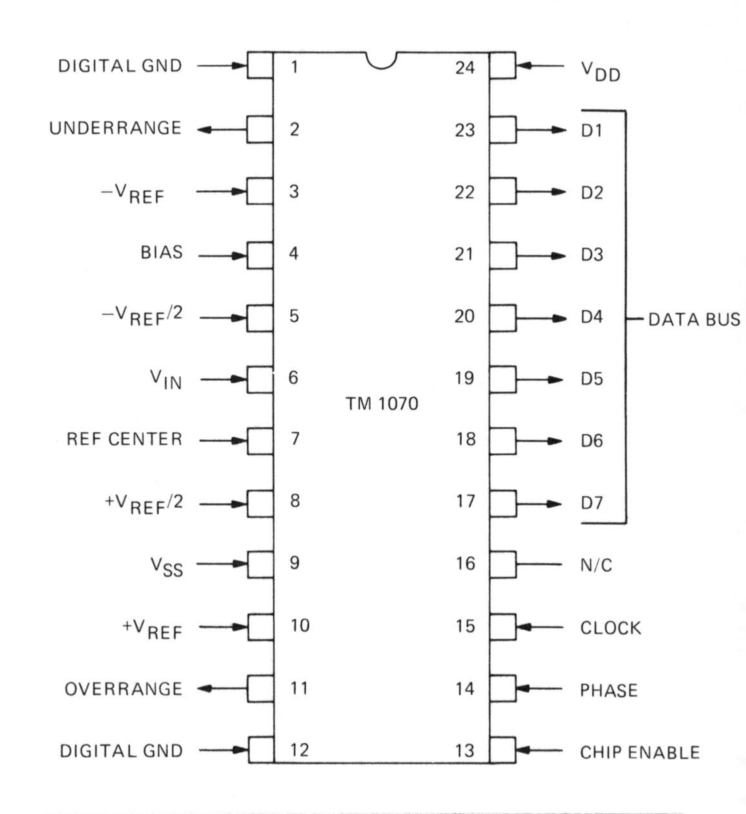

The two control functions, CLOCK and CHIP ENABLE, can be inverted using the PHASE (pin 14) input. The PHASE line determines the polarity of these two inputs, and when it is LOW, all input logic to them is reversed. However, the PHASE inversion has no effect on the binary output.

If a reading is taken where V_{IN} (analog input) is greater than $+V_{REF}$ or less than $-V_{REF}$, the OVERRANGE (pin 11) or UNDERRANGE (pin 2) outputs will go HIGH, respectively. An OVERRANGE condition drives all digital outputs HIGH.

Power Supply

The chip is powered using two supplies: a positive voltage to V_{DD} (pin 24) and a negative voltage toV_{SS} (pin 9). The two voltages don't have to be equal, but their sum must not exceed 12-volts. The analog ground is derived from their common junction.

The DIGITAL GND (pins 1, 12) is common to the output driver transistors only, and separate from the analog ground. This allows the open-drain outputs to interface directly with TTL, CMOS, and ECL logic—regardless of the analog voltages used—by returning the pull-up resistors to their respective logic levels instead of TM1070's V_{DD}. The DIGITAL GND is tied to the logic common.

Although the nominal power dissipated by the chip is 100-mW, the device has been designed with a programmable current source. The operating current is established by the BIAS (pin 4) input, and is equal to approximately 100 times the bias value. In other words, a 100-uA bias current will operate the chip at 10-mA. At lower conversion rates (under 1-MHz), the operating current can be reduced, with a considerable power savings. However, as CLOCK rates increase, the supply current must increase proportionally.

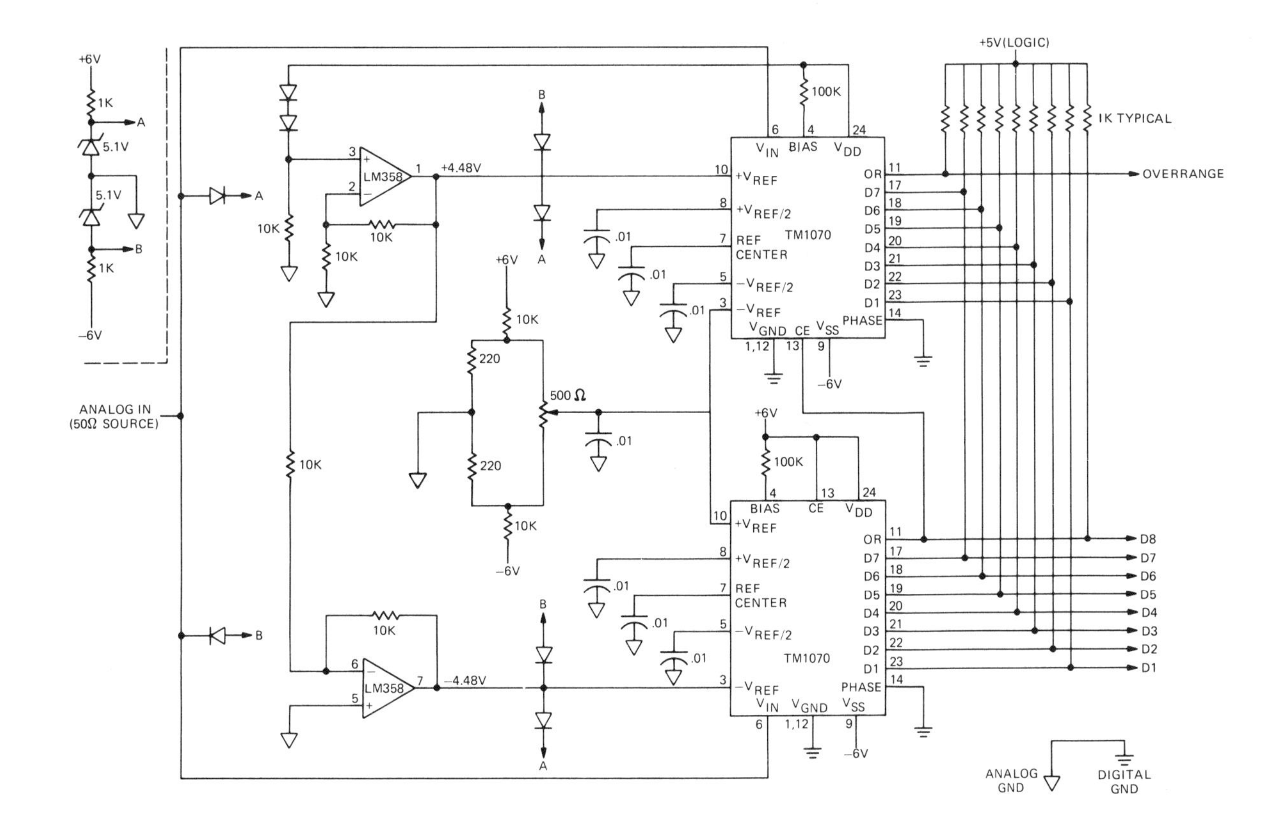

Figure 1

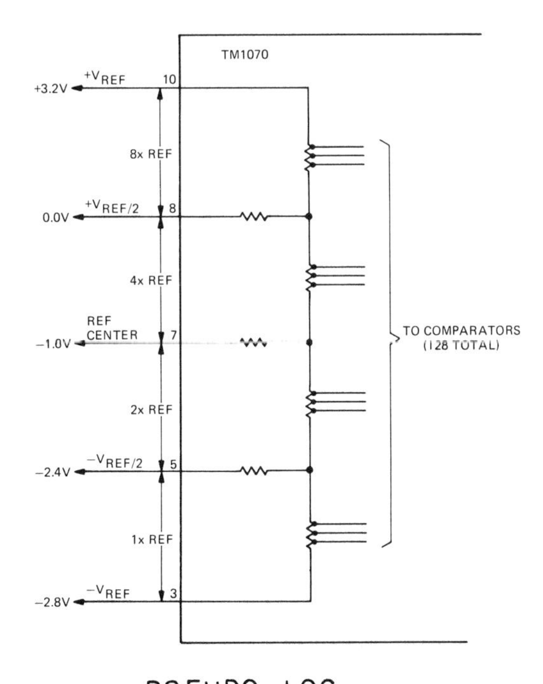

Figure 2

MC10315

Seven-Bit Parallel A/D Flash Converter

The MC10315, by Motorola, is a 7-bit high-speed A/D converter employing ECL logic. The device consists of 128 parallel latched comparators which are individually referenced to a tap on a resistive voltage divider. As the input voltage exceeds the comparison voltage, the comparators output a signal. The signals are fed into a 7-bit encoder, where they are translated into binary code and output through ECL emitter followers.

More significant, however, is the fact that no sample and hold requirements are needed for the conversion, allowing the MC10315 to attain sampling frequencies up to 15-MHz.

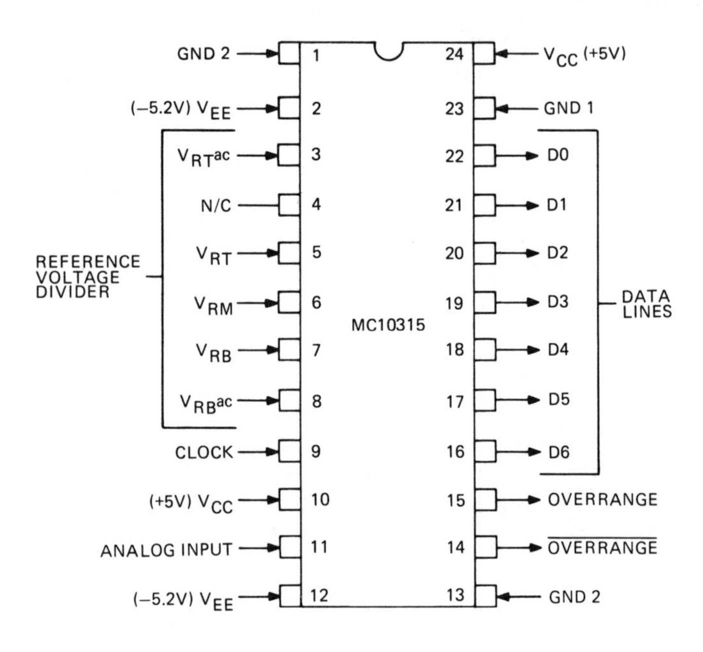

Clocking Requirements

Conversion is performed within a single CLOCK cycle (pin 9). The rising edge of the clock pulse holds the analog input stable by latching all the input comparators. The output latches are subsequently released by the same edge, and updated. The falling edge of the clock latches the output DATA LINES (pins 22 through 16) anew, while releasing the comparators. To minimize the chance of clocking errors and uncertainty, rise and fall times of the clock edge should be in the range from 2-ns to 7-ns.

Analog Input

The analog signal to be converted is presented to the device on pin 11, the ANALOG INPUT. The input current required by the ANALOG INPUT is a function of the input voltage, and is directly proportional to the number of working gates. The more comparators in operation, the higher the input current. Input capacitance is also affected in the same way. Maximum input current is 400-uA (all comparators ON); large signal capacitance ranges from 40-pf to 70-pf.

To overcome gain errors due to current and capacitive loading on the input, it is recommended you drive the comparators with a low output impedence buffer amplifier. An output impedence of 10-ohms will easily drive 200-MHz with little error.

Reference Inputs

The top of the resistance divider ladder, V_{RT}, and the bottom of the ladder, V_{RB}, are accessible through pins 5 and 7, respectively. The working voltage across the resistive ladder is between 1.0 and 2.0-volts, where V_{RT} is always positive with respect to V_{RB}. Voltages lower than 1.0-volts will degrade linearity due to comparator offsets.

Establishing the input range is a matter of preference and input voltage conditions. Grounding V_{RB} will give a positive unipolar range to the input, while grounding V_{RT} will measure a negative input signal. For bipolar operation, the reference source is proportionally split and grounded. V_{RT} must not exceed +2.5 volts (referenced to GND2), and V_{RB} must remain above −2.5 volts.

Additional taps on the reference ladder are also pinned out. The V_{RT}ac (pin 3) and V_{RB}ac (pin 8) are intended for ac decoupling through bypass capacitors. The V_{RM} (pin 6) input is the center tap of the resistor chain, and can be used as a voltage reference input to improve linearity if necessary, as can be V_{RT}ac and V_{RB}ac.

Power Supply

Two power supplies are needed to power the chip. Driving the ECL logic requires a −5.2-volt source (pins 2, 12) while a +5.0-volt input (pins 10, 24) is demanded by the comparators. Care must be taken to prevent digital ground currents from flowing into the analog ground. As is the practice with ECL devices, the logic V_{CC} line is referenced to ground; and as usual, two grounds are provided: GND1 (pin 23) and GND2 (pins 1, 13). The digital output transistors are sourced through GND1, thereby isolating them from the analog ground, which is referenced to GND2.

Outputs

Since the outputs are configured as ECL emitter followers, they must comply with MECL design rules. This means that each output line must be terminated with a pulldown resistor. A 510-ohm resistor to V_{EE} (−5.2-volts) will provide good fall times in most applications, while holding down the power dissipation. To further minimize power and decrease pulse reflections, the outputs can also be returned to an established V_{TT} (−2.0-volts) line.

Additional Information

The MC10315 is virtually identical to the MC10317, and the reader will find more information concerning the MC10315 by turning to that section.

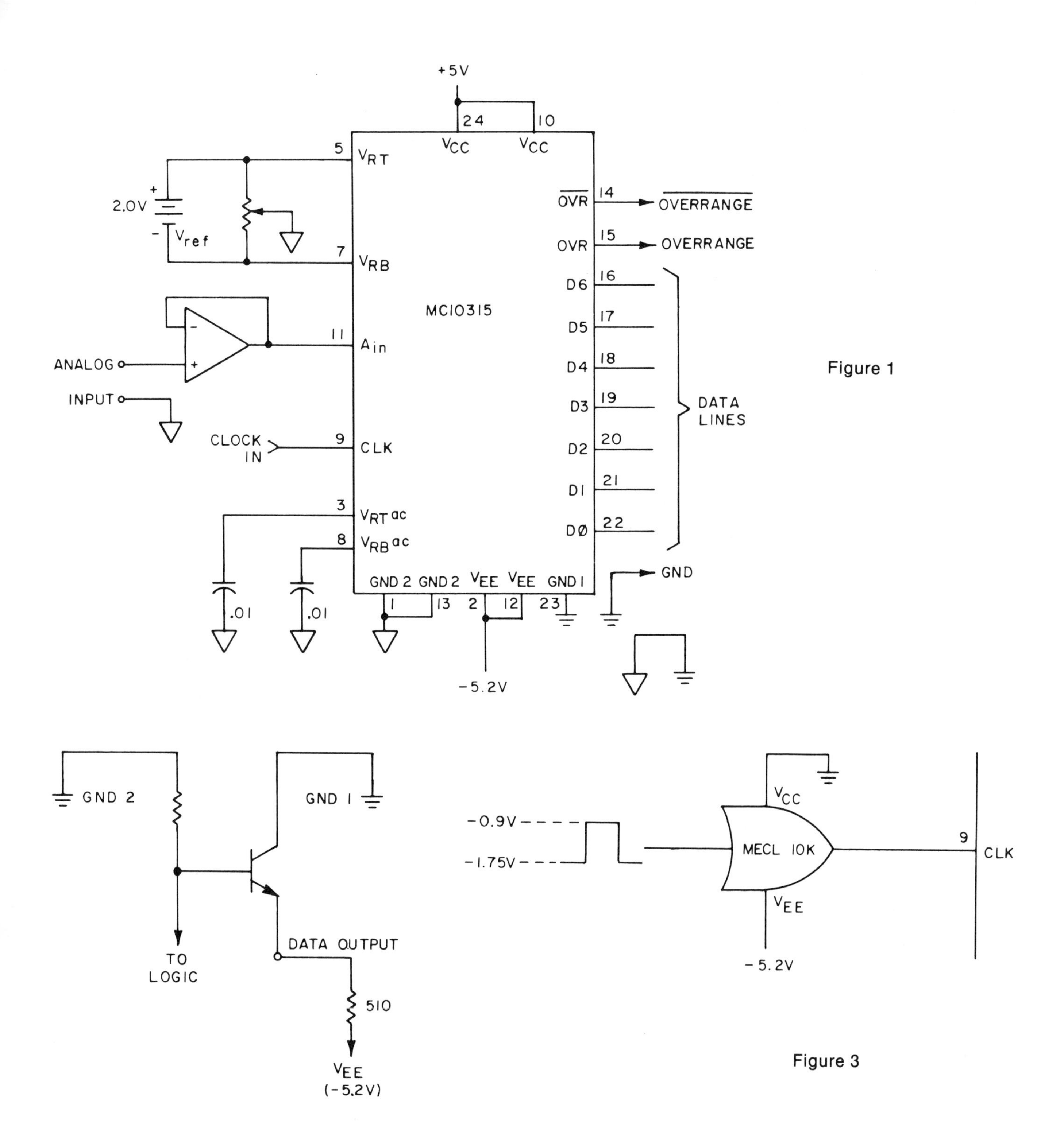

Figure 1

Figure 2

Figure 3

All digital outputs must be treated according to MECL 10K design rules. Gnd 1 is equivalent to MECL 10K V_{CC1}. Gnd 2 is equivalent to MECL 10K V_{CC2}.

MC10317

Seven-Bit Parallel A/D Flash Converter

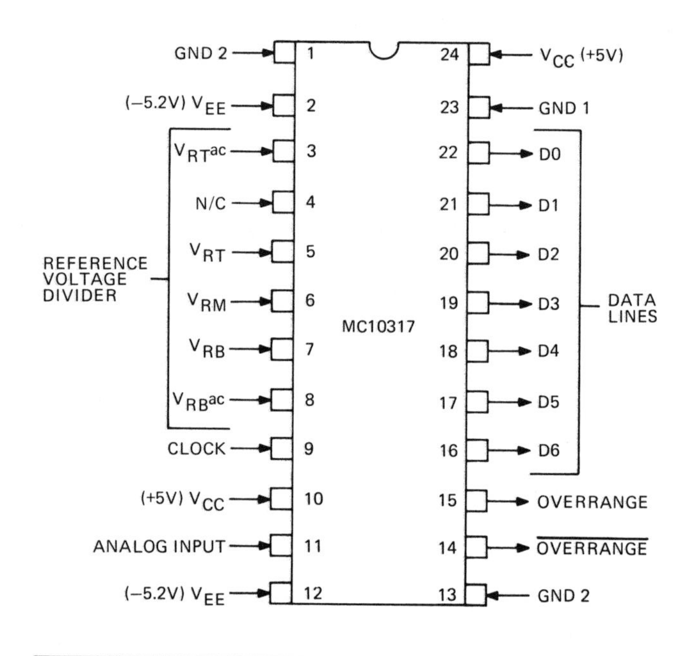

The MC10317, by Motorola, is a 7-bit high speed A/D converter employing ECL logic. The MC10317 and MC10315 are identical devices except for the method of overranging. This difference allows the two chips to be used in conjunction to extend the 7-bit output.

Although the 7-bit converter will suffice for most applications, 8-bits of resolution and accuracy can be obtained by stacking the MC10315 and MC10317 together and wire-ORing the data outputs. In view of the fact that the individual chip has been thoroughly reviewed in the section on MC10315, we will confine this discussion to the interaction of the two devices.

Overranging

It is best to first define the difference between these chips. When the MC10315 is driven into an overrange condition by the input voltage exceeding the input range, all output data bits will remain HIGH. In addition, an OVERRANGE indicator (pin 15) goes HIGH to notify the user of the fact. The MC10317, on the other hand, forces all outputs LOW during an overrange.

Since the output transistors are ECL emitter followers, it's possible to parallel the output DATA lines (pins 22-16) of these two chips and achieve a wire-OR function. The overrange output now represents the eighth bit.

Circuit Expansion

The schematic in fig. 1 shows how this is done, and has been set up to measure unipolar negative voltages (between −V and ground). The bottom converter is an MC10317, and will measure the lower end of the scale, with the MC10315 stacked above it. The first A/D will convert the input voltage until it exceeds the input limits, after which all the outputs go LOW and the OVERRANGE goes HIGH. The OVERRANGE output of the MC10317 serves as the eighth bit.

The input to the MC10315 now does the conversion, with the outputs ORed to the data lines. Should the top chip run out of input range, all outputs will end up HIGH, providing a true termination of the digital word.

Reference Voltage

The top of the MC10315 resistance ladder, V_{RT} (pin 5), is grounded, establishing a negative unipolar range. Next, the bottom of the ladder, V_{RB} (pin 7) connects to the top of the MC10317 resistor chain. This point is referenced to $V_{ref}/2$ to ensure that this node is midscale. Unit to unit variation in the resistance ladder can shift this point if a reference is not used, causing midscale linearity errors. The bottom pinout of MC10317 (pin 7) is returned to V_{ref}.

Care should be taken when interconnecting V_{RB} and V_{RT} of the MC10315 and MC10317 respectively. Reference ladder current flow through the resistance of the printed circuit board, sockets, and even the pin leads can manifest a significant IR drop of several millivolts, which can lead to midscale errors.

Clocking

The CLOCK inputs (pin 9) are driven by a common clock using standard ECL format. Depending upon the frequency to be encoded, it may be necessary to skew the rising clock edge from one chip to the other to compensate for slight differences in aperture delay time (latching) which may occur between the two devices.

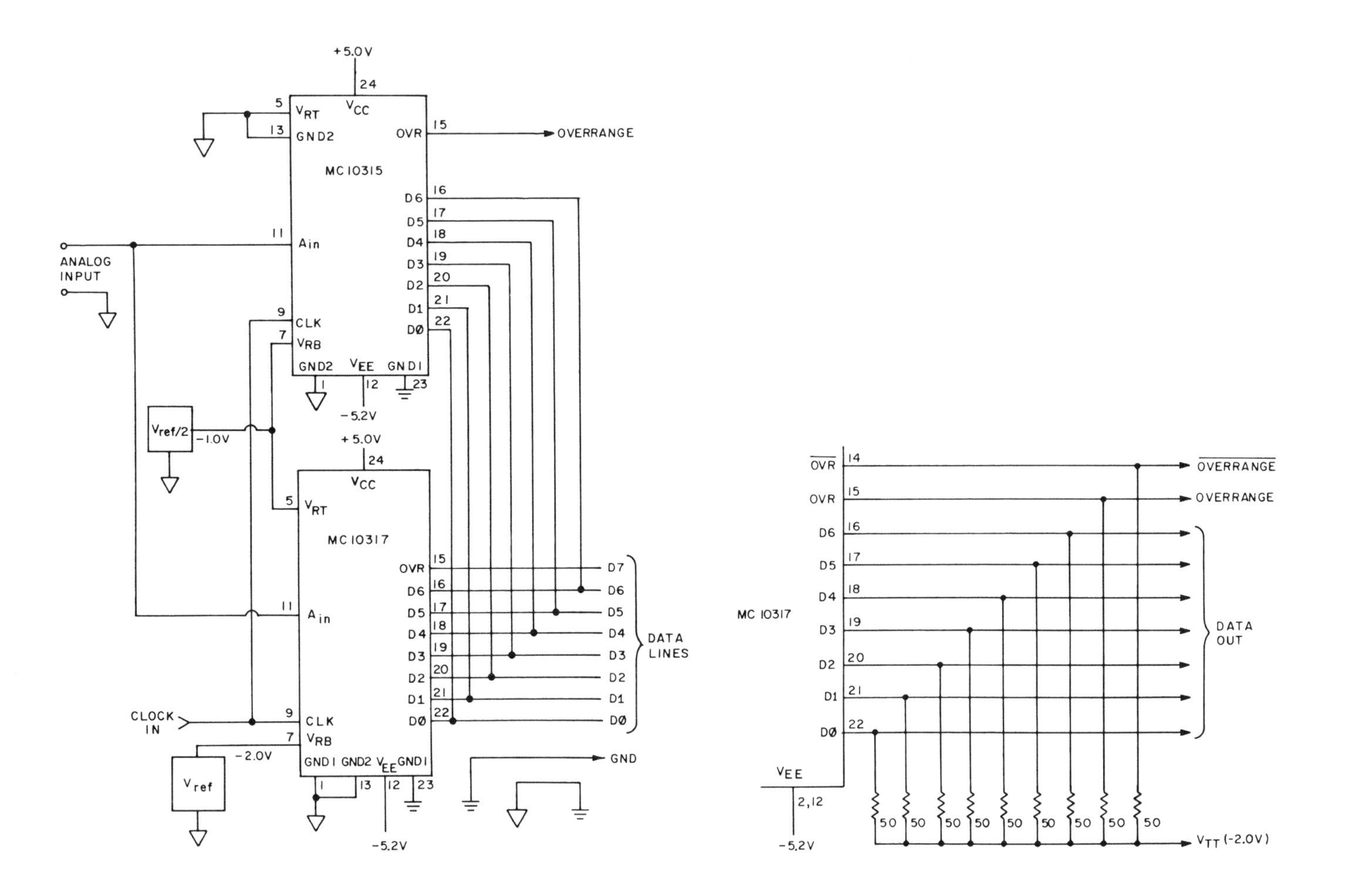

Figure 1

Figure 2

All outputs must be terminated following MECL 10K design rules. Figure 2 illustrates the preferred method.

ADDITIONAL INFORMATION

The MC10317 is virtually identical to the MC10315, and the reader will find more information concerning the MC10317 by turning to that section.

APPENDIX

Bibliography

Am6108/Am6148
Product Specification ABI-3067
Advanced Micro Devices

The Am7990 Family Ethernet Node
November 1982
Advanced Micro Devices

The AmZ8000 Family Data Book
Advanced Micro Devices

Data-Acquisition Databook 1982
Integrated Circuits
Analog Devices

Components Data Catalog
January 1982
Intel Corp.

Peripheral Design Handbook
August 1981
Intel Corp.

The Complete VLSI LAN Solution
Order Number: 210783-001
Intel Corp.

Microelectronic Data Book 1982/1983
Mostek Corp.

Motorola Microprocessors Data Manual
Series B, 1981
Motorola, Inc.

DP8340 Serial Bi-Phase Transmitter/Encoder
B-F-1819
National Semiconductor

DP8341 Serial Bi-Phase Receiver/Decoder
B-F-1820
National Semiconductor

NEC 1982 Catalog
NEC Electronic USA, Inc.
Microcomputer Division

The CRT Handbook
by Gerry Kane
OSBORNE/Mc-GRAW HILL

PMI 1982 Product Catalog
Linear Integrated Circuits
Precision Monolithics

COS/MOS Memories, Microprocessors, and Support Systems
SSD-260
RCA Solid State

Rockwell 10937 Data Sheet
Document No. 29000D85
Order No. D85
Rockwell International

Intelligent Display Controllers Data Sheet
Part Numbers: 10938 and 10939
Document No. 29000D96
Order No. D96
Rockwell International

EDLC Ethernet Data Link Controller
Preliminary Data Sheet, November, 1982
SEEQ Technology, Inc.

MOS Microprocessor Data Manual
Signetics Corp.

Data Catalog 1982/1983
Standard Microsystems Corp.

Synertek 1981-1982 Data Catalog
Synertek, A Division of Honeywell

TMS 9918A Video Display Processor
Data Manual
Texas Instruments

Semiconductor Databook 1982-1983
Unitrode Corp.

The Selection of Local Area Computer Networks
NBS Special Publications 500-96
U.S. Department of Commerce

1983 Components Handbook
Western Digital Corp.

1983 Network Products Handbook
Western Digital Corp.

WD1001 Winchester Disk Controller
OEM Manual
Document 80-031003-00A1
Western Digital Corp.

Microcomputer Components Data Book
1981 Data Book
Zilog

Microprocessor Applications Reference Book
Volume 1
Zilog

Manufacturers

Advanced Micro Devices
901 Thompson Place
Sunnyvale, CA 94086

Analog Devices
One Technology Way
PO Box 280
Norwood, MA 02062

Cybernetics Micro Systems
PO Box 3000
San Gregorio, CA 94074

Datel-Intersil
11 Cabot Blvd.
Mansfield, MA 02048

Intel
3065 Bowers
Santa Clara, CA 95051

Mostek Corp.
1215 West Crosby Road
Carrollton, TX 75006

Motorola, Inc.
Marketing Communications Dept.
MOS Integrated Circuits Group
3501 Ed Bluestein Blvd.
Austin, TX 78721

National Semiconductor
2900 Semiconductor Drive
Santa Clara, CA 95051

NEC Electronics U.S.A. Inc.
One Natick Exclusive Park
Natick, MA 01760

Precision Monolothics, Inc.
1500 Space Park Drive
Santa Clara, CA 95050

RCA Solid State Division
Route 202
Somerville, NJ 08876

Rockwell International
4311 Jamboree Road
Newport Beach, CA 92660

SEEQ Technology, Inc.
1849 Fortune Drive
San Jose, CA 95131

Signetics Corps.
PO Box 409
Sunnyvale, CA 94086

Standard Microsystems Corp.
35 Marcus Blvd.
Hauppauge, NY 11788

Synertek
PO Box 552
Santa Clara, CA 95052

Telmos, Inc.
740 Kifer Road
Sunnyvale, CA 94086

Texas Instruments, Inc.
PO Box 1443
Houston, TX 77001

Unitrode Corporation
5 Forbes Road
Lexington, MA 02173

Western Digital Corp.
2445 McCabe Way
Irvine, CA 92714

Zilog
1315 Dell Ave.
Campbell, CA 95008

Index